From Great Discoveries in Number Theory
to Applications

Michal Křížek · Lawrence Somer · Alena Šolcová

From Great Discoveries
in Number Theory
to Applications

 Springer

Michal Křížek
Institute of Mathematics
Czech Academy of Sciences
Prague, Czech Republic

Lawrence Somer
Department of Mathematics
Catholic University of America
Washington, DC, USA

Alena Šolcová
Department of Applied Mathematics
Czech Technical University in Prague
Prague, Czech Republic

ISBN 978-3-030-83898-0   ISBN 978-3-030-83899-7   (eBook)
https://doi.org/10.1007/978-3-030-83899-7

Mathematics Subject Classification (2010): 11-XX, 05Cxx, 11Dxx, 11B39, 28A80

This Springer imprint is published by the registered company Springer Nature Switzerland AG
The registered company address is: Gewerbestrasse 11, 6330 Cham, Switzerland

Preface

*For us, mathematical theorems and their proofs
are like gold nuggets to a prospector.*
THE AUTHORS

We encounter integer numbers daily, and they are literally everywhere around us. It is not possible to avoid them, ignore them, or to be indifferent to them. So let us take together a journey through the world of integers, to get acquainted with their fascinating and sometimes magic properties. We will discover some surprising connections between number theory and geometry (see, e.g., Chaps. 1, 4, 7, and 8). We shall see which laws are followed by integers. We will also show that number theory has many practical applications without which we could not imagine the modern technical world. It has a big influence on everything we do.

This treatise on integer numbers is based on our more than 70 works on elementary and algebraic number theory that we published between the years 2001 and 2021 mostly in prestigious international journals such as *Journal of Number Theory, Integers, The Fibonacci Quarterly, Discrete Mathematics, Journal of Integer Sequences, Proceedings of the American Mathematical Society, and Czechoslovak Mathematical Journal* (see, e.g., dml.cz). Most of our results were reported at many international conferences on number theory and also the regular Friday seminar *Current Problems in Numerical Analysis,* which takes place at the Institute of Mathematics of the Czech Academy of Sciences in Prague [426].

The book is intended for a general mathematical audience—especially for those who can appreciate the beauty of both abstract and applied mathematics. We only assume that the reader is familiar with the basic rules of arithmetic and has no problem with adjustments of algebraic formulas. Only very rarely it is necessary to understand some relationships from linear algebra or calculus. Most chapters can be read independently from one another. Some parts are quite simple, others more complicated. If some part is too difficult, there is no problem in skipping it.

At the end of the book, there are several tables and a fairly extensive bibliography to attract attention to some important works in number theory. For inspiration, there are also several links to websites, although we are well aware that they are not subjected to any review and change quite frequently. Newly defined terms are highlighted in italics in the text for the convenience of the reader. They can also be found in the Index.

In order to read the individual chapters, it is not necessary that the reader understands all theorems. There are 230 of them. Mathematicians formulate their ideas in the form of mathematical theorems that contain only what is relevant in the problem in question. We provide proofs of most statements so that one can verify their validity. For more complicated proofs, we only give a reference to the corresponding literature. The most beautiful feature of number theory is that the main ideas of proofs of every statement usually differ from each other. Formulations of mathematical theorems presented in this book often take only one or two lines, which makes it relatively easy to understand what a particular theorem says.

Mathematical theorems are valid forever. They are independent of position and time. Parliament does not decide about their validity by voting, nor the religious or political system in some country, nor does it depend on cultural customs. For example, the famous Pythagorean Theorem is valid on Earth as well as on the distant Andromeda galaxy M31, and it will also be valid after millions of years. Definitions of mathematical terms do not allow a double meaning. Also absolutely accurate formulations of mathematical problems do not allow more interpretations. The vague expressions we witness in daily life lead to a number of misunderstandings. Only a small percentage of our population is able to express their ideas accurately and perceive the beauty of mathematics. This was aptly stated by the well-known Hungarian mathematician Cornelius Lanczos (1893–1974) as follows:

> Most of the arts, as painting, sculpture, and music, have emotional appeal to the general public. This is because these arts can be experienced by one or more of our senses. Such is not true of the art of mathematics; this art can be appreciated only by mathematicians, and to become a mathematician requires a long period of intensive training. The community of mathematicians is similar to an imaginary community of musical composers whose only satisfaction is obtained by the interchange among themselves of the musical scores they compose.

There are many books on number theory. Let us mention, e.g., [6, 9, 28, 56, 82, 85, 91, 127, 132, 137, 138, 176, 235, 284, 291, 321–324, 327, 333, 344, 347, 350, 395, 413, 424]. However, our book contains some nonstandard topics. For instance, we will see how triangular numbers are related to the bell-work machinery of the Prague Astronomical Clock, what kind of mathematics is hidden in the traditional Chinese calendar, how the Fundamental Theorem of Arithmetic was used to design a message to extraterrestrial civilizations, how number theory is related to chaos, fractals, and graph theory. We will construct a $3 \times 3 \times 3$ magic cube containing only prime numbers. We will also get acquainted with the latest results from the hunt for the largest prime numbers and what prime numbers are good for. We shall present a number of their various real-life technical applications in completely different areas. We shall see how identification numbers of Czech organizations or bank account numbers are protected against possible errors with the help of prime numbers and error-detecting codes. We also discuss the so-called error-correcting codes, which automatically correct errors and we shall see how they are constructed. Further, we show how large prime numbers are used to transmit secret messages, to generate pseudorandom numbers, and what is their significance for digital signatures. We also demonstrate how congruences can be applied in scheduling sport tournaments. We give other examples, where number theory is useful and charming at the same time.

In a number of discussions, many researchers helped us to improve the content of this book, in particular, L'ubomíra Balková, Jan Brandts, Karel Břinda, Yann Bugeaud, Pavel Burda, Walter Carlip, Antonín Čejchan, Karl Dilcher, Petr Golan, Václav Holub, Jan Chleboun, František Katrnoška, Petr Klán, Martin Klazar, Michal Kliment, Kurt Koltko, Sergey Korotov, Pavel and Filip Křížek, František Kuřina, Florian Luca, Attila Mészáros, Karel Micka, Jaroslav Mlýnek, Vladimír Novotný, Pavla Pavlíková, Edita Pelantová, Jan Pernička, Štefan Porubský, Andrzej Schinzel, Bangwei She, Ladislav Skula, László Szalay, Bedřich Šofr, Jakub Šolc, Pavel Trojovský, Jiří Tůma, Tomáš Vejchodský, and Václav Vopravil. We really appreciate their help, and they deserve our great thanks. Furthermore, we are deeply grateful to Hana Bílková and Eva Ritterová for their technical assistance in the final typesetting of the manuscript.

Finally, we are indebted to Ms. Elena Griniari, Ms. Tooba Shafique, Mrs. Kay Stoll, Mr. Vijayakumar Selvaraj, and Ms. Sindhu Sundararajan from Springer-Verlag for helpful cooperation in the preparation of this book. Our great thanks go to the Springer Publishing House for its care in the graphical design of the book and to the referees for valuable suggestions. We are also grateful to our families for patience and understanding.

The work on this book was supported by RVO 67985840 of the Czech Republic. Chapter 12 was partly supported also by Grant No. 20-01074S of the Grant Agency of the Czech Republic. These supports are gratefully acknowledged.

Prague, Czech Republic   Michal Křížek
Washington, DC, USA   Lawrence Somer
Prague, Czech Republic   Alena Šolcová
May 2021

Contents

Glossary of Symbols

$\mathbb{N} = \{1, 2, 3, \dots\}$	Set of natural numbers
$\mathbb{Z} = \{\dots, -2, -1, 0, 1, 2, \dots\}$	Set of integer numbers
$\mathbb{P} = \{2, 3, 5, 7, 11, \dots\}$	Set of prime numbers
$\mathbb{Q} = \{\frac{m}{n} \mid m, n \in \mathbb{Z}, n \neq 0\}$	Set of rational numbers
$\mathbb{R}$	Set of real numbers
$\mathbb{C}$	Set of complex numbers
B_n	Bernoulli numbers
C_n	Cullen numbers
F_m	Fermat numbers
K_n	Fibonacci numbers
L_n	Lucas numbers
M_p	Mersenne numbers
$P_{k,n}$	Polygonal numbers
P_n	Pentagonal numbers
S_n	Square numbers
T_n	Triangular numbers
W_n	Woodall numbers
$(n_1, n_2, \dots, n_k)$	Greatest common divisor of $n_1, n_2, \dots, n_k$
$[n_1, n_2, \dots, n_k]$	Least common multiple $n_1, n_2, \dots, n_k$
$\{n_1, n_2, \dots, n_k\}$	Set of k numbers $n_1, n_2, \dots, n_k$ (the order does not matter)
$\langle n_1, n_2, \dots, n_k \rangle$	Ordered k-tuple of numbers $n_1, \dots, n_k$ (dependent on the order)
(a_i)	Sequence
$\lfloor a \rfloor$	Integer part of a real number a
$\lvert S \rvert$	Number of elements of the set S (also the absolute value)
$\approx$	Approximate equality
$n \equiv k \pmod{m}$	n is congruent to k modulo m
$n \not\equiv k \pmod{m}$	n is not congruent to k modulo m
$m \mid n$	m divides n
$m \nmid n$	m does divides n

$m^j \,\|\, n$	m^j exactly divides n for $1 < m \leq n$
$\max(m, n)$	Maximum of numbers m and n
$\min(m, n)$	Minimum of numbers m and n
sgn	Signum
$\text{ord}_d n$	Order of n modulo d
$n!$	Product $1 \cdot 2 \cdots n$, n-factorial
det	Determinant
$\log_b$	Logarithm to the base b
log	Natural logarithm
e	Euler number 2.718 281 828...
$G(n)$	Directed graph (digraph) with n vertices
$\tau(n)$	Number of all positive divisors of n
$\omega(n)$	Number of all different prime divisors of n
$\sigma(n)$	Sum of all positive divisors of n
$s(n)$	Sum of all positive divisors of n less than n
ϕ	Euler totient function
λ	Carmichael lambda function
$\left(\dfrac{a}{p}\right)$	Legendre symbol for an odd prime p
$\left(\dfrac{a}{m}\right)$	Jacobi symbol for an odd number m
$\dbinom{n}{k}$	Binomial coefficient n over k
Re z	Real part of a complex number z
Im z	Imaginary part of a complex number z
$\bar{z}$	Complex conjugate number to z
i	Imaginary unit
i, j, k	Integer indices (subscripts)
$\exists$	There exist(s)
$\forall$	For all
$\mathcal{O}(\cdot)$	Landau symbol: $f(\alpha) = \mathcal{O}(g(\alpha))$ if there exists $C > 0$ such that $|f(\alpha)| \leq C|g(\alpha)|$ for $\alpha \to 0$ or $\alpha \to \infty$
$\emptyset$	Empty set
π	Ludolph number 3.141 592 653...
$\pi(x)$	Number of primes less than or equal to x
$\prod$	Product
$\sum$	Sum
$\cap$	Intersection
$\cup$	Union
$\setminus$	Set subtraction
$\subset$	Subset
$\in$	Is element of
$\notin$	Is not element of
$\{x \in A; \mathcal{P}(x)\}$	Set of all elements x from A which possess property $\mathcal{P}(x)$

$f : A \rightarrow B$	Mapping (function) f from the set A to the set B
$\Rightarrow$	Implication
$\Leftrightarrow$	Equivalence
$:=$	Assignment
$\square$	Halmos symbol

Chapter 1
Divisibility and Congruence

1.1 Introduction

According to ancient Chinese philosophy,
all phenomena arose from the dusty axis,
which split into two complete opposites,
yin *and* **yang**.

In one of the oldest Chinese books *I-Ching (Book of Changes)*, which dates approximately from the 8th century BC, there is a picture (so-called hexagram) containing 8×8 boxes. Each box contains 6 broken or full horizontal lines (see Fig. 1.1). The broken line indicates the old Chinese principle *yin* and the full principle of *yang*, which are in opposition. *Yin* is associated with the Moon, humidity, darkness, Earth, woman, and passivity, *yang* on the other hand with the Sun, drought, light, heaven, man, and activity.

The prominent German mathematician Gottfried Wilhelm Leibniz (1646–1716) associated this hexagram with the discovery of a binary system. Considering zero instead of the broken line and one instead of the full line, the symbols in particular boxes from left to right (starting from the top line) can be interpreted as the numbers 0, 1, 2, 3, ... written in the binary system. The first number in the upper left corner is therefore zero, even though this notation was not used for operations with numbers in the 8th century BC. The last number in the lower right corner corresponds to 63, which is written as 111111 in the binary system. The use of zero nowadays seems completely natural, but its discovery and in particular, its symbolic representation signified great progress in mathematics over the entire world (cf. Fig. 1.2).

Although the ancient Chinese did not perform with the symbols *yin–yang* any arithmetic operations, we cannot deny they were the first to represent numbers by the binary system. The discovery of the binary system found practical application

M. Křížek et al., *From Great Discoveries in Number Theory to Applications*, https://doi.org/10.1007/978-3-030-83899-7_1

Fig. 1.1 The first depiction
of a binary system from the
8th century BC

Fig. 1.2 Symbols *yin–yang*
can be found on the current
South Korean flag

only in today's computer age, i.e. almost three thousand years later. Computers display and process all information (including numbers) in the binary system. This is the easiest way in electronic circuits of a computer to process data. Thus, the functioning of e-mail, scanners, copiers, digital cameras, compact disks CD and DVD, cell phones, and the worldwide network of Internet is actually based on the ancient Chinese principles of *yin* ($= 0$) and *yang* ($= 1$).

However, nature has discovered the binary (or if you wish quartic base 4) system in the course of evolution more than three billion years ago. On the double helix of deoxyribonucleic acid (DNA), which is contained in each cell, there are four bases: adenine A, cytosine C, guanine G, and thymine T. If we replace them by the pairs 00, 01, 10, and 11, then each strand of DNA will correspond to a sequence of zeros and ones that actually represents genetic information recorded in the binary system. Note that the genetic code is nearly universal for animals and plants [290] with a few exceptions, see [193].

By means of the replication $R(A) = T$, $R(C) = G$, $R(G) = C$, and $R(T) = A$ one gets the second strand of DNA, thereby forming a double helix (see [162]). Nature thus actually discovered a simple logical operation: negation. For example,

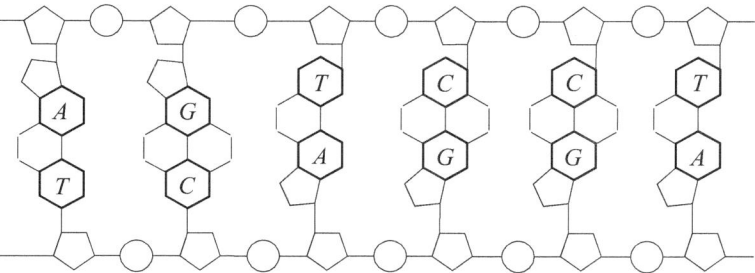

Fig. 1.3 Schematic illustration of DNA structure. Nucleotides are denoted by A, C, G, and T. Molecules of deoxyribose sugar (marked by pentagons) and phosphoric acid (marked by circles) are connected by strong covalent bonds. In this way, they protect genetic information against damage. The whole DNA molecule is actually twisted into a double helix

the nucleotides $\ldots AGTCCT\ldots$ on the upper strand (see Fig. 1.3) corresponding to the sequence of bits

$$\ldots 001011010111 \ldots$$

pass during DNA replication on $\ldots TCAGGA\ldots$ corresponding to the complementary sequence

$$\ldots 110100101000 \ldots,$$

where $A = 00$, $C = 01$, $G = 10$, and $T = 11$. The human genome in one cell is 750 MB (1 Byte = 8 bits).

Putting $A = 0$, $C = 1$, $G = 2$, and $T = 3$, the entire genetic information is transferred to the base 4 system. The sum of these numerical values is for all permissible pairs $A - T$, $C - G$, $G - C$, and $T - A$ always equal to 3. According to George Gamow, number theory could be used to elucidate the functioning of genes.

Here allow us a small detour. When designing the model for the structure of DNA in 1953, the fact that one of the authors Francis H. C. Crick was a physicist with abstract mathematical-physical thinking, played an important role. Already in 1950 Crick realized that if we connect the same molecules exactly in the same way, they will lie on a space helix (or especially on a circle or a straight line), see [73]. He received the Nobel Prize together with James D. Watson for the discovery of the double helix of DNA. Similarly, the founder of genetics Johann Gregor Mendel was a mathematician. His precise work with statistical data from crossing peas allowed him to discover laws of heredity (see [163, 269]). We further note that there are deep connections between the structure and function of DNA and topology and other areas of mathematics (see Benham et al. [29]).

In antiquity and the Middle Ages, number systems of different bases were used. For example, it is documented that the ancient Babylonians used a system with a base of 60. The origin of the words dozen (12) and pile (60) also illustrates that not only the decimal system was not used.

Recall that a positive integer n in a system of base b can be uniquely written in the form

$$n = c_k b^k + c_{k-1} b^{k-1} + \cdots + c_1 b + c_0,$$

where $c_i \in \{0, 1, \ldots, b - 1\}$ are its digits, i.e., the set of digits has exactly b elements and $c_k \neq 0$. (The existence of such an expression can be proved by induction and the uniqueness can be obtained similarly as in Theorem 2.2.) If $b \geq 10$, then letters can play the role of other digits. For instance, in the hexadecimal system the following digits are used: 0, 1, 2, ..., 9, A, B, C, D, E, F. In this book, however, we will mostly use the decimal or binary system.

We set out on a journey to find secrets of integer numbers and will introduce some of their literally magical and unexpected features. In particular, we want to convey that numbers are not just for fun, but that number theory also has a huge number of practical applications. In the past, Euclid, Fermat, Euler, and many others actually only "played" with numbers, proved various statements about them, without knowing what a huge amount of their results would appear and what practical use their mathematical theorems would have. Moreover, the same result can often be used in a number of completely different situations. In this book, we will see this, for example, in the Chinese Remainder Theorem or Fermat's Little Theorem. Most statements from number theory wait for their practical use and many of them will never find applications. But this does not diminish their beauty.

Humanity has been dealing with investigation of prime numbers and their sometimes surprising properties for several millennia. But not until the 20th century was it discovered that prime numbers also have a number of useful applications. For example, since 1986 birth numbers ($\approx$ social security numbers) in the Czech Republic are formed to be divisible by the prime number 11. This is because the computer immediately detects an error as soon as you type a given birth number wrongly in one of its digits. If we are wrong in more than one digit, then there is a large probability that the computer is also able to detect an error. This is one example of a so-called error-detecting code. A somewhat more complicated code based on the prime 11 protects against a possible error appearing in bank account numbers, identification numbers of organizations, ISSN numbers of journals, and ISBN numbers of publications.

Larger primes having more than a hundred of digits, are used in modern cryptographic systems with public encryption key (for instance in the RSA method for the transmission of secret messages). Also, a digital signature is based on large primes. Efficient pseudo-random generators or algorithms for very fast multiplication of large numbers can be constructed using prime numbers. Prime numbers have a number of applications in signal analysis as well as in image processing using various theoretical transformations, such as when filtering data obtained from radars, sonar, modems, and radiotelescopes.

With the publication of the fundamental work *Disquisitiones arithmeticae* [118] by Gauss at the beginning of the 19th century, number theory established itself as a systematic mathematical discipline whose main subject is the study of properties of integers. Then in the 20th century it has found wide application in various fields of human activities.

It is used in a number of software products, in information security, in data compression, mathematical genetics, in physics, astronomy, robotics, and crystallography. Bar codes are commonly encountered in the commercial sphere today thanks to the enormous advances of optoelectronics. Their introduction in stores increases the speed of sales up to 400%. Error-correcting codes are used in data transmission from interplanetary probes or on railways to ensure their reliable operation. Also the functioning of digital cameras, telecommunication satellites, and music players is based on number theory. These modern achievements of civilization would never have been possible without number theory. Although they were not here in the 19th century, we would hardly give them up at present. Unfortunately, the general public does not realize this and considers them to be obvious. It is often underestimated how much human ingenuity, intellectual effort, and mathematical results are hidden in these technical equipments. For example, computer tomographs would not work without complex numbers. The reason is that the fast Fourier transform relies on complex arithmetic and it is needed for a fast calculation of the Radon inverse transformation in real time (see [185]).

In the following chapters, we will focus on geometric imagination, which can to a large extent facilitate the understanding of some algebraic statements, relations and basic concepts of number theory, such as the famous algebraic identities $(a \pm b)^2 = a^2 \pm 2ab + b^2$, $a^2 - b^2 = (a + b)(a - b)$, the Pythagorean Theorem (see Fig. 2.1) and the relation for the sum of an arithmetic sequence (see Fig. 1.4). In total, there are about 70 pictures in this book and it contains notes on the historical background of some concepts and methods.

Finally, let us mention that in solving equations we must exactly specify the set of admissible solutions. For example, the equation

Fig. 1.4 Geometric interpretation of the well-known Gaussian relation for the sum of the first n members of an arithmetic sequence $a + (a + d) + (a + 2d) + \cdots + b = \frac{1}{2}n(a + b)$, where d is the difference of two neighboring terms and $b = a + (n - 1)d$

$$x^2 + y^2 = 2$$

has no solution in the set of even numbers. It has only one solution $(x, y) = (1, 1)$ in the set of natural numbers $\mathbb{N}$ and exactly four solutions $(\pm 1, \pm 1)$ in the set of integer numbers $\mathbb{Z}$. There are countably many solutions in the set of rational numbers $\mathbb{Q}$, e.g., $(\frac{1}{5}, \frac{7}{5})$, $(\frac{7}{13}, \frac{17}{13})$, $(\frac{7}{17}, \frac{23}{17})$, and there are uncountably many solutions in the set of real numbers $\mathbb{R}$ or complex numbers $\mathbb{C}$.

1.2 Natural Numbers

From ancient times people used the numbers $1, 2, 3, \ldots$ to express the number of some objects. The oldest use of zero was recorded in India. For a long time, zero was not even considered to be a number. Moreover, at present historians still do not have a year zero (but it is used by astronomers).

Sometimes we encounter the question of whether zero is or is not a natural number. Unfortunately, it is not possible to give a clear answer to this question of the YES/NO type, since whether or not we consider zero to be a natural number is a matter of definition. It is advisable to include zero in the set of natural numbers, for example, when determining the number of elements of finite sets, because the number of elements of the empty set is zero.

On the other hand, there are good reasons why it is sometimes advantageous not to include zero in the set of natural numbers. This is, for example, to avoid division by zero or when raising natural numbers to a natural power. In particular, the symbol 0^0 cannot be unambiguously assigned to one value that would naturally correspond to standard arithmetical operations with real numbers. For example, for $n = 1, 2, \ldots$ we have $0^n = 0$, while $n^0 = 1$. Archimedes' axiom presented below could not be applied if 0 would be a natural number. It is also not possible to define reasonably the least common multiple of, for example, the numbers 0 and 3, as we shall see in Sect. 1.4. Therefore, more often zero is not considered to be a natural number.

The *set of natural numbers (positive integers)* will be denoted by

$$\mathbb{N} = \{1, 2, 3, \ldots\}.$$

It took a long time for mathematicians to figure out how in fact, natural numbers should be introduced. Among several options, the following four axioms formulated around 1891 by the Italian mathematician Giuseppe Peano (1858–1939) were defined. They use a special function "successor", truthfully characterize the set of natural numbers and are called *Peano's axioms* after him:

(A1)   *There exists a unique natural number that is not a successor of any natural numbers.* We will denote this number by the symbol 1.

(A2)   *Each natural number has exactly one successor.*

(A3) *Each natural number is a successor of at most one natural number.*
(A4) *Any set that contains the natural number 1 and for each natural number also*
      contains its successor, is the set of natural numbers.

The main idea of the principle of mathematical induction is based on these axioms. If we want to prove that some property $V(n)$ holds for all natural numbers n, then first we prove that $V(n)$ is valid for $n = 1$. Then we prove that if the property $V(n)$ holds for some natural number n, then $V(n + 1)$ also holds for the successor $n + 1$ of the number n.

The *set of integer numbers* is denoted by

$$\mathbb{Z} = \{\ldots, -2, -1, 0, 1, 2, \ldots\}.$$

This set therefore consists of natural numbers, numbers opposite to them (i.e. with minus sign), and zero. It is closed under the addition and multiplication operations, i.e., for any $m, n \in \mathbb{Z}$ we have $m + n \in \mathbb{Z}$ and $m \cdot n \in \mathbb{Z}$. The following relations $=$, $<$, $>$, $\leq$, and $\geq$ are also established on the set $\mathbb{Z}$.

Convention. Integer numbers in this book will mostly denoted by i, j, k, ℓ, m, n, p, q, r, s, $t, \ldots$, unless otherwise specified.

The set of natural numbers $\mathbb{N}$ is *well ordered,* which means that an arbitrary nonempty subset has a least element. The sets of integers $\mathbb{Z}$, rational numbers $\mathbb{Q}$, and real numbers $\mathbb{R}$ do not have a similar property.

The fact that the set of natural numbers is well ordered is equivalent to the principle of mathematical induction (see e.g. [395, p. 40]). It is actually an axiom, i.e., a statement which is accepted without proof, because it does not contradict our intuition.

Already in antiquity, Archimedes (287–212 BC) realized that the set of natural numbers is well ordered.

Archimedes' axiom. *For any natural numbers j and k there exists a natural number n such that $nj \geq k$.*

Archimedes' axiom can be proved. So it is a mathematical theorem, but for historical reasons, it is called an axiom. If it were not true, then there would exist $j, k \in \mathbb{N}$ such that $nj < k$ for each $n \in \mathbb{N}$. Since the set $\mathbb{N}$ is well ordered, there exists a smallest element $k - mj$ of the subset $M = \{k - nj \; ; \; n \in \mathbb{N}\} \subset \mathbb{N}$ However, the element $k - (m + 1)j$ is also in M and satisfies

$$k - (m + 1)j = (k - mj) - j < k - mj,$$

which contradicts the minimality of the element $k - mj$ from the set M.

Archimedes' axiom has a nice geometric interpretation. It says, in fact, how many line segments of length j cover a line of length k.

1.3   Simple Criteria of Divisibility

We say that d *divides* (without remainder) a natural number n, if there exists $k \in \mathbb{N}$ such that $n = d \cdot k$. In this case we shall write

$$d \mid n,$$

for instance, $3 \mid 6$. The number d is called a *divisor* of the number n and the numbers 1 and n are called *trivial divisors* of n. If $1 < d < n$, then d is called a *nontrivial divisor*, and if $d < n$, then d is called a *proper divisor* of n. If m does not divide n, we shall write $m \nmid n$, for instance, $5 \nmid 6$. Similar definitions can also be introduced for integer numbers when $m \neq 0$. An integer divisible by 2 is called *even*, otherwise *odd*.

Theorem 1.1 *A natural number n written in the decimal system is divisible by*

(a) *two, if its last digit is even,*
(b) *three, if the sum of all its digits is divisible by* 3,
(c) *four, if the number formed by the last two digits of n is divisible by* 4,
(d) *five, if its last digit is 0 or* 5,
(e) *six, if it is even and divisible by* 3,
(f) *seven, if twice the number of hundreds increased by the number formed by the last two digits is divisible by* 7,
(g) *eight, if the number formed by the last three digits of n is divisible by* 8,
(h) *nine, if the sum of all its digits is divisible by* 9,
(i) *ten, if its last digit is* 0.

Proof Let n be an arbitrary natural number. In the decimal system it can be uniquely written in the form

$$n = c_k 10^k + \cdots + c_2 10^2 + c_1 10 + c_0, \tag{1.1}$$

where its digits $c_k, \ldots, c_2, c_1, c_0$ are from the set $\{0, 1, 2, \ldots, 9\}$ and $c_k \neq 0$. From (1.1) we immediately get (a), (c), (d), (g), and (i).

Denote by s the sum of all digits of the number n, i.e.,

$$s = c_k + \cdots + c_2 + c_1 + c_0.$$

Then

$$n - s = c_k(10^k - 1) + \cdots + c_2 99 + c_1 9,$$

where each term on the right-hand side is divisible by nine. That is why, n is divisible by three (respectively nine) exactly when s is divisible by three (respectively nine). Hence, (b) and (h) hold.

We observe that (e) follows immediately from (a) and (b). It remains to prove (f). The next proof of this criterion comes from Václav Holub. Twice the number of hundreds of n increased by the number formed by the last two digits is equal to

$$m = 2c_k 10^{k-2} + \cdots + 2c_2 + 10c_1 + c_0.$$

By (1.1) we see that the difference

$$n - m = 98c_k 10^{k-2} + \cdots + 98c_2$$

is divisible by seven, since $7 \mid 98$. Now if 7 divides m, then 7 also divides the sum $m + (n - m) = n$.   □

Example By Theorem 1.1 we have $7 \mid 1239$, since 7 divides $2 \cdot 12 + 39 = 63$.

Example For larger numbers it is usually necessary to apply an appropriate rule several times. For instance,

$$3 \mid 188887777788885,$$

whose sum of digits 105, which is divisible by 3, since $3 \mid (1 + 5)$.

We will deal with divisibility by 11 in Theorem 11.1. Let us further introduce rules for divisibility by the numbers 13, 17, and 19.

Remark Let n be a given natural number, let k be the number of tens in n, and let $c_0 \in \{0, 1, \ldots, 9\}$ be the last digit of n. Then

$$n = 10k + c_0.$$

A natural number n is divisible by 13 if four times the last digit added to the number of tens is divisible by 13. To see this we set $m = k + 4c_0$. Then

$$13 \mid n = 10k + c_0 = 10(m - 4c_0) + c_0 = 10m - 39c_0 \Leftrightarrow 13 \mid 10m \Leftrightarrow 13 \mid m.$$

For instance, $13 \mid 507$, since 13 divides $4 \cdot 7 + 50 = 78$.

A natural number n is divisible by 17, if five times the last digit subtracted from the number of tens is divisible by 17. To prove this we put $m = k - 5c_0$. Then

$$17 \mid n = 10k + c_0 = 10(m + 5c_0) + c_0 = 10m + 51c_0 \Leftrightarrow 17 \mid 10m \Leftrightarrow 17 \mid m.$$

For example, $17 \mid 357$, since 17 divides $35 - 5 \cdot 7 = 0$.

A natural number n is divisible by 19, if double the last digit added to the number of tens is divisible by 19. We set $m = k + 2c_0$. Then

$$19 \mid n = 10k + c_0 = 10(m - 2c_0) + c_0 = 10m - 19c_0 \Leftrightarrow 19 \mid 10m \Leftrightarrow 19 \mid m.$$

For instance, $19 \mid 1026$, since 19 divides $2 \cdot 6 + 102 = 114$ and 19 divides $11 + 2 \cdot 4 = 19$.

The divisibility tests by 13 and 19 are applications of a theorem by Carl Fredrik Liljevalch in 1838 (see [176, p. 283]).

Recall that the *binomial coefficient* $\binom{n}{m}$ (read n over m) is defined by

$$\binom{n}{m} = \frac{n!}{m!(n-m)!} \quad \text{for integers } n \geq m \geq 0,$$

where

$$n! = 1 \cdot 2 \cdot \ldots \cdot (n-1) \cdot n \quad \text{for } n \in \mathbb{N}, \quad 0! = 1.$$

The symbol $n!$ is called n *factorial*.

Remark Let $k = n - m$ be a natural number. Then we have

$$k!\binom{n}{m} = k!\frac{n!}{m!k!} = \frac{n!}{m!} = (m+1)(m+2)\cdots(m+k).$$

Thus we see that the product of k consecutive numbers on the right-hand side is always divisible by $k!$. Moreover, it can be proved that $(m+1)(m+2)\cdots(m+k)$ for $k \geq 2$ and $m \geq 2$ is never equal to the power ℓ^j for some $j \geq 2$ and $\ell \geq 2$ (see Erdős and Selfridge [105]).

1.4 The Least Common Multiple and the Greatest Common Divisor

Let m and n be arbitrary natural numbers. Denote by $M \subset \mathbb{N}$ a subset of all common multiples of m and n. The set M is clearly nonempty, because it contains e.g. the product mn. Since $\mathbb{N}$ is well ordered, M must contain a smallest element, which we denote by $[m, n]$ and call the *least common multiple* of the numbers $m, n \in \mathbb{N}$. Thus, it is the smallest natural number divisible by both m and n.

Similarly, the *greatest common divisor* of two integer numbers m and n, which are not zero at the same time, is the largest integer that divides both m and n. The greatest common divisor of numbers m and n will be denoted by (m, n).

A basic property of the largest common divisor and least common multiple is obviously

$$(m, n) = (n, m), \quad [m, n] = [n, m].$$

For $k, m, n \in \mathbb{N}$ the following distributive properties hold

$$[k, (m, n)] = ([k, m], [k, n]) \quad \text{and} \quad (k, [m, n]) = [(k, m), (k, n)].$$

Theorem 1.2 *For any natural numbers m and n we have*

$$mn = (m, n)[m, n]. \tag{1.2}$$

Proof Denote by $d \geq 1$ an arbitrary common divisor of m and n. Then $\frac{m}{d}$ and $\frac{n}{d}$ are natural numbers, $\frac{m}{d}n$ is an integer multiple of the number n, and $\frac{n}{d}m$ is an integer multiple of the number m. Therefore, $\frac{mn}{d}$ is a common multiple of the numbers m and n. Now if d is the greatest common divisor of the numbers m and n, then $\frac{mn}{d}$ has to be the least common multiple m and n. □

Example For $m = 18$ and $n = 27$ we have $(m, n) = 9$, $[m, n] = 54$, and hence $18 \cdot 27 = 9 \cdot 54$.

The greatest common divisor and the least common multiple of more than two numbers can be defined by induction similarly as for two numbers. For $k > 2$ and integer numbers $n_1, \ldots, n_k$ we set

$$(n_1, \ldots, n_{k-1}, n_k) = \big((n_1, \ldots, n_{k-1}), n_k\big) \quad \text{if } n_1 \neq 0,$$
$$[n_1, \ldots, n_{k-1}, n_k] = \big[[n_1, \ldots, n_{k-1}], n_k\big] \quad \text{if } n_1 n_2 \cdots n_k \neq 0.$$

1.5   Coprime Numbers

Natural numbers m and n are called *coprime,* if $(m, n) = 1$. An interesting real-life technical application of coprime numbers is depicted in Fig. 1.5. It shows two gear ratios. In the left part of the figure, the larger wheel has 20 teeth and the smaller one 10 teeth. If there is one tooth slightly damaged on the larger wheel (it is marked with a dot in Fig. 1.5), then it fits into exactly the same gap in the smaller wheel after each turn of the larger wheel. Just at this gap, the smaller wheel will be very quickly worn out. In the right part of Fig. 1.5 we see two wheels with 25 and 12 teeth. Since $(25, 12) = 1$, there will be completely uniform wear. Let us still note that the ratio of the teeth is actually almost the same in both cases: 2 and 2.083.

Here is another practical use of coprime numbers. The fixed part of the caliper is equipped with a scale in which each centimeter is divided into 10 mm. On the moving part, the so-called vernier is divided into 10 equally long pieces, which together have 9 mm (see Fig. 1.6). The fact that 9 and 10 are coprime allows us to determine the dimensions of small objects with precision to the nearest tenth of a millimeter. When measuring, we determine which line of the vernier merges with some line on the millimeter scale of the caliper. So many tenths of a millimeter is then added to the measured millimeters (i.e. to their largest integer value). If we were to choose instead of 9 mm another length that divides 10 mm, then several lines could merge so we would not know which data applies.

Fig. 1.5 For the number of teeth on the left and right gears we have $(20, 10) = 10$ and $(25, 12) = 1$, respectively

Fig. 1.6 Vernier on a caliper

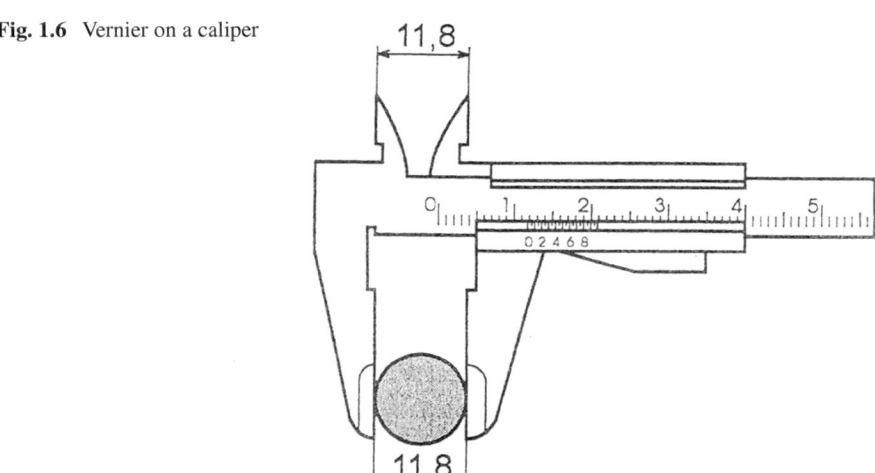

The author of this elegant idea is a Portuguese royal mathematician and cosmographer Pedro Nunes (1502–1578), who first used it for accurate angle measurements. At present, a vernier-like device is used also in micrometers and is called the *nonius* in his honor.

1.6 Euclidean Algorithm

To calculate the greatest common divisor (m, n) of two large natural numbers $m \geq n$ the well-known *Euclidean algorithm* is often used. It can be briefly characterized as follows:

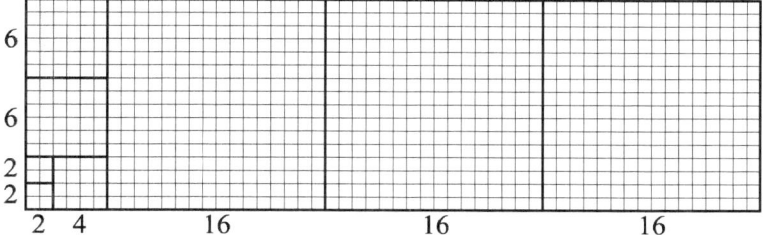

Fig. 1.7 Geometric illustration of the reduction of input data during the use of the Euclidean algorithm for calculating the greatest common divisor (54, 16)

If n divides m, then $(m, n) = n$, otherwise we have

$$(m, n) = (n, z),$$

where $z \geq 1$ is the remainder when dividing the number m by the number n. Since $z < m$, larger problem is thus converted to a smaller one. The next steps of the algorithm then proceed similarly. The original problem is thus reduced to smaller and smaller parts until we get the remainder 0.

For instance, if $m = 54$ and $n = 16$, then by the Euclidean algorithm we get

$$(54, 16) = (16, 6) = (6, 4) = (4, 2) = 2.$$

Now let us imagine that we have a squared paper with dimensions 54×16 (see Fig. 1.7). From this we will gradually cut off squares as large as possible and we will perform this as long as possible (see [188]), i.e., in the first step we cut off 3 squares 16×16, in the next step we cut 2 squares 6×6, etc. The length of the side of the square that we have left, is the result of the Euclidean algorithm, i.e. the largest common divisor of numbers 54 and 16.

If the numbers m and n are coprime, the Euclidean algorithm ends with the least possible square 1×1. For example, two consecutive natural numbers are always coprime.

To calculate the least common multiple of $[m, n]$, it is also useful to apply the Euclidean algorithm first, because $(m, n) \leq [m, n]$, and then use the relation (1.2). We return to the Euclidean algorithm in Theorem 7.7.

1.7 Linear Diophantine Equations

The name Diophantine equation is derived from the name of the Greek mathematician *Diophantus*, who lived in Alexandria in the 3rd century AD and dealt with solving various problems in number theory. Diophantine equations are equations with integer coefficients, whose solution is sought in integers. In solving systems of Diophantine equations we usually have more unknowns than equations. In this section we shall deal only with one linear Diophantine equation with two integer unknowns.

Theorem 1.3 *Let $k = (m, n)$ for some integers m and n, which are not simultaneously zero. Then there exist integers x and y such that*

$$mx + ny = k. \tag{1.3}$$

Proof Let S be a set of all integers of the form $ma + nb$, where a and b are integers. The number m or n is not zero, and thus the set S contains nonzero integers. Since $t = ma + nb$ is in S, the number $-t = m(-a) + n(-b)$ is also in S. Hence, S contains natural numbers. Since $\mathbb{N}$ is well ordered (see Sect. 1.2), there exists a smallest natural number d in S of the form $d = mx + ny$. We claim that $d = (m, n)$.

First we show that d is a common divisor of m and n. Let $u = ma_0 + nb_0$ be an arbitrary element in S. By division we find that $u = qd + r$, where $0 \le r < d$ is the remainder. Thus we have

$$ma_0 + nb_0 = q(mx + ny) + r,$$

i.e.,

$$r = m(a_0 - qx) + n(b_0 - qy)$$

and $r \in S$. Since $r \ge 0$ and $r < d$, it follows that $r = 0$ due to the choice of d. Therefore, d divides u for all $u \in S$. However, $m = m \cdot 1 + n \cdot 0 \in S$ and $n = m \cdot 0 + n \cdot 1 \in S$, which means that d divides both m and n.

Finally, let e be an arbitrary common divisor of m and n. Then e divides $mx + ny = d$, and thus $d = (m, n) = k$. □

Theorem 1.3 has a very nice geometric interpretation (see Burton [56, p. 22]). Each line $mx + ny = k$ passes through the grid points (x, y), which are solutions of the linear Diophantine equation (see Fig. 1.8 for $m = 2$, $n = -3$, and $k = 1$). Relation (1.3) is called *Bézout's identity*.

Example Let us show how to solve the following Diophantine equation

$$8x - 27y = 1$$

by a method similar to the Euclidean algorithm. Since $(8, 27) = 1$, this equation has by Theorem 1.3 a solution and we have

$$x = 3y + \frac{3y + 1}{8}.$$

Fig. 1.8 A straight line corresponding to the linear Diophantine equation $2x - 3y = 1$ passes through points with coordinates $\ldots, (-1, -1), (2, 1), (5, 3),$ $(8, 5), (11, 7), \ldots$

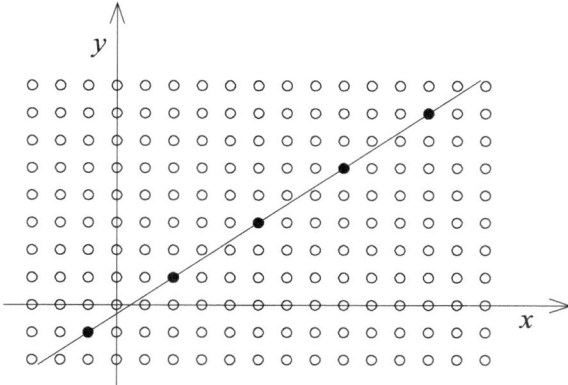

To get an integer solution, $3y + 1$ must be a multiple of 8, i.e., there exists an integer v such that

$$3y + 1 = 8v, \quad \text{and thus } y = 2v + \frac{2v - 1}{3}.$$

Hence, we can choose $v = 2$ and by backward substitution we get that the pair $y = 5$ and $x = 17$ is a solution. For $v = 5, 8, 11, \ldots$ and $v = -1, -4, -7, \ldots$ we obtain further pairs of solutions.

1.8 Congruence

In this section we will discuss the concept of congruence, introduced by the German mathematician Carl Friedrich Gauss. He used it for various calculations, such as which day falls on Easter Sunday (see Remark below). Congruences have many other practical applications in cryptography, in astronomy when creating calendars (see [56, p. 122]), in generating pseudorandom numbers, etc., as we shall see in Sects. 11.2–11.5.

Let n, z be integers and $m \in \mathbb{N}$. Then we say that n *is congruent to z modulo m* and write

$$n \equiv z \quad (\text{mod } m),$$

if $n - z$ is divisible by m. The number m is called the *modulus*.

The notion congruence modulo 12 can be clearly demonstrated on the dial of a classical clock.

We now derive some practical rules for calculating congruences. Obviously $a \equiv a$ (mod m) for any integer a. If $a \equiv b$ (mod m), then for arbitrary integers a, b, c and $k \geq 0$ we easily find that

$$b \equiv a \quad (\text{mod } m),$$
$$a \pm c \equiv b \pm c \quad (\text{mod } m),$$
$$ac \equiv bc \quad (\text{mod } m),$$
$$a^k \equiv b^k \quad (\text{mod } m), \tag{1.4}$$

where the last congruence follows from the equality

$$a^k - b^k = (a - b)\left(a^{k-1} + a^{k-2}b + \cdots + b^{k-1}\right), \quad k > 1.$$

From the above relationships, it is clear that the relation "$\equiv$" modulo m is reflexive and symmetric. Since transitivity holds as well, it is actually an equivalence relation on the set of integers.

If $(c, m) = 1$, then the congruence $ac \equiv bc$ (mod m) can be canceled by c, i.e., $a \equiv b$ (mod m). If $a \equiv b$ (mod m) and $c \equiv d$ (mod m), then obviously

$$a + c \equiv b + d \quad (\text{mod } m),$$
$$a - c \equiv b - d \quad (\text{mod } m),$$

and moreover,

$$ac \equiv bd \quad (\text{mod } m). \tag{1.5}$$

For $a = b + im$ and $c = d + jm$ it indeed holds that $ac = bd + (jb + id + ijm)m$, and thus congruence (1.5) is satisfied. Moreover, by (1.4) we also get that

$$f(a) \equiv f(b) \quad (\text{mod } m)$$

for an arbitrary polynomial f with integer coefficients.

Remark The Gaussian algorithm on which day of the year y is Easter Sunday proceeds as follows:

For the period 1900–2099 we set $m = 24$ and $n = 5$. Let a, b, c, d, e be the smallest nonnegative numbers satisfying the congruences

$$a \equiv y \quad (\text{mod } 19),$$
$$b \equiv y \quad (\text{mod } 4),$$
$$c \equiv y \quad (\text{mod } 7),$$
$$d \equiv (m + 19a) \quad (\text{mod } 30),$$
$$e \equiv (n + 2b + 4c + 6d) \quad (\text{mod } 7).$$

Here $a + 1$ is called *golden number* (the ordinal number of the year in the Metonic cycle having a period of 19 years in which there are 235 lunations). If $d + e < 10$ then the Easter Sunday will be on the $(22 + d + e)$th March, if $d + e = 35$ it will be on the $(d + e - 16)$th April and otherwise it will be on the $(d + e - 9)$th April.

However, this algorithms has two exceptions, when $d + e = 34$ and $a > 10$. In 1954 and 2049, the Easter Sunday is not on the 25th April, but one week earlier (for details see [427]).

Let us introduce some other interesting congruences.

Example It is easy to verify that

$$n^5 \equiv n \pmod{10} \quad \forall\, n \in \mathbb{N},$$

i.e., the fifth power of any natural number n ends at the same digit as n itself.

Example Choose two digits $a, b \in \{1, 2, \ldots, 9\}$ such that $a + b = 10$. Then for any $n \in \mathbb{N}$ the numbers a^{2n} and b^{2n} have the same last digit. This follows from the congruence $b \equiv -a \pmod{10}$, since by (1.4) we have

$$b^{2n} \equiv (-a)^{2n} \equiv a^{2n} \pmod{10},$$

which proves the result.

Example Choose two digits $a, b \in \{1, 2, \ldots, 9\}$ such that $a + b = 10$. Then for any $n \in \mathbb{N}$ the last digit of $a^{2n+1} + b^{2n+1}$ is zero. This follows from the congruence $b \equiv -a \pmod{10}$, since by (1.4) we obtain

$$b^{2n+1} \equiv (-a)^{2n+1} = -a^{2n+1} \pmod{10},$$

and thus $a^{2n+1} + b^{2n+1} \equiv 0 \pmod{10}$. From this we observe, e.g., that the cube of a number that ends in $0, 1, 2, 3, 4, 5, 6, 7, 8, 9$, will end in $0, 1, 8, 7, 4, 5, 6, 3, 2, 9$, respectively.

1.9 The Chinese Remainder Theorem

The ancient Chinese mathematician Sun Zi around the 4th–5th century formulated in his book on arithmetic the following problem:

Let x denote the number of certain objects. If we count them in triplets, there will be two left, if we count them in fives, there will be three left, and finally if we count them by seven, there will be two left. How much is x?

This problems can be obviously converted into a system of three congruences for an unknown value x:

$$x \equiv 2 \pmod{3}, \tag{1.6}$$

$$x \equiv 3 \pmod{5}, \tag{1.7}$$

$$x \equiv 2 \pmod{7}. \tag{1.8}$$

The problem (1.6)–(1.8) can also be formulated as follows:

I have a certain number of yuans. If I bought a hen in threes yuans, I would have
2 yuans left, if I bought rabbits in five yuans, I would have 3 yuans left, and if I
bought turkeys in seven yuans, I would have 2 yuans left. How many yuans do I
have? Therefore, Chinese call a similar type of problem the "money problem". In
his book of 1275, Yang Hui gives five other similar examples (see Martzloff [260,
p. 311]).

Let us deal with a simpler problem:

*There is a staircase in front of the house. If I go up three steps, two steps will left.
If I go up five steps, three stairs will left. How many stairs does the staircase have?*

We see that this problem can be written using congruences (1.6) and (1.7), i.e.,

$$x \equiv 2 \quad (\text{mod } 3),$$
$$x \equiv 3 \quad (\text{mod } 5),$$

where x is the number of stairs. Subtracting the corresponding equations $x = 3i + 2$
and $x = 5j + 3$, we get $0 = 3i - 5j - 1$, i.e., $j = \frac{3i-1}{5}$ and the choice $i = 2 + 5n$
yields the solution $x = 8 + 15n$ for $n \geq 0$, since the staircase cannot have a negative
number of stairs.

Figure 1.9 shows a square whose side length $x = 8$ we are looking for. It corresponds to the smallest solution $n = 0$. We see that the square is subdivided into
several rectangles. At the bottom left is a rectangle whose side lengths correspond to
the residues 2 and 3. In the upper right part there are two rectangles whose side lengths
correspond to the moduli 3 and 5. Dimensions of the remaining rectangles are com-
bined from residues and moduli. The number of segments marked on the horizontal
and vertical sides of the square are contained in the first and second congruences,
respectively.

Now we introduce one of the most used theorems in number theory, which allows
us to solve problems of these types that lead to a systems of linear congruences of
one unknown (see e.g. (1.6)–(1.8)). It is very similar to the Euclidean algorithm. We

Fig. 1.9 Geometric
interpretation of problem
(1.6)–(1.7)

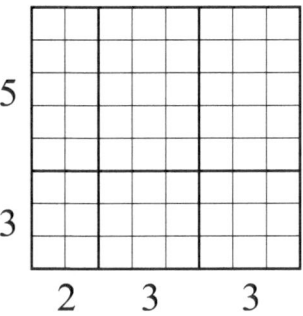

will see its usefulness in many other places of this book. Because its proof will be constructive, the corresponding algorithm can be used to solve a given system of congruences.

Theorem 1.4 (Chinese Remainder Theorem) *Let $m_1, m_2, \ldots, m_k$ be pairwise coprime natural numbers. Then for the system of simultaneous congruences*

$$
\begin{aligned}
x &\equiv r_1 \pmod{m_1}, \\
x &\equiv r_2 \pmod{m_2}, \\
&\vdots \\
x &\equiv r_k \pmod{m_k},
\end{aligned}
\tag{1.9}
$$

where the r_i are integers, there exists one and only one solution x modulo m, where

$$ m = m_1 m_2 \cdots m_k. $$

Proof First we prove the existence of a solution x. Define n_i by the equalities

$$ m = m_1 n_1 = m_2 n_2 = \cdots = m_k n_k. \tag{1.10} $$

Since m_i and n_i are coprime, there exist by Theorem 1.3 integers $y_i, i = 1, 2, \ldots, k$, such that
$$ n_i y_i \equiv 1 \pmod{m_i}. \tag{1.11} $$

We claim that the general solution (1.9) has the form

$$ x \equiv r_1 n_1 y_1 + r_2 n_2 y_2 + \cdots + r_k n_k y_k \pmod{m}. \tag{1.12} $$

To check that such a number x solves system (1.9), we choose $i \in \{1, \ldots, k\}$. From (1.10) we see that all terms, except for the ith term on the right-hand side of (1.12), contain the factor m_i, and thus

$$ x \equiv r_1 n_1 y_1 + r_2 n_2 y_2 + \cdots + r_k n_k y_k \equiv r_i n_i y_i \equiv r_i \pmod{m_i}, $$

where the last equivalence is obtained by multiplication of the congruence (1.11) by r_i.

In order to prove uniqueness, assume that x_1 and x_2 are two solutions of system (1.9). Then $x_1 \equiv x_2 \pmod{m_i}$ for each $i = 1, \ldots, k$. Since the m_i are pairwise coprime, we have $x_1 \equiv x_2 \pmod{m}$. Hence, the solution (1.9) is uniquely determined modulo m.   □

Using Theorem 1.4 on the system (1.6)–(1.8), where $(3, 5, 7) = 1$, we find the solution $x = 23$, which is the smallest in positive integers.

Table 1.1 Sino-representation

z_2	0	1	2	3	4
$z_1 = 0$	0	6	12	3	9
$z_1 = 1$	10	1	7	13	4
$z_1 = 2$	5	11	2	8	14

In the thirteenth century Qin Jiushao solved congruence (1.11) using an analogue of the Euclidean algorithm (see [260, p. 317] or example in Sect. 1.7). This is because the problem of finding y_i satisfying congruence (1.11) can be converted to the solution of the Diophantine equation

$$n_i y_i - m_i v_i = 1$$

for unknowns y_i and v_i.

The Chinese Remainder Theorem actually states that any natural number x not greater than $m = m_1 m_2 \cdots m_k$ can be uniquely determined by the k-tuples of remainders $\langle r_1, r_2, \ldots, r_k \rangle$ after division by the numbers $m_1, m_2, \ldots, m_k$, where

$$0 \le r_i < m_i, \quad i = 1, 2, \ldots, k.$$

This is called sino-representation (see Schroeder [350]). For instance, if $k = 2$, $m_1 = 3$, and $m_2 = 5$, we get Table 1.1.

We see that the inside of Table 1.1 contains each of the numbers from 0 to 14 just once. Notice further that these numbers increase by one in the direction of the diagonals. A similar structure based on sino-representation also played an important role in the creation of the traditional Chinese calendar, which has a sixty-year cycle.

The rows correspond to 12 earth branches and columns to 10 heavenly tribes. Each line is dedicated to some animal and each double column corresponds to one of the five elements that make up the universe (see Table 1.2). For more details about the Chinese calendar, see the articles [195–197]. The Chinese Remainder Theorem has a number of other practical applications (see Schroeder [350]) in calculating convolutions, Fourier transform, quadratic congruences, etc.

1.10   Dirichlet's Pigeonhole Principle

Dirichlet's pigeonhole principle, is usually used to prove the existence of such objects which cannot be constructed or their construction is very difficult.

Theorem 1.5 (Dirichlet's pigeonhole principle) *Let $n \in \mathbb{N}$. If more than n items are distributed into n groups, then there are at least two items in at least one group.*

Table 1.2 Traditional Chinese calendar. The year of the mouse (rat) is in 2020, 2032, 2044, 2056,... In more details, 2020 is a metal rat year, 2032 is a water rat year, etc.

	ť'ia	i	ping	ting	wu	ť'i	keng	sin	žen	kuej	
Mouse	1		13		25		37		49		shǔ
Cow		2		14		26		38		50	niú
Tiger	51		3		15		27		39		hǔ
Rabbit		52		4		16		28		40	tù
Dragon	41		53		5		17		29		lóng
Snake		42		54		6		18		30	shé
Horse	31		43		55		7		19		mǎ
Sheep		32		44		56		8		20	yáng
Ape	21		33		45		57		9		hóu
Chicken		22		34		46		58		10	jī
Dog	11		23		35		47		59		gǒu
Pig		12		24		36		48		60	zhū
	Wood		Fire		Ground		Metal		Water		

Proof Let k_i be the number of items in the ith group for $i = 1, 2, \ldots, n$. If there were at most one item in each group, i.e. $k_i \leq 1$ for all $i = 1, 2, \ldots, n$, then the number of all items would be

$$k_1 + k_2 + \cdots + k_n \leq 1 + 1 + \cdots + 1 = n,$$

which is a contradiction. □

Dirichlet's pigeonhole principle can also be proved by mathematical induction (see [344, p. 150]). At first glance, it looks to be a very simple statement. However, using it properly, very strong statements can be obtained as we shall see e.g. in Fermat's proof of the Christmas Theorem 3.12.

With it, for example, it is easy to prove that in Prague there exist two inhabitants having the same number of hairs. No person has more than $n = 100\,000$ hairs and Prague has more than one million inhabitants. Its inhabitants can be divided into groups so that the ith group contains all inhabitants of Prague who have just i hairs, where $i = 0, 1, \ldots, n$. Since there are more inhabitants in Prague than the number of groups, there must be by Dirichlet's pigeonhole principle at least one group with at least two inhabitants (in fact many more) with the same number of hairs.

This surprising example nicely illustrates how Dirichlet's pigeonhole principle is used. However, to find two concrete citizens who have the same number of hairs, would be practically extremely difficult.

Theorem 1.6 *Given $n + 1$ natural numbers, there are at least two of them such that their difference is divisible by n.*

Proof Let us distribute the given numbers into n groups according to which remainder they yield when they are divided by n, i.e., in the first group we include those numbers that produce the remainder 0, in the second group those numbers that produce the remainder 1, etc. By Theorem 1.5 there exists at least one group containing at least two numbers that yield the same remainder after division by n. Their difference is thus divisible by n.   □

Example We prove that there exists a power of 37 that ends with 00001. Consider the numbers
$$37, \ 37^2, \ 37^3, \ldots, \ 37^{100001}.$$

We distribute them into $n = 10^5$ groups so that in the ith group we put those powers of 37^k which after division by the number n yield the remainder i for $i = 0, 1, 2, \ldots, n - 1$. (We cannot get the zero remainder, since the numbers 37^m and n are coprime.) By Theorem 1.5 there exist two natural numbers $j > m$ such that

$$37^j - 37^m = 37^m(37^{j-m} - 1)$$

is divisible by n. Since $(37^m, n) = 1$, the difference $37^{j-m} - 1$ is divisible by n, and thus the division of 37^{j-m} by n yields the remainder 1.

Example Assume that five points A, B, C, D, E lie on the sphere $\mathbb{S}^2 = \{(x, y, z) \in \mathbb{R}^3; \ x^2 + y^2 + z^2 = 1\}$. We show that there exists a closed hemisphere $H \subset \mathbb{S}^2$ which contains at least four of these points. Let h be the great circle passing through A and B. Each of the remaining three points C, D, E belongs to at least one of the two hemispheres whose boundary is h. By Theorem 1.5, at least two of the remaining three points must lie in the same hemisphere which also contains A and B.

Let us still note that the assumption H is closed (i.e. $h \subset H$) is necessary. To see this it is enough to consider five points uniformly distributed on the equator of $\mathbb{S}^2$.

A number of similar examples can be found, for example, in the book [54]. A somewhat stronger version of Dirichlet's pigeonhole principle states:

Theorem 1.7 *Let $k, n \in \mathbb{N}$. If more than kn items are distributed into n groups, then there are at least $k + 1$ items in at least one group.*

Chapter 2
Prime and Composite Numbers

2.1 The Fundamental Theorem of Arithmetic

A positive integer p is called a *prime,* if p has exactly two different positive divisors. Every prime p is thus greater than 1 and it is divisible by itself and 1. Positive integers greater than 1 are called *composite* if they are not primes.

The set of positive integers $\mathbb{N}$ can be thus divided into three disjoint groups: the number 1, prime numbers from the set

$$\mathbb{P} = \{2,\ 3,\ 5,\ 7,\ 11,\ 13, \dots \}$$

and composite numbers $4,\ 6,\ 8,\ 9,\ 10,\ 12,\ 14, \dots$

Theorem 2.1 (Euclid) *If p is a prime and $p \mid ab$ for $a, b \in \mathbb{N}$, then $p \mid a$ or $p \mid b$.*

Proof If $p \mid a$, we are done. If $p \nmid a$, then $(a, p) = 1$ and by Theorem 1.3 there exists integer numbers x and y such that $ax + py = 1$, and thus

$$abx + pby = b.$$

Since $p \mid ab$ and $p \mid p$, we see that $p \mid b$. $\qquad\square$

As a consequence we obtain the following Fundamental Theorem of Arithmetic, which claims that every integer $n > 1$ can be uniquely written as a product of positive integer powers of primes $p_1 < p_2 < \cdots < p_r$:

$$n = p_1^{\alpha_1} p_2^{\alpha_2} \cdots p_r^{\alpha_r} = \prod_{i=1}^{r} p_i^{\alpha_i}, \qquad (2.1)$$

where $\alpha_i \in \mathbb{N}$ for $i \in \{1, \dots, r\}$ and the symbol $\prod$ denotes the product. The numbers p_i are called *prime number factors.*

© The Author(s), under exclusive license to Springer Nature Switzerland AG 2021
M. Křížek et al., *From Great Discoveries in Number Theory to Applications*,
https://doi.org/10.1007/978-3-030-83899-7_2

Theorem 2.2 (Fundamental Theorem of Arithmetic) *If*

$$p_1^{\alpha_1} p_2^{\alpha_2} \cdots p_r^{\alpha_r} = q_1^{\beta_1} q_2^{\beta_2} \cdots q_s^{\beta_s},$$

where $p_1 < p_2 < \cdots < p_r, q_1 < q_2 < \cdots < q_s$ *are primes and* $r, s, \alpha_i, \beta_i \in \mathbb{N}$, *then*

$$r = s, \quad p_i = q_i, \quad \alpha_i = \beta_i$$

for every $i = 1, \ldots, r$.

Proof Let $m = p_1^{\alpha_1} p_2^{\alpha_2} \cdots p_r^{\alpha_r}$, $n = q_1^{\beta_1} q_2^{\beta_2} \cdots q_s^{\beta_s}$, and let $m = n$. If some prime p divides m, then by Euclid's Theorem 2.1 the prime p has to divide p_k for some $k \in \{1, \ldots, r\}$. From the definition of a prime it follows that $p = p_k$. Since $m = n$, the prime p has to divide some q_ℓ for $\ell \in \{1, \ldots, s\}$ and by a similar manner we get that $p = q_\ell$. From this we have $p_k = q_\ell$. Similarly, if $q \mid m$, then $q = q_i = p_j$ for some i and j. Since the primes p_i and q_j are ordered by size, it holds that $p_1 = q_1, \ldots, p_r = q_r$ and $r = s$.

Assume that $\alpha_i > \beta_i$ for some $i \in \{1, \ldots, r\}$. Dividing the equality $m = n$ by p^{β_i}, we get

$$p_1^{\alpha_1} \cdots p_i^{\alpha_i - \beta_i} \cdots p_r^{\alpha_r} = p_1^{\beta_1} \cdots p_i^0 \cdots p_r^{\beta_r},$$

which is a contradiction by Theorem 2.1, since the left-hand side is divisible by p_i, whereas the right-hand side not. Similarly we obtain a contradiction also for the case when $\alpha_i < \beta_i$. Hence, $\alpha_i = \beta_i$ for all $i \in \{1, \ldots, r\}$.   □

Example Let us show that $\log_{10} 3$ is an irrational number. Suppose to the contrary that $\log_{10} 3 = \frac{m}{n}$ for some $m, n \in \mathbb{N}$. Then $10^{m/n} = 3$. This implies that

$$10^m = 2^m \cdot 5^m = 3^n$$

which contradicts the Fundamental Theorem of Arithmetic 2.2.

The Fundamental Theorem of Arithmetic has a lot of practical applications. An unusual use in the message to extraterrestrial civilizations will be discussed in Sect. 11.6.

Remark Analogues to Theorem 2.2 may not apply at all in other algebraic systems. For instance, in the field $Q(\sqrt{-5})$ (see [138, p. 211] for details) we can factor the number 21 by two different manners into two irreducible numbers which cannot be factored further:

$$21 = 3 \cdot 7 \quad \text{and} \quad 21 = 1 + 2^2 \cdot 5 = (1 + 2\sqrt{-5})(1 - 2\sqrt{-5}).$$

Note that the imaginary unit $i \notin Q(\sqrt{-5})$.

Remark One of the reasons why do not wish the number 1 to be a prime is that we would not have unambiguous exponents in (2.1), e.g. $1 = 1^2 = 1^3$. Another reason is the expression of the so-called Riemann's ζ-function using the product over all primes (see (3.19) below).

According to the Fundamental Theorem of Arithmetic 2.2, for any $m, n \in N$ we can write

$$m = \prod_{i=1}^{\infty} p_i^{k_i}, \quad n = \prod_{i=1}^{\infty} p_i^{\ell_i},$$

where p_i is the ith prime and $k_i, \ell_i \geq 0$ for all $i \in \mathbb{N}$ (in particular, if $m = 1$, then all $k_i = 0$). The greatest common divisor and the least common multiple of two positive integers m and n can be calculated as follows:

$$(m, n) = \prod_{i=1}^{\infty} p_i^{\min(k_i, \ell_i)}, \quad [m, n] = \prod_{i=1}^{\infty} p_i^{\max(k_i, \ell_i)}.$$

Remark The number $n = 30$ has the following remarkable property. It is the largest integer for which all integers greater than 1, coprime to n, and less than n are primes. The proof is based on Bonse's inequality

$$p_1 p_2 \cdots p_n \geq p_{n+1}^2 \quad \text{for } n \geq 4,$$

see [35] and [315, pp. 187–192]. The other integers with this property are 3, 4, 6, 8, 12, 18, 24.

2.2   Euclid's Theorem on the Infinitude of Primes

It is interesting that we can find arbitrarily long intervals of consecutive composite numbers. Let us illustrate it on a simple example.

Example For any $n > 1$ consider a finite progression

$$n! + 2, \; n! + 3, \ldots, \; n! + n$$

which contains $n - 1$ consecutive numbers. We observe that the first term of this progression is divisible by two, the second term by three, etc. Finally, the last term is divisible by n. For instance, if $n = 1001$, we obtain 1000 consecutive composite numbers. On the other hand, the following theorem was already known to the ancient Greek mathematician Euclid (see the ninth Book, Prop. 20, of his *Elements*).

Theorem 2.3 (Euclid) *There are infinitely many primes.*

Proof Assume to the contrary that there exist only a finite number of primes $p_1, p_2, \ldots, p_n$ and set

$$m = p_1 p_2 \cdots p_n + 1. \tag{2.2}$$

Since dividing m by p_i always yields the remainder 1, no p_i divides m. According to the Fundamental Theorem of Arithmetic 2.2, m is another prime or m is composite and divisible by a prime different from $p_1, p_2, \ldots, p_n$, which is a contradiction. $\square$

There are also infinitely many composite numbers, since all even numbers greater than 2 are composite.

Remark In some publications, it is incorrectly stated that m from the previous theorem must be a new prime number. However, this is not true. To see this it is enough to consider a finite sequence of primes 2, 3, 5, 7, 11, and 13. Then we observe that

$$m = 2 \cdot 3 \cdot 5 \cdot 7 \cdot 11 \cdot 13 + 1 = 59 \cdot 509. \tag{2.3}$$

Hence, the number m is composite and is not divisible by any prime from the sequence 2, 3, 5, 7, 11, 13. We will return to this topic in Sect. 4.7.

Let us recall that all primes in a given bounded interval which starts by 1 can be found by the famous *Eratosthenes sieve*. We just cross out the number 1 and all composite numbers from this interval in turn. That is, we cross out all numbers divisible by 2, then by 3, then by 5 etc. In this way we get primes:

$$\cancel{1},\ 2,\ 3,\ \cancel{4},\ 5,\ \cancel{6},\ 7,\ \cancel{8},\ \cancel{9},\ \cancel{10},\ 11,\ \cancel{12},\ 13, \ldots \tag{2.4}$$

This sieve is named after the famous Greek mathematician and astronomer Eratosthenes of Cyrene (276–194 BC) who also estimated the circumference of the Earth.

Remark The Russian mathematician Pafnuty Lvovich Chebyshev (1821–1894) proved that for any positive integer n there always exists a prime p such that

$$n \le p \le 2n.$$

2.3 Pythagorean Triples

First we recall a well-known theorem.

Theorem 2.4 *Let a, b, c be the lengths of sides of a triangle. Then this triangle is right with hypotenuse c if and only if*

$$a^2 + b^2 = c^2.$$

Fig. 2.1 A visual proof of
the Pythagorean Theorem by
the ancient Chinese
mathematician Liu Hui (3rd
century AD): The areas of the
squares with sides $a + b$ and
c satisfy the equations
$(a + b)^2 = c^2 + 2ab$ and
$2ab + (a - b)^2 = c^2$. Each
of these equalities yields
$a^2 + b^2 = c^2$

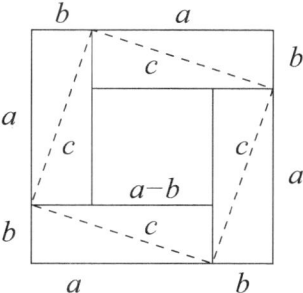

The implication $\Rightarrow$ is the classical Pythagorean Theorem (see Fig. 2.1). The converse implication $\Leftarrow$, which is more important from a practical point of view, is unfortunately omitted in many textbooks.

Let $a^2 + b^2 = c^2$ for an ordered triple $\langle a, b, c \rangle$ of positive integers. Then this triple is called a *Pythagorean* and the corresponding triangle from Theorem 2.4 is called a *Pythagorean triangle*. Moreover, if a, b, c have no common divisor $d > 1$, then $\langle a, b, c \rangle$ is called a *primitive Pythagorean triple*.

The following theorem is presented in Diophantus's Arithmetic whose Latin version was published in 1621 by Claude-Gaspard Bachet de Méziriac (1581–1638). It shows Euclid's fundamental formulae for generating primitive Pythagorean triples.

Theorem 2.5 *An ordered triple $\langle a, b, c \rangle$ of positive integers is a primitive Pythagorean triple if and only if there exists coprime positive integers $m > n$, one odd and one even, such that*

$$a = m^2 - n^2, \quad b = 2mn, \quad c = m^2 + n^2, \tag{2.5}$$

or

$$a = 2mn, \quad b = m^2 - n^2, \quad c = m^2 + n^2. \tag{2.6}$$

The numbers m and n are determined uniquely.

Proof $\Rightarrow$: Let $\langle a, b, c \rangle$ be a primitive Pythagorean triple. Then at most one of these numbers can be even. Assume for a moment that both a and b are odd. Then a^2 and b^2 are both one larger than some multiple of 4, since

$$(2j + 1)^2 = 4j^2 + 4j + 1.$$

Hence, $a^2 + b^2$ is of integer 2 greater than some multiple of 4. From this it follows that c^2 is divisible by 2, but is not divisible by 4, which is impossible. Therefore, one of the numbers a or b has to be even. Without loss of generality we may assume that $b = 2k$ and that the numbers a and c are odd. Then

$$k^2 = \frac{c^2 - a^2}{4} = \frac{c+a}{2} \cdot \frac{c-a}{2}.$$

Since a and c are odd, the numbers $\frac{1}{2}(c+a)$ and $\frac{1}{2}(c-a)$ are integers. They have to be coprime, because each of their divisors divides their sum (which is c) and also their difference (which is a) and we assumed that these numbers are coprime.

Every prime p that divides $(c-a)/2$, also divides k^2, and therefore, p^2 divides k^2. Since p does not divide $(c+a)/2$, the square p^2 has to divide $(c-a)/2$. The factorization of $(c-a)/2$ into primes thus has only even exponents which means that $(c-a)/2$ is a square. Analogously we find that $(c+a)/2$ is also a square. So we can write

$$\frac{c+a}{2} = m^2 \quad \text{and} \quad \frac{c-a}{2} = n^2.$$

We observe that m and n are coprime, they are determined uniquely and $m > n$. Since $c = m^2 + n^2$ is odd, one of the numbers m and n has to be odd and the second even. Further we see that $a = m^2 - n^2$, and thus

$$b = \sqrt{c^2 - a^2} = 2mn.$$

$\Leftarrow$: Conversely let $m > n$ be coprime positive integers such that one is odd and the second is even. First we set

$$a = m^2 - n^2, \quad b = 2mn, \quad c = m^2 + n^2.$$

Then clearly a and c are odd, b is even and

$$a^2 + b^2 = (m^2 - n^2) + (2mn)^2 = (m^2 + n^2)^2 = c^2.$$

Consequently, $\langle a, b, c \rangle$ is a Pythagorean triple. If a, b, c would be divisible by some odd prime p, then also $c + a = 2m^2$ and $c - a = 2n^2$ would be divisible by p, which contradicts the fact that m and n are coprime. From this it follows that $\langle a, b, c \rangle$ is a primitive Pythagorean triple.

The case when a is even and b is odd can be investigated similarly. $\square$

The most famous primitive Pythagorean triples are:

$$\langle 3, 4, 5 \rangle, \quad \langle 5, 12, 13 \rangle, \quad \langle 7, 24, 25 \rangle, \quad \langle 8, 15, 17 \rangle, \quad \langle 9, 40, 41 \rangle.$$

Note that for all positive integers d the following choices

$$a = d(m^2 - n^2), \quad b = d(2mn), \quad c = d(m^2 + n^2)$$

and

$$a = d(2mn), \quad b = d(m^2 - n^2), \quad c = d(m^2 + n^2)$$

yield all Pythagorean triples (see Sierpiński [353]).

Using the *Lebesgue identity*

$$(m^2 + n^2 - r^2 - s^2)^2 + (2ms - 2nr)^2 + (2mr + 2ns)^2 = (m^2 + n^2 + r^2 + s^2)^2,$$

we can look for quadruples satisfying the Diophantine equality

$$a^2 + b^2 + c^2 = d^2.$$

For instance, taking $m = n = s = 2$ and $r = 1$, we get

$$3^2 + 4^2 + 12^2 = 13^2.$$

At the end of this section, let us prove one more theorem whose author is Pierre de Fermat.

Theorem 2.6 (Fermat) *No integer of the form $4k - 1$ for $k \in \mathbb{N}$ is a sum of two squares of integers.*

Proof The square of an even number is of the form $4k$ and the square of an odd number $2j + 1$ has the form $4j^2 + 4j + 1 = 4k + 1$. (In particular, $4k - 1$ is never the square of an integer.) Thus the sum of two squares has the form $4k$ or $4k + 1$ or $4k + 2$, but it is never of the form $4k + 3 = 4(k + 1) - 1$.   □

2.4   Fermat's Method of Infinite Descent

In this chapter we will introduce Fermat's method of infinite descent, which is close to the principle of mathematical induction. It relies on the property that a set of positive integers $\mathbb{N}$ is well ordered, which means that any of its non-empty subsets has a smallest element.

Fermat's method of infinite descent is based on the following theorem.

Theorem 2.7 *Let M be a subset of $\mathbb{N}$ and suppose that for an arbitrary $m \in M$ there exists $n \in M$ such that $n < m$. Then the set M is empty.*

Proof Assume to the contrary that M is not empty. Since $\mathbb{N}$ is well ordered, there exists its smallest element m of the set M. Then by the assumption of the theorem there exists an element $n \in M$, which is smaller than m. This is a contradiction of the minimality of m. Hence, $M = \emptyset$.   □

Theorem 2.7 is used mainly in proofs of the non-existence of positive integers with certain properties. We will show its usefulness on two statements, which Pierre de Fermat himself dealt with around 1640. The proofs of both statements will illustrate how Fermat's infinite descent method is actually used. The first statement concerns Pythagorean triangles.

Since a or b in Theorem 2.5 is even, the area

$$P = \frac{ab}{2} \tag{2.7}$$

of a Pythagorean triangle is always a positive integer. Now we will apply Theorems 2.5 and 2.7 to the proof of the following property.

Theorem 2.8 *There is no Pythagorean triangle whose area is a square of a natural number.*

Proof Suppose to the contrary that there exist $a, b, c \in \mathbb{N}$ such that $a^2 + b^2 = c^2$ and that the area P of a Pythagorean triangle is the square of some positive integer. We may assume that $\langle a, b, c \rangle$ is a primitive Pythagorean triple. This means that the greatest common divisor of a, b, c is one, which is equivalent to the property that a and b are coprime. If this were not so, then we could set $a' = a/d$, $b' = b/d$ and $c' = c/d$, where d is the greatest common divisor of a, b, c. The triple $\langle a', b', c' \rangle$ would be again Pythagorean and the area of the corresponding Pythagorean triangle would again be a square as follows from (2.7). By Theorem 2.5 it is clear that exactly one of the numbers a and b is even. Without loss of generality we may assume that b is even. Hence, there exist positive integers $m > n$ such that

$$a = m^2 - n^2, \quad b = 2mn, \quad c = m^2 + n^2. \tag{2.8}$$

Since a, b, c are coprime, from relation (2.8) we immediately get that m and n are coprime as well. Therefore, $m, n, m + n$, and $m - n$ are also coprime in pairs, since m and n are not both odd.

By (2.7) and (2.8) the area of the Pythagorean triangle is

$$P = mn(m + n)(m - n). \tag{2.9}$$

Since P is a square, all particular factors in (2.9) have to also be squares, i.e., there exist $u, v, k, \ell \in \mathbb{N}$ such that

$$m = u^2, \quad n = v^2, \quad m + n = u^2 + v^2 = k^2, \quad m - n = u^2 - v^2 = \ell^2. \tag{2.10}$$

Summing and subtracting the last two equations, we obtain

$$2m = 2u^2 = k^2 + \ell^2, \quad 2n = 2v^2 = k^2 - \ell^2 = (k + \ell)(k - \ell). \tag{2.11}$$

Since we assume that $a = m^2 - n^2$ is odd (see (2.8)), the numbers $k^2 = m + n$ and $\ell^2 = m - n$ from relation (2.10) are also odd. Hence, k and ℓ are odd, too.

Denote by j the greatest common divisor of $k + \ell$ and $k - \ell$. Clearly $j \geq 2$, since $k + \ell$ and $k - \ell$ are even. We see that j divides $2k$ and also 2ℓ. Hence, j^2 divides

$4(k^2 + \ell^2) = 8m$ and also $4(k^2 - \ell^2) = 8n$. Since m and n are coprime, we have $j \le 2$. The numbers $k + \ell$ and $k - \ell$ thus have no common divisor except for 2. Therefore, by (2.11) there exist $r \in \mathbb{N}$ and an even number $s \in \mathbb{N}$ such that either

$$k + \ell = 2r^2 \quad \text{and} \quad k - \ell = s^2,$$

or

$$k + \ell = s^2 \quad \text{and} \quad k - \ell = 2r^2.$$

From this and (2.11) we get

$$(r^2)^2 + \left(\frac{s^2}{2}\right)^2 = \left(\frac{k + \ell}{2}\right)^2 + \left(\frac{k - \ell}{2}\right)^2 = \frac{k^2 + \ell^2}{2} = u^2.$$

We observe that the triple $\langle r^2, s^2/2, u \rangle$ is a new Pythagorean triple. The area of the corresponding triangle is $P' = r^2 s^2/4$, which is evidently a square of a positive integer, since s is even. However, the area P' is smaller than P, i.e.,

$$P' = \frac{r^2 s^2}{4} = \frac{k^2 - \ell^2}{8} < k^2 < u^2 v^2 k^2 \ell^2 = P,$$

where the last equality follows from (2.9) and (2.10).

Consequently, we have verified the assumptions of Theorem 2.7 for the set

$$M = \{P \in \mathbb{N}; \ \exists a, b, c, i \in \mathbb{N}: \ P = i^2 = \frac{ab}{2}, \ a^2 + b^2 = c^2\}.$$

Hence, $M = \emptyset$. $\qquad\square$

Note that the right triangle with sides 3, 4, and 5 has an integer area equal to 6. On the other hand, the following statement is valid.

Theorem 2.9 *There does not exist a right triangle with rational lengths of its sides whose area is* 1.

Proof Assume to the contrary that there exist $a, b, c, q, r, s \in \mathbb{N}$ such that

$$\left(\frac{a}{q}\right)^2 + \left(\frac{b}{r}\right)^2 = \left(\frac{c}{s}\right)^2 \quad \text{and} \quad \frac{1}{2}\frac{a}{q}\frac{b}{r} = 1.$$

Then $(ars)^2 + (bqs)^2 = (cqr)^2$ and $\frac{1}{2}(ars)(bqs) = (qrs)^2$, which by Theorem 2.8 cannot happen. $\qquad\square$

Fermat also used the method of infinite descent in the proof of the following statement, see e.g. [361].

Theorem 2.10 *The biquadratic equation* $x^4 + y^4 = w^2$ *has no solution for positive integers* $x, y,$ *and* w.

Proof Suppose to the contrary that there exists a solution of the equation $x^4 + y^4 = w^2$. Then

$$(x^2)^2 + (y^2)^2 = w^2. \tag{2.12}$$

Obviously, we can assume that the greatest common divisor of numbers x^2, y^2, and w is 1. If their greatest common divisor d were to be greater than 1, then we could divide the Eq. (2.12) by d^2 and obtain another coprime numbers satisfying (2.12).

Since $\langle x^2, y^2, w \rangle$ is a Pythagorean triple of coprime numbers, exactly one of the numbers x or y has to be even. Without loss of generality, we may assume that x is odd and y is even. Hence, by Theorem 2.5 there exist positive integers $m > n$ such that

$$x^2 = m^2 - n^2, \quad y^2 = 2mn, \quad w = m^2 + n^2. \tag{2.13}$$

If the numbers m and n were not coprime, then x^2, y^2, and w in (2.13) would also not be coprime, which is a contradiction. Thus m and n are coprime. From (2.13) we see that

$$x^2 + n^2 = m^2.$$

Since x is odd, by Theorem 2.5 there exist positive integers $r > s$ such that

$$x = r^2 - s^2, \quad n = 2rs, \quad m = r^2 + s^2. \tag{2.14}$$

The numbers r and s are again coprime. If they were not coprime, then the greatest common divisor of x, m and n would be greater than 1, which contradicts the fact that m and n are coprime. By (2.13), $2mn$ is the square of a positive integer. Since m and n are coprime, there exist positive integers u and v such that

$$n = 2u^2, \quad m = v^2. \tag{2.15}$$

From this and (2.14) we have $2u^2 = 2rs$, and thus there exist positive integers g and h such that $r = g^2$ and $s = h^2$. Using the relation $m = v^2$ and (2.14), we get

$$g^4 + h^4 = v^2.$$

This equation has the same shape as the original equation (2.12), but with $v < w$, since by (2.13) and (2.15)

$$v = \sqrt{m} < m^2 + n^2 = w.$$

Again, the assumptions of Theorem 2.7 are satisfied for the set

$$M = \{w \in \mathbb{N}; \ \exists x, y \in \mathbb{N}: \ x^4 + y^4 = w^2\},$$

and thus $M = \emptyset$. $\qquad\qquad\qquad\qquad\qquad\qquad\qquad\qquad\qquad\qquad\qquad\qquad\qquad$ □

In a manner similar to Theorem 2.10, it can be proved that the difference of two fourth powers of positive integers is never the square of a positive integer.

Fermat used the method of infinite descent with enthusiasm to also prove other hypotheses. This method is similar to that which is known today as the method of mathematical induction.

2.5 Elliptic Curves

An *elliptic curve* is a set of all solution $x, y \in \mathbb{R}^2$ of the equation

$$y^2 + a_1 xy + a_3 y = x^3 + a_2 x^2 + a_4 x + a_6, \tag{2.16}$$

where the coefficients a_i are rational numbers (sometimes x, y are taken to be in $\mathbb{Z}$ or $\mathbb{Q}$ or $\mathbb{C}$). It is obvious that no elliptic curve can be an ellipse. Their name is only related to the use of these curves for calculation of the length of elliptic arcs, e.g. in the determination of trajectories of planets [215, p. 186]. Equations of type (2.16) with integer coefficients were already solved by the Greek mathematician Diophantus (and also Niels Henrik Abel) because of their interesting properties. In the proof of Fermat's Last Theorem, the following special elliptic curve

$$y^2 = x(x - a^p)(x + b^p)$$

was employed, where $p \geq 5$ is a prime.

By means of the linear substitution $y \mapsto y - a_1 x/2 - a_3/2$, the Eq. (2.16) can be simplified as follows

$$y^2 = x^3 + b_2 x^2 + 2b_4 x + b_6, \tag{2.17}$$

where b_i are suitable rational numbers. By another linear substitution $x \mapsto x - b_2/3$ representing only a shift, we can reduce (2.17) to

$$y^2 = x^3 + Ax + B. \tag{2.18}$$

The associated *discriminant*

$$\Delta = -\left(\frac{A}{3}\right)^3 - \left(\frac{B}{2}\right)^2 = -\frac{4A^3 + 27B^2}{108}$$

has a very nice geometric interpretation. If $\Delta = A = 0$, then the curve $y^2 = x^3$ has the so-called *cusp* singularity and the corresponding graph is connected (see the left graph of Fig. 2.2). If $\Delta = 0$ and $A \neq 0$, then the elliptic curve (2.18) crosses itself (see the middle graph of Fig. 2.2). If $\Delta \neq 0$, then the polynomial $p(x) = x^3 + Ax + B$ has three different roots; otherwise it has a multiple root. If $\Delta < 0$, then there are one real and two complex conjugate roots and the corresponding elliptic curve is

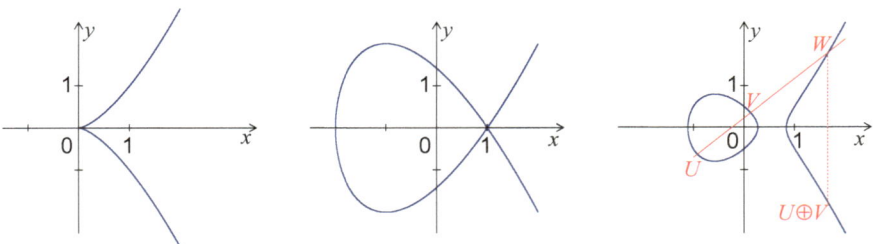

Fig. 2.2 Elliptic curves $y^2 = x^3$, $y^2 = x^3 - 3x + 2 = (x-1)^2(x+2)$, and $y^2 = x^3 - x + \frac{1}{4}$. On the last curve we can define a group operation $\oplus$ as sketched on the right graph, i.e., the element W is inverse to $U \oplus V$

connected. If $\Delta > 0$, then there are only real roots and the elliptic curve has two components (see the right graph of Fig. 2.2).

Now we shall describe a specific group of points on an elliptic curve which was used to prove Fermat's Last Theorem 2.11. Recall that a *group* G is a set on which an associative binary operation is defined $\circ : G \times G \to G$ with neutral element n and such that for any $g \in G$ there exists exactly one inverse element $g^{-1} \in G$ for which $g \circ g^{-1} = g^{-1} \circ g = n$.

On points of the elliptic curve (2.18) it is possible to define a group with operation $\oplus$ when $\Delta \neq 0$ (see the right graph of Fig. 2.2) and with neutral element at infinity. The inverse element to a given point of an elliptic curve is defined as its mirror image with respect to the axis x. To get acquainted with such a group we present some concrete examples.

Example Consider the elliptic curve $\mathcal{C}$ given by the relation

$$y^2 = x^3 - x + \tfrac{1}{4} \tag{2.19}$$

whose graph consists of two components (see Fig. 2.3). For each point $U = (x, y) \in \mathcal{C}$ we first introduce the inverse element defined by the relation

$$\ominus U = (x, -y), \tag{2.20}$$

which again lies on $\mathcal{C}$ due to (2.19).

Next we describe how to define the binary group operation $\oplus$. Let $U, V \in \mathcal{C}$ be two distinct points lying on the straight line $y = kx + q$. Then from (2.19) we get the cubic equation

$$(kx + q)^2 = x^3 - x + \tfrac{1}{4}, \tag{2.21}$$

which has two different real solutions (x-coordinates of the points U and V). Thus, the third root must also be real. We have to distinguish two cases:

Fig. 2.3 Commutative group of points on the elliptic curve (2.19)

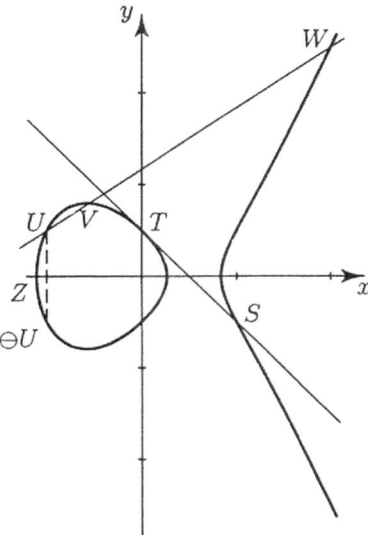

(a) If this root is simple, then the point $W \in C$, whose x coordinate is just the third root of Eq. (2.21), also lies on the straight line $y = kx + q$, and $U \neq W \neq V$.

(b) If this root is double, then the straight line $y = kx + q$ is tangent at one point to the curve C, i.e. either $W \equiv U$ or $W \equiv V$.

The points $U, V, W \in C$ are therefore collinear (i.e., they lie on one straight line) and at least two of them are different from each other. On the set of such points C we define the group operation $\oplus$ by

$$U \oplus V = \ominus W. \tag{2.22}$$

We immediately observe that this operation is commutative. By (2.19), (2.20), (2.21), and (2.22) one can easily verify that e.g. for the point $T = (0, \frac{1}{2})$ on Fig. 2.3 we have

$$T \oplus T = \left(1, \tfrac{1}{2}\right) = \ominus S,$$
$$T \oplus T \oplus T = \left(-1, \tfrac{1}{2}\right) = \ominus U,$$
$$T \oplus T \oplus T \oplus T = \left(2, -\tfrac{5}{2}\right) = \ominus W.$$

In (2.20) we defined inverse elements to any point of the curve C. Elements that are inverse to themselves are all intersections of the curve C with the x-axis (e.g. the point $Z = \ominus Z$ in Fig. 2.3).

Concerning the neutral element of this group, it must be such a point N that for any $U \in C$ we have

$$N \oplus U = U \oplus N = U. \tag{2.23}$$

That is, the points U, $\ominus U$ and N lie on one straight line which is parallel with the y-axis. But such a line is not of the form $y = kx + q$, and thus (2.21) cannot be employed. However, because (2.23) holds for any point $U \in \mathcal{C}$, the neutral element N should lie on each line parallel to the y-axis. This means that N is an improper point located at infinity, i.e., it does not actually belong to the curve $\mathcal{C}$. Now it is necessary to add the equality $N \oplus N = N$ to the relations (2.22) and (2.23) defining the group operation $\oplus$. Note that the Eq. (2.22) can be equivalently written as follows:

$$U \oplus V \oplus W = N.$$

It remains to prove the associativity of the group operation $\oplus$. However, such a proof is technically complicated and is beyond the scope of this presentation.

Example Consider another elliptic curve of type (2.18)

$$y^2 = x^3 - 43x + 166. \tag{2.24}$$

It can be shown that it has exactly six rational solutions (x, y): $(3, \pm 8)$, $(-5, \pm 16)$, and $(11, \pm 32)$ which are by chance all integers. Notice that they lie on the two straight lines

$$y = 3x - 1 \quad \text{and} \quad y = -3x + 1.$$

Adding the neutral element to these six points, we get a finite group that is isomorphic to the ordinary cyclic group C_7.

On the other hand, the equation

$$y^2 = x^3 - 2$$

has infinitely many rational solutions (e.g. $(3, \pm 5)$). Thus, one of the key questions in solving equations of type (2.18) is: *Which of these equations have a finite number of rational solutions and which have an infinite number of solution?* John Tate in [394] developed a sophisticated method that enables us to verify which elliptic curves are finitely generated.

Only a few people know that the arithmetic of elliptic curves is implemented in mobile phones, payment cards, traffic control systems, etc. For example, the number of a credit card can be encoded by converting this number to a point on an elliptic curve. To encrypt the information using elliptic curve cryptography an ingenious transformation is used that moves this point to another point of a given elliptic curve, see Hendrik W. Lenstra, Jr. [228].

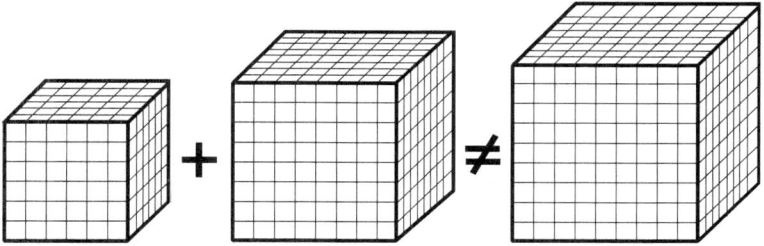

Fig. 2.4 Graphical illustration of Fermat's Last Theorem for the case $n = 3$ and $728 = 6^3 + 8^3 \neq 9^3 = 729$

2.6 Fermat's Last Theorem

In 1670 Samuel de Fermat (1630–1690), the son of Pierre de Fermat, published Diophantos's *Arithmetic* extended by notes that his father wrote in the margins of his copy from 1621. In one of these notes, P. Fermat states without any proof the sentence that is today generally called Fermat's Last Theorem: *Cubem autem in duos cubos, aut quadratoquadratum in duos quadratoquadratos, et generaliter nullam in infinitum ultra quadratum potestatem in duos eiusdem nominis fas est dividere ...*

Using modern terminology, the paragraph above can be rewritten in English as follows (see Fig. 2.4): *It is impossible to separate a cube into two cubes, or a fourth power into two fourth powers, or in general, any power higher than the second, into two like powers ...*

Theorem 2.11 (Fermat's Last Theorem) *There are no integers $x, y, z \in \mathbb{N}$ and $n \geq 3$ such that*

$$x^n + y^n = z^n. \tag{2.25}$$

First let us introduce a simple and amusing application of this theorem. We shall prove that the number $\sqrt[n]{2}$ is irrational for $n \geq 3$. Assume to the contrary that

$$\sqrt[n]{2} = \frac{z}{y}$$

for some positive integers y and z. Then $2 = z^n/y^n$. From this

$$2y^n = y^n + y^n = z^n$$

which is a special case of (2.25) that has no solution. This is a contradiction.

Fermat's Last Theorem is one of the most famous mathematical problems of all times. The exponent $n \geq 3$ is either divisible by an odd prime, or n is a power of 2. Now we show that it is enough to prove the above theorem only for an arbitrary odd prime exponent or for $n = 4$. Assume for a moment that (2.25) has a solution for some $n = pq$, where p is an odd prime and $q > 1$. Then

$$x^{pq} + y^{pq} = \left(x^q\right)^p + \left(y^q\right)^p = \left(z^q\right)^p = z^{pq}. \tag{2.26}$$

Hence, it is enough to set $x' = x^q$, $y' = y^q$, $z' = z^q$ and we see that Eq. (2.25) has a solution also for $n = p$. Analogously it can be demonstrated that if there exists some solution of (2.25) for $n = 2^k$, where $k > 2$, then there must exist a solution also for $n = 4$,

$$x^{2^k} + y^{2^k} = \left(x^{2^{k-2}}\right)^4 + \left(y^{2^{k-2}}\right)^4 = \left(z^{2^{k-2}}\right)^4 = z^{2^k}.$$

Fermat himself proved that there is no solution of (2.25) for $n = 4$. He used his favorite method of infinite descent. Equation (2.25) thus has no solution also for $n = 8, 16, 32, \ldots$.

Theorem 2.12 *Fermat's Last Theorem holds for $n = 4$, i.e., the equation*

$$x^4 + y^4 = z^4 \tag{2.27}$$

has no solution in the set of positive integers.

Proof Assume to the contrary that there exist positive integers x, y, z satisfying (2.27). Setting $w = z^2$, we immediately get a contradiction with Fermat's Theorem 2.10. □

A century later, Leonhard Euler used the infinite descent method for $n = 3$ (cf. Sect. 2.4). However, in his original proof from 1770 a mistake was found that was corrected later. Also Christian Huygens treated the case $n = 3$ (which is proved e.g. in [138], [215, p. 183]). In 1825 Peter Gustav Lejeune Dirichlet and three years later also Adrien-Marie Legendre proved Fermat's Last Theorem for $n = 5$. The case $n = 7$ was demonstrated by Gabriel Lamé and Victor-Amédée Lebesgue in 1840. In 1983, Gerd Faltings [109] proved the Mordell hypothesis which states that for any exponent $n > 3$ the Eq. (2.25) has only finitely many solutions such that $(x, y, z) = 1$. A detailed historical survey containing various special statements concerning the solution of (2.25) is given in [101, 320, 324, 398]. Guy Terjanian [397] proved by elementary methods that if p is an odd prime and x, y, z are integers such that $2p$ does not divide either x or y, then $x^{2p} + y^{2p} = z^{2p}$ does not hold.

It took more than 350 years until Fermat's Last Theorem was proved for all prime exponents. The English mathematician Andrew Wiles from Princeton University found a way to prove it. In June 1993 he presented his proof during three lectures at the Newton Institute in Cambridge in front of about fifty specialists in number theory. Later a gap in his proof was found, but it was removed relatively soon. In 1995 he published together with Richard Taylor a complete proof, from which Fermat's Last Theorem follows (see [396, 418]). The method used by these authors could not have been known by Fermat, since many specific concepts (e.g. elliptic curves, groups, modular forms) were not invented yet.

Now we will briefly introduce some other basic ingredients of Wiles' proof. Modular forms are much more abstract mathematical objects than elliptic curves. These forms are certain mappings with many symmetries (see e.g. (2.29) and (2.30) below). Denote by $SL_2(\mathbb{Z})$ the multiplicative group of 2×2 matrices

$$g = \begin{pmatrix} a & b \\ c & d \end{pmatrix},$$

where a, b, c, d are integers and

$$\det g = ad - bc = 1.$$

The inverse element of this nonabelian group is obviously given by

$$g^{-1} = \begin{pmatrix} d & -b \\ -c & a \end{pmatrix}.$$

For $g \in SL_2(\mathbb{Z})$ and $z \in \mathbb{C}$ define the so-called *action* gz by means of a complex rational function (see e.g. [172, p. 98])

$$gz = \frac{az + b}{cz + d}. \tag{2.28}$$

Further, let $H = \{z \in \mathbb{C}; \ \mathrm{Im}\ z > 0\}$. Then we easily find that for an arbitrary $g \in SL_2(\mathbb{Z})$ the upper part of the Gaussian complex halfplane H is mapped on H, i.e., $\mathrm{Im}\ z > 0$ implies $\mathrm{Im}\ gz > 0$. Namely,

$$\mathrm{Im}\ gz = \mathrm{Im}\frac{az + b}{cz + d} = \mathrm{Im}\frac{(az + b)(c\bar{z} + d)}{|cz + d|^2} = |cz + d|^{-2}\mathrm{Im}(adz + bc\bar{z}),$$

where $\bar{z}$ is the complex conjugate of z. Since

$$\mathrm{Im}(adz + bc\bar{z}) = (ad - bc)\mathrm{Im}\ z = \det g\ \mathrm{Im}\ z = \mathrm{Im}\ z,$$

we get

$$\mathrm{Im}\ gz = |cz + d|^{-2}\mathrm{Im}\ z \quad \forall g \in SL_2(\mathbb{Z}).$$

Hence, to any group element $g \in SL_2(\mathbb{Z})$ we can assign by (2.28) a transformation $H \to H$.

Recall that a complex function is called *meromorphic*, if it is holomorphic on an open connected subset of the complex plane $\mathbb{C}$ up to a set of isolated poles. Let $k \in \mathbb{Z}$ and let a meromorphic function f defined in the upper halfplane H satisfy

$$f(gz) = (cz + d)^k f(z) \quad \forall g \in SL_2(\mathbb{Z}). \tag{2.29}$$

Moreover, assume that f is meromorphic at infinity (this means that the Fourier series $f(z) = \sum_{m \in \mathbb{Z}} a_m q^m$, where $q = \exp(2\pi i z)$, has only finitely many nonzero coefficients a_m for $m < 0$, and f has a pole at ∞, if $f(1/z)$ has a pole at 0). Then f is called a *modular form with weight k of the group* $SL_2(\mathbb{Z})$.

From (2.29) we observe a high degree of symmetry of modular forms. For instance, choosing $g = \left(\begin{smallmatrix} 1 & 1 \\ 0 & 1 \end{smallmatrix}\right)$, we find that

$$f(z+1) = f(z). \tag{2.30}$$

Similarly for $g = \left(\begin{smallmatrix} 0 & 1 \\ -1 & 0 \end{smallmatrix}\right)$ we get

$$f(-1/z) = (-z)^k f(z). \tag{2.31}$$

We will now briefly describe what it means for an elliptic curve to be modular. Let $E(\mathbb{Q})$ be an elliptic curve over rational numbers $\mathbb{Q}$ with conductor N. The conductor N is a certain invariant which divides the discriminant of the curve $E(\mathbb{Q})$. Let N_p denote the number of points lying on the elliptic curve E over a finite field F_p with p elements, where p is a prime number. Put $a_p = p + 1 - N_p$ if $p \nmid N$, and $a_p = \pm 1$ if $p \mid N$ and $p^2 \nmid N$. The sign of $a_p \in \mathbb{Z}$ depends on the behavior of the elliptic curve over F_p. Further let $a_1 = 1$ and for coprime integer numbers a_m and a_n we set $a_{mn} = a_m a_n$. For $s \in \mathbb{C}$ and $E(\mathbb{Q})$ define the function $L(s, E)$ by

$$L(s, E) = \prod_p L_p(s, E)^{-1} = \sum_{n=1}^{\infty} \frac{a_n}{n^s},$$

where for a given prime number p we have

$$L_p(s, E) = \begin{cases} 1 - a_p p^{-s} + p^{1-2s} & \text{for } p \nmid N, \\ 1 - a_p p^{-s} & \text{for } p \mid N \text{ and } p^2 \nmid N, \\ 1 & \text{for } p^2 \mid N. \end{cases}$$

An elliptic curve is said to be *modular*, if the function

$$f(q, E) = \sum_{n=1}^{\infty} a_n q^n$$

is modular, where $q = e^{2\pi i \tau}$ and $\tau \in \mathbb{C}$.

In the 1950s, a Japanese mathematician Yutaka Taniyama stated a hypothesis that there is a connection between elliptic curves and modular forms. Another Japanese mathematician, Goro Shimura, then helped him to refine it. That is why today it is usually called the *Taniyama–Shimura Conjecture*.

Taniyama–Shimura Conjecture. *Every rational elliptic curve is modular.*

According to this conjecture, any rational elliptic curve can be identified with a certain modular form. The connection between the Taniyama–Shimura Conjecture and Fermat's Last Theorem originated when German mathematician Gerhard Frey at a conference in Oberwolfach in 1985 declared:

If it were to happen that

$$a^p + b^p = c^p$$

for some natural numbers a, b, c and a prime number exponent $p \geq 5$ (see (2.26)), then the rational elliptic curve of the form (2.17)

$$y^2 = x(x - a^p)(x + b^p) \tag{2.32}$$

(the so-called *Frey curve*) would yield a counter-example to the Taniyama–Shimura Conjecture.

In 1990, Kenneth Ribet in [325] proved that Frey was right.

Theorem 2.13 (Ribet) *If the Taniyama–Shimura Conjecture is true, then Fermat's Last Theorem holds for any prime exponent $p \geq 5$.*

Due to this theorem, mathematicians gained a new approach to attack Fermat's problem. And it was Andrew Wiles who found how to handle this problem, although he proved only a restricted Taniyama–Shimura Conjecture for the so-called semistable elliptic curves. These curves have a square-free conductor. Fortunately, this special case contains Fermat's Last Theorem,

Theorem 2.14 (Wiles–Taylor) *Every semistable rational elliptic curve is modular.*

The proof uses an elliptic curve over complex numbers with the set $E(p)$ of the so-called p-torsion points, where p is a prime number. Furthermore, a homomorphism from $E(p)$ to the multiplicative group $GL_2(F_p)$ of 2×2 matrices over a finite field F_p with characteristic p is investigated. In 1995, Wiles published his proof in Annals of Mathematics [418] and a part of this proof is also in the common paper [396] with Richard Taylor. However, their proof is based on a number of other non-trivial statements published elsewhere. Methods used by Wiles are explained also in [142, 285, 355].

A complete proof of the Taniyama–Shimura Conjecture also for nonsemistable elliptic curves was later given by Christophe Breuil, Brian Conrad, Fred Diamond, and Richard Taylor in the comprehensive paper [44].

Theorem 2.15 *Every rational elliptic curve is modular.*

We shall return to Fermat's Last Theorem 2.11 in Sects. 4.3 and 4.5. Only few mathematical results have such a rich and dramatic history as this theorem. However, this story does not end, since there exist a series of open problems similar to Fermat's Last Theorem. For instance, the American mathematician Andrew Beal formulated the following conjecture for whose solution a large financial reward is announced (see [264]).

Beal's Conjecture. *The equation*

$$x^k + y^m = z^n$$

has no solution in positive integers k, m, n, x, y and z, where k, m and n are greater than 2 and the numbers x, y and z are mutually coprime.

The assumption that all three exponents are greater than 2 is essential, as follows from the identities $2^5 + 7^2 = 3^4$, $7^3 + 13^2 = 2^9$, etc.

During the 1980s, Masser, Oesterlé and Szpiro introduced a Diophantine inequality known as the abc Conjecture (see e.g. [125, p. 1227]). Its solution would have many applications (e.g. to prove Beal's Conjecture). Let a, b and c be positive integers and let the symbol $N(a, b, c) = r$ denote the product of the prime divisors of the numbers a, b and c, where each divisor appears only once.

The abc Conjecture. *For any real $\varepsilon > 0$ there exists $\mu > 1$ such that for any two coprime numbers a and b with sum $c = a + b$ we have*

$$\max(|a|, |b|, |c|) \leq \mu N(a, b, c)^{1+\varepsilon}.$$

The meaning of the previous inequality can be explained as follows. For instance, if $a = 2^m$, $b = 3^n$ for some large m and n, then the abc Conjecture states that c must have a big prime number divisor or a great amount of prime divisors so that r is large. There exist infinitely many cases when $r < c$ (e.g. for $a + b = 3 + 5^3 = 2^7 = c$ we have $r = 2 \cdot 3 \cdot 5 = 30 < 128 = c$). Another version of the abc Conjecture states that for any $\varepsilon > 0$ there exist only finitely many cases when $r^{1+\varepsilon} < c$.

In the case of Eq. (2.25), it is enough to choose $\varepsilon = \frac{1}{2}$. Then by the abc Conjecture we have $z^n < Cr^{3/2}$, where

$$r = \prod_{p|x^n y^n z^n} p = \prod_{p|xyz} p \leq xyz \leq z^3,$$

and thus $z^n < Cz^{9/2}$. From this follows the existence of n_0 such that Fermat's Last Theorem holds for all exponents $n > n_0$.

Below we give further consequences that would follow from the truth of the abc Conjecture below (see [409]):

1. Beal's Conjecture given above has at most finitely many counterexamples.
2. Pillai's Conjecture holds. Pillai's Conjecture states that the equation

$$x^m - y^n = k$$

has only finitely many solutions for any fixed $k \in \mathbb{N}$, where the unknowns x, y, m, and n all take integral values greater than or equal to 2. Catalan's Conjecture given

below treats the case in which $k = 1$. It is noteworthy that there is no value of $k \geq 2$ for which it is known whether Pillai's Conjecture holds.
3. There are infinitely many primes which are not Wieferich primes. A Wieferich prime is a prime number p such that $p^2 \mid 2^{p-1} - 1$. The only known Wieferich primes p are $p = 1093$ and $p = 3511$. These primes are discussed in Sect. 4.3.

Another important and long unsolved conjecture which was recently proved is the following:

Catalan's Conjecture. *There are no consecutive positive integers except for* 8 *and* 9 *that are both powers.*

In the 14th century, the astronomer Levi Ben Gerson (1288–1344) proved that there are no other consecutive powers of 2 and 3. Catalan's Conjecture was finally proved by the Romanian mathematician Preda Mihăilescu [272] in 2004 (see also [49]). He showed that the equation

$$x^m - y^n = 1, \quad \text{where } m, n, x, y \geq 2,$$

has the only solution $m = y = 2$ and $n = x = 3$. Finally we note that in the 7th century the Indian mathematician Brahmagupta found the smallest solution $x = 151$ and $y = 120$ which is a special case of the so-called *Pell's equation*: $x^2 - ny^2 = 1$ when $n = 94$. More about this equation can be found in Burton [56].

2.7   Fermat's Little Theorem

The number $n^3 - n$ for $n \in \mathbb{N}$ is divisible by 3 (and also by 2), since it can be written as the product of three consecutive integers

$$n^3 - n = (n - 1)n(n + 1).$$

Similarly the number $n^5 - n$ is divisible by 5, since

$$n^5 - n = (n^2 - 1)n(n^2 + 5 - 4) = 5(n^2 - 1)n + (n - 2)(n - 1)n(n + 1)(n + 2).$$

We generalize these simple statements below in Fermat's Little Theorem 2.16 which belongs among the most used tools in number theory, as we shall see throughout this book (cf. Index). To prove it, we need the following implication, which follows directly from Euclid's Theorem 2.1:

If p is a prime and $ab \equiv 0 \pmod{p}$, then

$$a \equiv 0 \pmod{p} \quad \text{or} \quad b \equiv 0 \pmod{p}. \tag{2.33}$$

Theorem 2.16 (Fermat's Little Theorem) *If $a \in \mathbb{N}$ and p is a prime, then $p \mid (a^p - a)$.*

Proof The case $p = 2$ is obvious. Let p be a prime greater than 2. If $p \mid a$, then p also divides the number $a^p - a = a(a^{p-1} - 1)$. So let the integers p and a be coprime, i.e.,

$$(p, a) = 1.$$

We will show that $p \mid (a^{p-1} - 1)$. Consider the finite progression

$$a, \ 2a, \ 3a, \ldots, \ (p-1)a. \tag{2.34}$$

Dividing the numbers ia and ja, $1 \le j < i < p$, by the prime p, we cannot get the same remainder, since we would then have that $p \mid (i - j)a$, which by (2.33) contradicts the fact that $(a, p) = 1$. Thus the progression (2.34) yields $p - 1$ different nonzero remainders when dividing by the prime p. We get the same remainders (up to order), when the progression $1, 2, \ldots, p - 1$ is divided by the prime p. Let us multiply these numbers and also the numbers in (2.34). From this, (1.5), and by induction we get the congruence $a^{p-1}(p - 1)! \equiv (p - 1)! \pmod{p}$. Hence,

$$(a^{p-1} - 1)(p - 1)! \equiv 0 \pmod{p}$$

and the first factor on the left-hand side is divisible by the prime p due to (2.33), since $p \nmid (p - 1)!$. $\qquad\qquad\qquad\qquad\qquad\qquad\qquad\qquad\qquad\qquad\qquad\qquad\qquad\qquad\quad\square$

There are many other proofs of Fermat's Little Theorem, see e.g. [129, 357]. One of the oldest from 1736 is due to Euler.

Another proof considers a one-sided bracelet with p beads of a colors (see [120]). The number of all colorings of the bracelet is a^p. The number of single color bracelets is clearly a (see Fig. 2.5). If all beads do not have the same color, then p rotations about the angle $360°/p$ produce p different colorings, since p is a prime. (For a composite number p we could get the same configuration for a suitable rotation.) Therefore, $a^p = pn + a$ for some $n \in \mathbb{N}$, i.e. $p \mid a^p - a$.

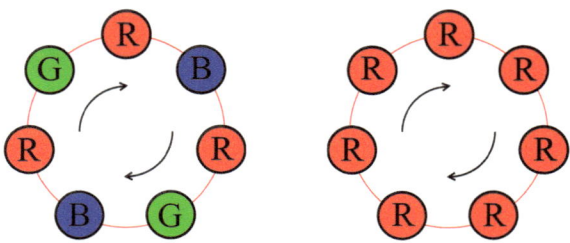

Fig. 2.5 A visual proof of Fermat's Little Theorem for $p = 7$ beads and $a = 3$ colors. Rotating the left one-sided bracelet about the angle $360°/7$, we obtain seven distinct strings: RBGBRRG, BGBRRGR, GBRRGRB, BRRGRBG, RRGRBGB, RGRBGBR, GRBGBRR. When rotating the right bracelet, we always get the same string RRRRRRR. Hence, the total number of all different colorings is $3^7 = 7n + 3$ for some number $n \in \mathbb{N}$, i.e., $7 \mid 3^7 - 3$

Fermat's Little Theorem is often formulated using a congruence as follows:

Theorem 2.17 (Fermat's Little Theorem) *If p is a prime number and $(a, p) = 1$, then*

$$a^{p-1} \equiv 1 \pmod{p}. \tag{2.35}$$

If $(a, p) = 1$, then the integer number

$$q(a) = \frac{a^{p-1} - 1}{p}$$

is called the *Fermat quotient*, see Karel Lepka [230].

Let d and a be coprime positive integers. The smallest positive exponent e for which $d \mid (a^e - 1)$ (i.e. $a^e \equiv 1 \pmod{d}$) is called the (*multiplicative*) *order of a modulo d*. We denote it as follows:

$$e = \mathrm{ord}_d a.$$

By Fermat's Little Theorem $p \mid (a^{p-1} - 1)$ for an arbitrary prime p not divisible by a. From this follows the existence of the order $e \le p - 1$. For instance, we can check that $\mathrm{ord}_7 2 = 3$ and $\mathrm{ord}_5 3 = 4$. In particular, $\mathrm{ord}_1 a = 1$ for any $a \in \mathbb{N}$.

Theorem 2.18 *If $e = \mathrm{ord}_d a$, then*

$$d \mid (a^n - 1) \tag{2.36}$$

for $n = ke$, $k \in \{1, 2, \dots\}$ and the relation (2.36) holds for these exponents only.

Proof If $n = ke$, then

$$a^n - 1 = a^{ke} - 1 = (a^e - 1)(a^{e(k-1)} + \cdots + a^e + 1),$$

and thus (2.36) holds due to the previous definition.

Assume now that $d \mid (a^{ke+h} - 1)$ for some $k \in \mathbb{N}$ and $0 < h < e$. Then

$$(a^{ke+h} - 1) - (a^{ke} - 1) = a^{ke}(a^h - 1).$$

Since both the numbers in parentheses on the left-hand side are by assumption divisible by d and by (2.36), $(d, a^{ke}) = 1$, we obtain that $a^h - 1$ is divisible by d. However, this contradicts the minimality of the exponent e. □

Let p be a prime. According to Fermat's Little Theorem, the order modulo p of a number a coprime to p is $p - 1$ or a smaller integer. The number $g \not\equiv 0 \pmod{p}$ is called a *primitive root modulo p*, if

$$\mathrm{ord}_p g = p - 1, \quad \text{i.e.} \quad g^{p-1} \equiv 1 \pmod{p}.$$

Usually we choose g so that $0 < g < p$. Such a g is thus a primitive root, if $g^k \not\equiv 1$ (mod p) for all $k \in \{1, \ldots, p - 2\}$ (see Table 13.4).

For example, 3 is a primitive root modulo 2, since $3^1 \equiv 1$ (mod 2). The number 3 is also a primitive root modulo 17 (see Fig. 2.6), since $3^{16} \equiv 1$ (mod 17) and $3^k \not\equiv 1$ (mod 17) for all $k = 1, \ldots, 15$. In this case, the order of 3 modulo 17 is 16, i.e. $\mathrm{ord}_{17} 3 = 16$.

Primitive roots have many applications e.g. in cryptography, in construction of pseudorandom number generators (see Sect. 11.5) and construction of regular polygons (see [199]). The existence of primitive roots is established by the following theorem (cf. Table 13.4).

Theorem 2.19 *If p is a prime, then there exists a primitive root modulo p.*

The proof is given e.g. in [118, Article 55] and [56, pp. 156–157]. We will generalize this result in Theorem 2.22.

If g is a primitive root modulo p, where p is a prime, then g is the generator of the set of all nonzero remainders modulo p, i.e., $\{g, g^2, g^3, \ldots, g^{p-1}\}$ consists of all $p - 1$ nonzero remainders modulo p. Hence, for any positive integer $a \not\equiv 0$ (mod p) there exists an exponent $n \in \{1, \ldots, p - 1\}$ such that $g^n \equiv a$ (mod p). This idea is further developed in [56, pp. 165–169].

Finally, note that in the 19th century primitive roots were used to multiply large numbers (for that time). For that reason books with such tables were also produced then, e.g. by Carl Gustav Jacobi.

2.8 Euler–Fermat Theorem

Leonhard Euler (1707–1783) generalized Fermat's Little Theorem 2.17 also for a non-prime modulus. To elucidate his idea, we first introduce the *Euler totient function* ϕ. For every $n \in \mathbb{N}$ its value $\phi(n)$ is defined as the number of all positive integers not exceeding n that are coprime to n, i.e.,

$$\phi(n) = |\{m \in \mathbb{N}; \ m \leq n, \ (m, n) = 1\}|,$$

where $| \cdot |$ denotes the number of elements. We easily find that

$$\phi(1) = 1, \ \phi(2) = 1, \ \phi(3) = 2, \ \phi(4) = 2, \ \phi(5) = 4, \ \phi(6) = 2, \ \phi(7) = 6, \ \ldots$$

Further values of ϕ are listed in Table 13.3. From the definition of the function ϕ we observe that its values are always even when $n > 2$, since $(n - m, n) = 1$ whenever $1 \leq m \leq n$ and $(m, n) = 1$. If p a prime, then obviously

$$\phi(p) = p - 1 \tag{2.37}$$

and

$$\phi\left(p^k\right) = (p-1)p^{k-1} \tag{2.38}$$

for any $k \in \mathbb{N}$.

The Euler totient function has the following important property:

$$(m, n) = 1 \quad \Longrightarrow \quad \phi(mn) = \phi(m)\phi(n). \tag{2.39}$$

A proof is given e.g. in [118, Article 38], [56, p. 125], or [291, p. 69]. Consequently, if the prime-power factorization of n is given by

$$n = \prod_{i=1}^{r} p_i^{k_i},$$

where $p_1 < p_2 < \cdots < p_r$, $k_i \in \mathbb{N}$, then by (2.38) and (2.39)

$$\phi(n) = \prod_{i=1}^{r}(p_i - 1)p_i^{k_i-1} = n \prod_{i=1}^{r} \frac{(p_i - 1)p_i^{k_i-1}}{p_i^{k_i}}$$

$$= n\left(1 - \frac{1}{p_1}\right)\left(1 - \frac{1}{p_2}\right)\cdots\left(1 - \frac{1}{p_r}\right). \tag{2.40}$$

Notice that the number on the right-hand side is an integer, since $\phi(n)$ on the left-hand side is an integer.

Carl Friedrich Gauss derived another important formula (see [118, Article 39], [56, p. 134], [284, p. 17])

$$\sum_{d|n} \phi(d) = n \quad \forall n \in \mathbb{N},$$

where summation is done over all divisors d of n. To see this consider the n fractions

$$\frac{1}{n}, \frac{2}{n}, \ldots, \frac{n}{n}.$$

Suppose that $d \mid n$ and look at all the fractions m/n for which $(m, n) = d$, where $1 \le m \le n$. Then there are $\phi(n/d)$ possibilities for m, namely,

$$m = di,$$

where $1 \le i \le n/d$ and $(i, n/d) = 1$. It now follows that

$$n = \sum_{d|n} \phi\left(\frac{n}{d}\right) = \sum_{d|n} \phi(d).$$

Remark According to [315, Chap. 27], $n = 30$ is the largest number n such that all numbers greater than 1 that are coprime to n and not greater than n are primes. The original proof given by the Russian mathematician Pafnuty Lvovich Chebyshev (1821–1894) was later simplified by the student H. Bonse.

Theorem 2.20 *Let $n > 1$. Then $\phi(n) < n - 1$ if and only if n is composite; and $\phi(n) = n - 1$ if and only if n is a prime.*

Proof The equivalences immediately follow from relations (2.37), (2.38), and (2.40).   □

Theorem 2.21 (Euler–Fermat) *Let $a, n \in \mathbb{N}$. Then*

$$a^{\phi(n)} \equiv 1 \pmod{n} \tag{2.41}$$

if and only if $(a, n) = 1$.

This theorem actually represents a necessary and sufficient condition for numbers a and n to be coprime. But we will not prove it now, as it is a direct consequence of Carmichael's Theorem 2.24 and relation (2.42), which provide stronger results.

From (2.37) and (2.35) we see that the Euler–Fermat Theorem 2.21 is a direct generalization of Fermat's Little Theorem 2.17.

Now we define primitive roots modulo an arbitrary positive integer $n \geq 2$ analogously to the situation as for primes. By the Euler–Fermat Theorem 2.21, the maximum possible order modulo n of any integer a coprime to n is equal to $\phi(n)$. If a is a positive integer such that $(a, n) = 1$, then a is called a *primitive root modulo n* if

$$\mathrm{ord}_n a = \phi(n).$$

The next theorem characterizes all positive integers $n \geq 2$ that have primitive roots.

Theorem 2.22 *Let $n \geq 2$. Then there exists a primitive root modulo n if and only if $n \in \{2, 4, p^k, 2p^k\}$, where p is an odd prime and $k \in \mathbb{N}$. Moreover, if n has a primitive root, then n has exactly $\phi(\phi(n))$ incongruent primitive roots.*

The proof is given e.g. in [56, pp. 160–164] or [291, pp. 102–104]. If p is a prime, then p has exactly $\phi(p - 1)$ incongruent primitive roots (cf. Fig. 2.6). For instance, the number 3 has only one primitive root, since $\phi(\phi(3)) = \phi(2) = 1$ and this root equals 2. The prime number 7 has two primitive roots, since $\phi(\phi(7)) = \phi(6) = 2$.

According to Theorem 2.22 there exist primitive roots modulo p^k, where p is a prime and $k \in \mathbb{N}$ except for the case $p = 2$ and $k \geq 3$. The next theorem states the maximum possible order modulo 2^k for $k \geq 3$.

Theorem 2.23 *Let a be an odd integer and $k \geq 3$. Then*

$$\mathrm{ord}_{2^k} a \mid 2^{k-2} \quad and \quad 2^{k-2} < \phi(2^k) = 2^{k-1}.$$

In particular,

$$\mathrm{ord}_{2^k} 5 = 2^{k-2}.$$

The proof is given in [291, pp. 103–105].

Let us further note that in 1875 H. J. Smith proved the following surprising relation (see [239])

$$\det\big((i,j)\big)_{i,j=1}^{n} = \prod_{k=1}^{n} \phi(k),$$

i.e., the determinant of the matrix of greatest common divisors is equal to the product of values of the Euler totient function.

By [243, p. 5], for any $n \in \mathbb{N}$ there exists k such that $\phi(k) = n!$. In [244], we deal with solutions of the congruence $n^2 \equiv 1 \pmod{\phi^2(n)}$.

2.9 Carmichael's Theorem

By the Euler–Fermat Theorem 2.21 and Theorems 2.18 and 2.22 we have: if a and n are coprime positive integers, then $\mathrm{ord}_n a \mid \phi(n)$ and there exists a positive integer b such that $\mathrm{ord}_n b = \phi(n)$ for $n = 2$ or 4 or p^k or $2p^k$, where p is an odd prime and $k \in \mathbb{N}$. As we shall see later, the maximum possible order modulo n can be significantly lower than $\phi(n)$, if n is not a power of a prime number (cf. Table 13.3). To this end we introduce the Carmichael lambda function $\lambda(n)$, which first appeared in [62]. Its definition is actually a mere modification of the Euler totient function $\phi(n)$.

For any positive integer n the *Carmichael lambda function* $\lambda(n)$ is defined as follows:

$$\lambda(1) = 1 = \phi(1),$$
$$\lambda(2) = 1 = \phi(2),$$
$$\lambda(4) = 2 = \phi(4),$$
$$\lambda(2^k) = 2^{k-2} = \tfrac{1}{2}\phi(2^k) \text{ for } k \geq 3,$$
$$\lambda(p^k) = (p-1)p^{k-1} = \phi(p^k) \text{ for any odd prime } p \text{ and } k \geq 1,$$
$$\lambda\big(p_1^{k_1} p_2^{k_2} \cdots p_r^{k_r}\big) = \big[\lambda\big(p_1^{k_1}\big),\ \lambda\big(p_2^{k_2}\big),\ \ldots,\ \lambda\big(p_r^{k_r}\big)\big],$$

where $p_1, p_2, \ldots, p_r$ are different primes and $k_i \in \mathbb{N}$ for $1 \leq i \leq r$.

The values of $\lambda(n)$ for $n \leq 32$ are listed in Table 13.3. By Theorems 2.22 and 2.23, if p is a prime and $k \in \mathbb{N}$, then $\lambda\big(p^k\big)$ is equal to the maximum possible order modulo p^k. From the definition of $\lambda(n)$ we also see that

$$\lambda(n) \mid \phi(n) \tag{2.42}$$

for all $n \in \mathbb{N}$ and that $\lambda(n) = \phi(n)$ if and only if $n \in \{1, 2, 4, p^k, 2p^k\}$, where p is an odd prime and $k \in \mathbb{N}$ (cf. Theorem 2.22).

Notice that $\lambda(n)$ can be much smaller than $\phi(n)$ if n has many prime factors. For instance, let $n = 2^6 \cdot 11 \cdot 17 \cdot 41 = 490688$. Then

$$\lambda(n) = \left[\lambda\left(2^6\right), \lambda(11), \lambda(17), \lambda(41)\right] = [16, 10, 16, 40] = 80,$$

while

$$\phi(n) = \phi\left(2^6\right)\phi(11)\phi(17)\phi(41) = 32 \cdot 10 \cdot 16 \cdot 40 = 204800.$$

The following theorem generalizes the Euler–Fermat Theorem 2.21. It actually states that $\lambda(n)$ is a universal order modulo n.

Theorem 2.24 (Carmichael) *Let $a, n \in \mathbb{N}$. Then*

$$a^{\lambda(n)} \equiv 1 \quad (\text{mod } n) \tag{2.43}$$

if and only if $(a, n) = 1$. Furthermore, there exists $b \in \mathbb{N}$ such that

$$\text{ord}_n b = \lambda(n). \tag{2.44}$$

Proof $\Rightarrow$: If $(a, n) = d > 1$, then $d \nmid \left(a^{\lambda(n)} - 1\right)$, and thus (2.43) does not hold.

$\Leftarrow$: Conversely, let $(a, n) = 1$. The congruence (2.43) obviously holds for $n = 1$. Suppose that $n \geq 2$. To prove that congruence (2.43) is valid, it suffices to show by Theorem 2.18 that

$$\text{ord}_n a \mid \lambda(n). \tag{2.45}$$

Consider the following prime-power factorization of n,

$$n = \prod_{i=1}^{r} p_i^{k_i}, \tag{2.46}$$

where $p_1 < p_2 < \cdots < p_r$, $k_i \in \mathbb{N}$, and let $q_i = p_i^{k_i}$ for $1 \leq i \leq r$. Since powers of different primes are coprime, we find that

$$\text{ord}_n a = \left[\text{ord}_{q_1} a, \text{ord}_{q_2} a, \dots, \text{ord}_{q_r} a\right]. \tag{2.47}$$

On the other hand, we have

$$\text{ord}_{q_i} a \mid \lambda(q_i) \tag{2.48}$$

for $1 \leq i \leq r$ due to (2.42), Theorems 2.18 and 2.23. According to the definition of the function $\lambda(n)$, (2.47), and (2.48), we see that relation (2.45) is true.

Further we prove that there exists $b \in \mathbb{N}$ such that (2.44) is satisfied. Consider the prime-power factorization (2.46). By the Chinese Remainder Theorem 1.4 and Theorems 2.22 and 2.23, there exists an integer b such that for $1 \leq i \leq r$,

$$b \equiv g_i \pmod{p_i^{k_i}}, \tag{2.49}$$

where g_1 is a primitive root modulo $p_1^{k_1}$, if p_1 is odd, or $p_1 = 2$ and $1 \leq k_1 \leq 2$, $g_1 = 5$, if $p_1 = 2$ and $k_1 \geq 3$, and g_i is a primitive root modulo $p_i^{k_i}$ for $2 \leq i \leq r$. Since $\left(g_i, p_i^{k_i}\right) = 1$ for $1 \leq i \leq r$, it follows that $(b, n) = 1$. We claim that $\mathrm{ord}_n b = \lambda(n)$. By (2.49), (2.42), and Theorem 2.23, we obtain

$$\mathrm{ord}_{q_i} b = \lambda(q_i) \tag{2.50}$$

for $1 \leq i \leq r$. Since

$$\mathrm{ord}_n b = \left[\mathrm{ord}_{q_1} b, \ \mathrm{ord}_{q_2} b, \ \ldots, \ \mathrm{ord}_{q_r} b\right],$$

we observe by (2.50) and the definition of λ that $\mathrm{ord}_n b = \lambda(n)$. $\square$

2.10 Legendre and Jacobi Symbol

The French mathematician Adrien-Marie Legendre (1752–1813) introduced a very useful symbol $\left(\frac{a}{p}\right) \in \{-1, 0, 1\}$ into number theory. Before we define it and discuss its properties, we define quadratic residues and nonresidues modulo a prime.

Let $n \geq 2$ and a be coprime integers. If the quadratic congruence

$$x^2 \equiv a \pmod{n}$$

has a solution x, then a is called a *quadratic residue modulo n*. Otherwise, a is called a *quadratic nonresidue modulo n*.

Let p be an odd prime. Then the *Legendre symbol* $\left(\frac{a}{p}\right)$ is defined as follows

$$1, \text{ if } a \text{ is a quadratic residue modulo } p,$$
$$-1, \text{ if } a \text{ is a quadratic nonresidue modulo } p,$$
$$0, \text{ if } p \mid a.$$

The quadratic character a with respect to a prime number p is thus expressed via the Legendre symbol. It is obvious that the sequence $\left(\left(\frac{a}{p}\right)\right)_{a=0}^{\infty}$ is periodic. For example,

Fig. 2.6 Black circles
denote the value (-1) of the
Legendre symbol for
$p = 17$. In this special case
they coincide with primitive
roots. The other circles
correspond to nonnegative
values of the Legendre
symbol $\left(\frac{a}{p}\right)$, while its zero
value is only at point 0

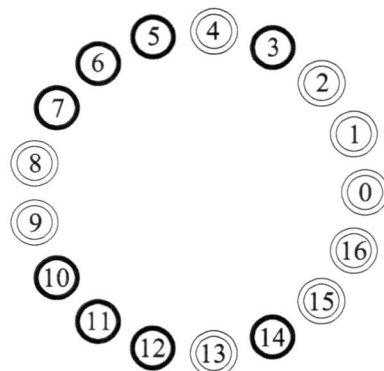

$$\left(\frac{0}{7}\right) = 0, \ \left(\frac{1}{7}\right) = 1, \ \left(\frac{2}{7}\right) = 1, \ \left(\frac{3}{7}\right) = -1, \ \left(\frac{4}{7}\right) = 1, \ \left(\frac{5}{7}\right) = -1, \ \left(\frac{6}{7}\right) = -1,$$

$$\left(\frac{7}{7}\right) = 0, \ \left(\frac{8}{7}\right) = 1, \ \left(\frac{9}{7}\right) = 1, \ \dots.$$

If p is an odd prime, then the number of quadratic residues is equal to the number of quadratic nonresidues, which is equal to $\frac{1}{2}(p - 1)$ (for a proof see e.g. [327, p. 279]). Values of the Legendre symbol for $p = 17$ are marked in Fig. 2.6.

The Legendre symbol plays an important role in testing prime numbers and in calculating primitive roots. In Sects. 5.2, 7.13, 11.5, etc., we shall see some of its practical applications. Leonhard Euler proposed a simple method how to calculate the Legendre symbol $\left(\frac{a}{p}\right)$.

Theorem 2.25 (Euler Criterion) *If p is an odd prime, then*

$$\left(\frac{a}{p}\right) \equiv a^{\frac{p-1}{2}} \quad (\text{mod } p). \tag{2.51}$$

Proof Let p be an odd prime. If $p \mid a$, then (2.51) is obviously valid. Now suppose that $p \nmid a$. By Fermat's Little Theorem 2.17,

$$a^{p-1} - 1 = \left(a^{(p-1)/2} - 1\right)\left(a^{(p-1)/2} + 1\right) \equiv 0 \quad (\text{mod } p).$$

Since p is a prime, then either $a^{(p-1)/2} \equiv 1 \pmod{p}$, or $a^{(p-1)/2} \equiv -1 \pmod{p}$, see (2.33). Hence, its suffices to prove that $a^{(p-1)/2} \equiv 1 \pmod{p}$ if and only if a is a quadratic residue modulo p.

Let a be a quadratic residue modulo p, i.e., $a \equiv b^2 \pmod{p}$ for some integer b such that $b \not\equiv 0 \pmod{p}$. Then by Fermat's Little Theorem 2.17 we have

$$a^{(p-1)/2} \equiv \left(b^2\right)^{(p-1)/2} \equiv b^{p-1} \equiv 1 \quad (\text{mod } p).$$

Conversely, assume that $a^{(p-1)/2} \equiv 1 \pmod{p}$ and let q be a primitive root modulo p. Then $a \equiv q^t \pmod{p}$ for some t such that $1 \le t \le p - 1$. Hence,

$$q^{t(p-1)/2} \equiv a^{(p-1)/2} \equiv 1 \pmod{p}.$$

However, $\text{ord}_p q = p - 1$, and thus $(p - 1) \mid t(p - 1)/2$. This implies that $2 \mid t$. So let $t = 2j$. Then

$$(q^j)^2 = q^t \equiv a \pmod{p}$$

and $\left(\frac{a}{p}\right) = 1$. $\square$

Below are some basic properties of the Legendre symbol, which will be employed in further chapters.

Theorem 2.26 *Let p be an odd prime and let a and b be integers. Then*

(i) *if $a \equiv b \pmod{p}$, then $\left(\dfrac{a}{p}\right) = \left(\dfrac{b}{p}\right)$;*

(ii) $\left(\dfrac{ab}{p}\right) = \left(\dfrac{a}{p}\right)\left(\dfrac{b}{p}\right)$;

(iii) *if $(a, p) = 1$, then $\left(\dfrac{a^2}{p}\right) = 1$, $\left(\dfrac{a^2 b}{p}\right) = \left(\dfrac{b}{p}\right)$;*

(iv) $\left(\dfrac{1}{p}\right) = 1$, $\left(\dfrac{-1}{p}\right) = (-1)^{\frac{p-1}{2}}$;

(v) $\left(\dfrac{2}{p}\right) = (-1)^{\frac{p^2-1}{8}}$.

Proofs of this theorem can be found in [56, pp. 178–183] and [291, pp. 132–135].

The celebrated Law of Quadratic Reciprocity was first proved by Gauss. He called it *Theorema aureum* (Golden Theorem). It is an invaluable tool in calculating the Legendre symbol $\left(\frac{p}{q}\right)$ when p and q are both odd primes (see [118, Articles 135–144]).

Theorem 2.27 (Law of Quadratic Reciprocity) *Let p and q be both odd primes. Then*

$$\left(\frac{p}{q}\right)\left(\frac{q}{p}\right) = (-1)^{\frac{p-1}{2}\frac{q-1}{2}}.$$

For proofs we refer to [56, pp. 188–190], [173, pp. 45–46], and [291, pp. 137–138].

Theorems 2.26–2.27 enable us to substantially simplify computation of the Legendre symbol. For example, if $p = 1777$, then by the property (ii), the Law of Quadratic Reciprocity, and properties (i), (iii), (v), and (iv) we obtain

$$\left(\frac{119}{1777}\right) = \left(\frac{7 \cdot 17}{1777}\right) = \left(\frac{7}{1777}\right)\left(\frac{17}{1777}\right) = (-1)^{888 \cdot 3}\left(\frac{1777}{7}\right) \times (-1)^{888 \cdot 8}\left(\frac{1777}{17}\right)$$

$$= \left(\frac{6}{7}\right)\left(\frac{3^2}{17}\right) = \left(\frac{2}{7}\right)\left(\frac{3}{7}\right) = (-1)^6\left(\frac{3}{7}\right) = (-1)^3\left(\frac{7}{3}\right) = -\left(\frac{1}{3}\right) = -1.$$

The Legendre symbol $\left(\frac{a}{p}\right)$, where p is a prime, was generalized by the German mathematician Carl Gustav Jacobi (1804–1851) as follows:

Let a be an integer and let $n \geq 3$ be odd. Let $n = p_1 p_2 \cdots p_r$, where the p_i's are odd primes, not necessarily distinct. Then the *Jacobi symbol* $\left(\frac{a}{n}\right)$ is defined by

$$\left(\frac{a}{n}\right) = \prod_{i=1}^{r}\left(\frac{a}{p_i}\right), \tag{2.52}$$

where $\left(\frac{a}{p_i}\right)$ is the Legendre symbol. We will use the Jacobi symbol in Proth's Theorem 3.2 and in definitions of several pseudoprimes.

Properties of the Jacobi symbol are similar to those of the Legendre symbol. For instance, the Law of Quadratic Reciprocity again holds. For odd primes, both symbols are equal.

Theorem 2.28 *Let $m > 1$ and $n > 1$ be odd integers and let a and b be integers. Then*

(i) *if $a \equiv b$ (mod n), then $\left(\frac{a}{n}\right) = \left(\frac{b}{n}\right)$,*

(ii) $\left(\frac{ab}{n}\right) = \left(\frac{a}{n}\right)\left(\frac{b}{n}\right)$,

(iii) $\left(\frac{a}{mn}\right) = \left(\frac{a}{m}\right)\left(\frac{a}{n}\right)$,

(iv) *if $(a, n) = 1$, then $\left(\frac{a^2}{n}\right) = \left(\frac{a}{n^2}\right) = 1$,*

(v) *if $(ab, mn) = 1$, then $\left(\frac{a^2 b}{m^2 n}\right) = \left(\frac{b}{n}\right)$,*

(vi) $\left(\frac{1}{n}\right) = 1$, $\left(\frac{-1}{n}\right) = (-1)^{\frac{n-1}{2}}$,

(vii) $\left(\frac{2}{n}\right) = (-1)^{\frac{n^2-1}{8}}$,

(viii) $\left(\frac{m}{n}\right)\left(\frac{n}{m}\right) = (-1)^{\frac{m-1}{2}\frac{n-1}{2}}$ *(Law of Quadratic Reciprocity).*

The proof can be found in [291, pp. 143–146].

If the Jacobi symbol $\left(\frac{a}{n}\right) = 1$, then a is not necessarily a quadratic residue modulo n, when n is composite. For instance,

$$\left(\frac{2}{15}\right) = \left(\frac{2}{3}\right)\left(\frac{2}{5}\right) = (-1)(-1) = 1,$$

but 2 is not a quadratic residue modulo 15.

Theorem 2.29 *If $n \geq 3$, then every primitive root is a quadratic nonresidue modulo n.*

Proof Let $n \geq 3$, $a \in \{1, \ldots, n-1\}$ be a quadratic residue modulo n, and let $(a, n) = 1$. We show that a is not a primitive root. By definition there exists an integer x such that

$$x^2 \equiv a \pmod{n}.$$

By (1.4) we can raise this congruence to $\phi(n)/2$, since $n \geq 3$ and $\phi(n)$ is even. Therefore, by the Euler–Fermat Theorem 2.21 we have

$$a^{\phi(n)/2} \equiv x^{\phi(n)} \equiv 1 \pmod{n}.$$

From this it follows that a is not a primitive root. $\qquad\square$

For completeness, let us point out that for $n = 2$ the number 1 is the primitive root modulo 2, but the set of quadratic nonresidues is empty. For $n = 1$ the sets of primitive roots and quadratic nonresidues modulo n are both empty.

2.11 Prime Factorization

Fermat's Little Theorem 2.17 can be used to show that a given number $q > 1$ is composite without knowing any non-trivial divisor. If we find some a coprime to q such that congruence (2.35) does not hold, then q is not a prime number. For example, if we show that q does not divide $2^{q-1} - 1$ for q odd, then q is composite.

It should be emphasized that the factorization of a given number n to prime factors is much more difficult than finding whether n is a prime or composite number. According to Gauss, finding the prime number factorization of a given number is a very important and fundamental role of arithmetic [118, Article 329].

The oldest method to factor the number n into prime factors gradually tries to divide n by all primes not exceeding $\sqrt{n}$. For instance, to show that 283 is a prime it is enough to verify that it is not divisible by 2, 3, 5, 7, 11, and 13. However, it must be emphasized that this method is not very effective. Suppose, for instance, we are able to perform one billion divisions per second. Then a factorization of $n = pq$, where p and q are unknown thirty digits primes, would take more time than the age of the universe ($\approx 13.7 \cdot 10^9$ years).

To explore this assertion further, let us denote by $\pi(x)$ the number of primes not exceeding x (see Fig. 2.7). Note that Carl Friedrich Gauss when he was a teenager conjectured that the probability that n is a prime number is $1/\log n$. By [138, 350] the value of $\pi(x)$ is approximately equal to $\frac{x}{\log x}$, where the error is less than 15 % for every $x \geq 3000$. In 1896 Jacques Hadamard (1865–1963) and independently also Charles-Jean de la Vallée Poussin (1866–1962) even proved this asymptotic equality for $x \to \infty$,

$$\pi(x) \approx \frac{x}{\log x} \qquad (Prime\ Number\ Theorem). \qquad (2.53)$$

Fig. 2.7 Graph of the function $\pi(x)$

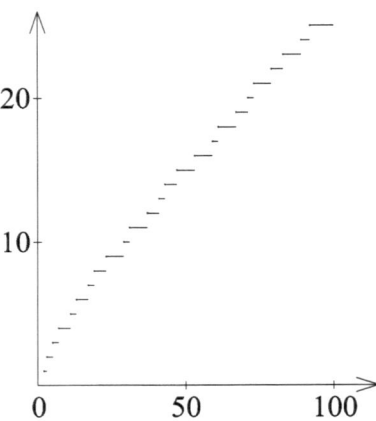

Since the integer part of $\sqrt{n}$ has 30 digits, there exist at least $10^{29}/(29 \log 10)$ $\approx 1.5 \cdot 10^{27}$ prime numbers less than $\sqrt{n}$. Because each year has about $3.2 \cdot 10^7$ s, we will perform $3.2 \cdot 10^{16}$ divisions during this time period. But to test all primes not exceeding $\sqrt{n}$, we would need at least $47 \cdot 10^9$ years to factor n which is more than three times the age of the universe. Mathematicians compare such an algorithm to an effort to smash an atom by a hammer. By having any additional information about the number to be factored or about its prime factors, the factorization can be done much more efficiently (see e.g. Theorems 4.5 and 4.17).

Further, we shall assume that n is odd and composite, because powers of two can be easily separated. One of the methods for factoring the number n is to try to express n as the difference of two squares. In this way we immediately get a factorization into nontrivial factors:

$$n = a^2 - b^2 = (a+b)(a-b), \tag{2.54}$$

when $a - b > 1$. If $n = qr$, where $q \geq r > 1$, then n can be written as the difference of two squares

$$n = \left(\frac{q+r}{2}\right)^2 - \left(\frac{q-r}{2}\right)^2,$$

where the numbers in parentheses are integer and nonnegative, since q and r are odd.

Using (2.54) Fermat proposed a method that is today called *Fermat's factorization method* after him. Note that his method is effective only if n is a product of two almost equally large factors. Fermat's algorithm proceeds as follows (see e.g. [30]):

If n is a square, we are finished. If it is not a square, we define

$$a = \lfloor \sqrt{n} \rfloor + 1,$$

where $\lfloor \sqrt{n} \rfloor$ denotes the integer part of $\sqrt{n}$. Further, we set

$$x = a^2 - n.$$

If x is a square, we are done, since $n = (a + \sqrt{x})(a - \sqrt{x})$. Otherwise we will calculate another possible candidate

$$(a + 1)^2 - n = a^2 + 2a + 1 - n = x + 2a + 1$$

and verify if it is a square or not. If so, we are again done, otherwise we increase a by two and we continue in the same way. After finitely many steps, we find two non-trivial divisors (see e.g. [56, pp. 97–99], [173, pp. 143–144]).

The following factorization method is attributed to Leonhard Euler. Let n can be expressed as the sum of two nonzero squares in two different ways, i.e.,

$$n = a^2 + b^2 = c^2 + d^2.$$

Without loss of generality we may assume that $a > c \geq d > b \geq 1$. Then n is composite, since

$$
\begin{aligned}
n &= \frac{(a^2 - d^2)(c^2 + d^2)}{a^2 - d^2} = \frac{a^2 c^2 - d^2(c^2 - a^2 + d^2)}{a^2 - d^2} \\
&= \frac{a^2 c^2 - d^2 b^2}{a^2 - d^2} = \frac{(ac + bd)(ac - bd)}{(a + d)(a - d)}.
\end{aligned}
$$

Since n is an integer, we can divide out all factors in the denominator and thus we obtain n as the product of two nontrivial factors. We will return to this method in Theorem 3.13.

The French mathematician Sophie Germain (1776–1831) found that each number of the form $a^4 + 4$ is composite for $a > 1$, because it can be split into two non-trivial factors,

$$a^4 + 4 = \left(a^2 + 2\right)^2 - 4a^2 = \left(a^2 + 2 + 2a\right)\left(a^2 + 2 - 2a\right). \tag{2.55}$$

The equality (2.55) is a special case of the formula

$$a^4 + 4b^4 = \left(a^2 + 2b^2\right)^2 - 4a^2 b^2 = \left(a^2 + 2b^2 + 2ab\right)\left(a^2 + 2b^2 - 2ab\right).$$

Another simple factorization for $m > 1$ was found by Léon François Antoine Aurifeuille in 1873,

$$2^{4m-2} + 1 = (2^{2m-1} + 2^m + 1)(2^{2m-1} - 2^m + 1). \tag{2.56}$$

Note that Fortuné Landry spent many years in searching factorization of the number $2^{58} + 1$, which he found in 1869. However, knowledge of the relationship (2.56) would have saved him a lot of time, because

$$2^{58} + 1 = \left(2^{29} + 2^{15} + 1\right)\left(2^{29} - 2^{15} + 1\right) = 536838145 \cdot 536903681,$$

where the first factor is divisible by 5.

Let us introduce some other known factorizations into two factors

$$a^m - b^m = (a - b)(a^{m-1} + a^{m-2}b + \cdots + ab^{m-2} + b^{m-1}).$$

If $m = 2k$ is even, then there is, in general, another factorization

$$a^{2k} - b^{2k} = (a + b)(a^{2k-1} - a^{2k-2}b + a^{2k-3}b^2 - \cdots - b^{2k-1}), \qquad (2.57)$$

Moreover, for an odd number $m = 2k + 1$ we have

$$a^{2k+1} + b^{2k+1} = (a + b)(a^{2k} - a^{2k-1}b + a^{2k-2}b^2 - a^{2k-3}b^3 + \cdots + b^{2k})$$

and as a special case we get

$$a^{2m} + b^{2m} = (a^2)^m + (b^2)^m = (a^2 + b^2)(a^{2m-2} - a^{2m-4}b^2 + \cdots + b^{2m-2}),$$

where m is odd.

The factorization into three factors can be performed for numbers of the form

$$3^{6k-3} + 1 = \left(3^{2k-1} + 1\right)\left(3^{2k-1} - 3^k + 1\right)\left(3^{2k-1} + 3^k + 1\right).$$

Other examples of special types of numbers, which can be a priori factored into smaller factors, are given in the book [416, p. 15].

As has already been said, it is much more difficult to factor a composite number n into two nontrivial factors than just find out if n is composite. The well-known RSA method for encryption of secret messages in which n is the product of two large prime numbers is based on this fact (see Sect. 5.2). The number of arithmetic operations necessary for a prime number factorization of such a large number grows exponentially. If n has, for example, 300 digits, then it is practically impossible to factor it by today's means. On the other hand, to verify whether a given number with 300 digits is a prime takes less than a few minutes on current computers.

An overview of the best algorithms for prime number factorization is given e.g. in [3, 82, 227, 307], see also Wagstaff [408, pp. 135–141, 158–163] for Pollard's rho method, Pollard's $p - 1$ method, and the continued fraction factoring algorithm. Lenstra's elliptic curve method (ECM) is especially suitable for finding prime factors up to about 40 digits (see [408, pp. 181–186]). Instead of using the group of integers modulo p, the ECM uses the group of points on an elliptic curve (see (2.18)). The two most effective methods for factoring numbers with a large number of digits are the quadratic sieve and the number field sieve, see [307] and [408, pp. 195–202, 207–217]. The quadratic sieve was developed by Carl Pomerance in 1981, while the idea of the number field sieve was introduced by John Pollard in 1988. The special number field sieve is especially suitable for the factorization of those numbers for which there exists a "close" number (± 10) composed only of small factors less than 20000. The general number field sieve is able to deal with numbers of a general form. The quadratic sieve is considered at present to be the fastest factorization method for numbers of up to about 100 digits, while the number field sieve is judged to the best method of factoring for numbers with more than 130 digits (see [307, p. 1483]).

Chapter 3
Properties of Prime Numbers

3.1 Criteria for Primality

Let j, m, n be positive integers. We say that m^j *exactly divides* n and write $m^j \parallel n$, if $m^j \mid n$, but $m^{j+1} \nmid n$. For $j = 0$ the symbol $m^0 \parallel n$ means that $m \nmid n$.

Theorem 3.1 *A positive integer n is a prime if and only if $n \mid \binom{n}{k}$ for all $k \in \{1, \ldots, n-1\}$.*

Proof $\Rightarrow$: For $k \in \{1, \ldots, n-1\}$ the number $n - k$ is also between 1 and $n - 1$. Since n is prime, $n \nmid k!$ and $n \nmid (n-k)!$. Hence, we have

$$n \left| \binom{n}{k} \right. = \frac{n!}{k!(n-k)!},$$

since $n \mid n!$.

$\Leftarrow$: Conversely, let n be composite and let p be the smallest prime which divides n. Hence, $1 < p < n$ and

$$\binom{n}{p} = \frac{n(n-1)\cdots(n-p+1)}{p(p-1)\cdots 2 \cdot 1}. \tag{3.1}$$

Assume that $p^j \parallel n$ for some positive integer j. Between p successive numbers $n, n-1, \ldots, n-p+1$ there exists exactly one divisible by p. Since $p \mid n$, we obtain

$$p \nmid (n-1)(n-2)\cdots(n-p+1).$$

Therefore, $p^j \parallel n(n-1)\cdots(n-p+1)$ and clearly $p \parallel p(p-1)\cdots 2 \cdot 1$. Thus by (3.1) we get $p^{j-1} \parallel \binom{n}{p}$. However, $n \nmid \binom{n}{p}$, since $p^j \parallel n$. $\qquad \square$

M. Křížek et al., *From Great Discoveries in Number Theory to Applications*, https://doi.org/10.1007/978-3-030-83899-7_3

As an important consequence of Theorem 3.1, we get the following rule, which makes it easier to manipulate with powers and congruences. If p is a prime, then

$$(a + b)^p \equiv a^p + b^p \pmod{p}. \tag{3.2}$$

From the $p + 1$ terms of $(a + b)^p$ modulo p by the Binomial Theorem

$$(a + b)^p = a^p + \binom{p}{1}a^{p-1}b + \binom{p}{2}a^{p-2}b^2 + \cdots + b^p$$

only a^p and b^p remain, since all others are divisible by p according to Theorem 3.1, i.e., $p \mid \binom{p}{k}$ for $1 \le k \le p - 1$, which can also be written as follows

$$\binom{p}{k} \equiv 0 \pmod{p}.$$

Theorem 3.1 has a number of further applications. For example, it was used in a pioneering article [4] by Agrawal, Kayal, and Saxena, where a new polynomial time deterministic algorithm for testing a prime number is described. For a given number n its computational complexity is $\mathcal{O}((\log n)^{12})$, where $\log n$ is proportional to the number of digits of n. According to Crandall and Pomerance [82, p. 207], the number of operations can be reduced to $\mathcal{O}((\log n)^{6+\varepsilon})$ for any $\varepsilon > 0$ by means of Gaussian periods. However, for the time being this algorithm is not used for practical purposes, because the exponent 6 is too large. We need to reduce the exponent to at most 3 with a small constant factor.

The French mathematician François Proth (1852–1879) during his short life published the following statement (see [313]).

Theorem 3.2 (Proth) *Let $q = k2^n + 1$, $k < 2^n$ and let $\left(\frac{a}{q}\right) = -1$. Then q is a prime if and only if*

$$a^{(q-1)/2} \equiv -1 \pmod{q}. \tag{3.3}$$

The proof is given e.g. in [199, p. 70]. Proth's Theorem has an important application in primality testing (see Theorem 4.13).

Theorem 3.3 *Let p be an odd prime. Then*

$$p \equiv 1 \pmod{4} \iff \left(\frac{-1}{p}\right) = 1,$$

$$p \equiv -1 \pmod{4} \iff \left(\frac{-1}{p}\right) = -1.$$

Proof By Theorem 2.26 for a prime of the form $p = 4k + 1$ we have

$$\left(\frac{-1}{p}\right) = (-1)^{(p-1)/2} = (-1)^{2k} = 1.$$

Analogously, for $p = 4k - 1$ we get

$$\left(\frac{-1}{p}\right) = (-1)^{(p-1)/2} = (-1)^{2k-1} = -1.$$

From this the converse implications also follow.   □

Theorem 3.4 *If p is a prime and a is a positive integer less than p, then there exists exactly one number $x \in \mathbb{N}$ less than p which is a solution of the congruence*

$$ax \equiv 1 \pmod{p}. \tag{3.4}$$

Proof By Fermat's Little Theorem 2.17

$$a^{p-1} = a(a^{p-2}) \equiv 1 \pmod{p},$$

since $(a, p) = 1$. Hence, we can choose x to be the positive integer less than p congruent to a^{p-2} modulo p.

Next, we prove the uniqueness of x. Let there exist two solutions of the congruence (3.4), i.e., $ax \equiv ay \equiv 1 \pmod{p}$. Since

$$ax - ay = a(x - y) \equiv 0 \pmod{p},$$

we see by Euclid's Theorem 2.1 that $p \mid a$ or $p \mid (x - y)$. However, the prime p cannot divide a, since $(a, p) = 1$, and consequently, we obtain $p \mid (x - y)$, i.e., $x \equiv y \pmod{p}$. Since both the positive integers x and y are less than p, we get $x = y$.   □

The unique positive integer x from Theorem 3.4 will be called the *inverse element* of a modulo a prime p.

Theorem 3.5 *If $p > 2$ is a prime and $a^2 \equiv 1 \pmod{p}$, then*

$$either \; a \equiv 1 \pmod{p}, \quad or \; a \equiv -1 \pmod{p}.$$

Proof Since $(a - 1)(a + 1) \equiv 0 \pmod{p}$, we have $p \mid (a - 1)(a + 1)$. By Euclid's Theorem 2.1 for a prime $p > 2$, either $p \mid (a - 1)$, or $p \mid (a + 1)$.   □

In the next section, we will deal with another important primality criterion.

3.2 Wilson's Theorem

John Wilson (1741–1793) was an outstanding British mathematician at the University of Cambridge. However, the following theorem that bears his name was not discovered by him.

Theorem 3.6 (Wilson) *A number* $p > 1$ *is prime if and only if*

$$(p - 1)! \equiv -1 \pmod{p}. \tag{3.5}$$

In 1770 Edward Waring first published in *Meditationes algebraicae* on p. 288 the implication $\Rightarrow$ without any proof and attributed it to John Wilson. He literally wrote that if p is a prime number, then the sum of $(p - 1)! + 1$ is divisible by p. At that time, the concept of congruences had not been introduced yet. According to Hardy and Wright [138, p. 81] the implication $\Rightarrow$ was already known to Gottfried Wilhelm Leibniz (1646–1716) in somewhat modified form. The converse implication $\Leftarrow$ was later proved by Joseph-Louis Lagrange in 1773. Therefore, sometimes Theorem 3.6 is called the *Wilson–Lagrange Theorem*.

Let us further note that the assumption of $p > 1$ is omitted in many textbooks. For $p = 1$ the congruence (3.5) is satisfied, but 1 is by definition not a prime number.

Proof of Theorem 3.6. $\Rightarrow$: If $p = 2$, then congruence (3.5) obviously holds. So let $p > 2$ be a prime and let a be an arbitrary positive integer less than p. By Theorem 3.4 there exists exactly one positive integer $b < p$ such that $ab \equiv 1 \pmod{p}$. From Theorem 3.5 we get that if $ab \equiv 1 \pmod{p}$ and $b \equiv a \pmod{p}$, then either $a \equiv 1 \pmod{p}$, or $a \equiv p - 1 \pmod{p}$. From this it follows that the integers $2, 3, \ldots, p - 2$ can be reordered as the progression $a_2, a_3, \ldots, a_{p-2}$ so that in pairs we have

$$a_i a_{i+1} \equiv 1 \pmod{p}$$

for $i = 2, 4, 6, \ldots, p - 3$. Between 2 and $p - 2$ there are exactly $p - 3$ numbers, which is an even number. Therefore,

$$(p - 1)! \equiv 1 \cdot (p - 1)a_2 \cdots a_{p-2} \equiv (p - 1)1^{(p-3)/2} \equiv -1 \pmod{p}.$$

$\Leftarrow$: We show that if n is composite, then $(n - 1)! \equiv 2 \pmod{n}$ for $n = 4$ and $(n - 1)! \equiv 0 \pmod{n}$ for $n > 4$. We immediately see that

$$(4 - 1)! = 6 \equiv 2 \pmod{4}.$$

Now assume that $n > 4$ is composite and that it is not a square. Then we can factor n so that

$$n = dk,$$

where $2 \leq d < k < n - 1$. Hence, $dk \mid (n - 1)!$ and $(n - 1)! \equiv 0 \pmod{n}$.

Finally assume that $n > 4$ and $n = k^2$. Then $3 \leq k < 2k \leq n - 1$, and thus $k(2k) \mid (n - 1)!$. Since $2k^2 \equiv 0 \pmod{n}$, in this case we also have $(n - 1)! \equiv 0 \pmod{n}$. $\square$

As a consequence of the previous proof of Wilson's Theorem, we get the following statement.

Theorem 3.7 *An integer $n > 4$ is composite if and only if*

$$(n - 1)! \equiv 0 \pmod{n}. \tag{3.6}$$

Wilson's Theorem 3.6 can be proved by several different methods (see e.g. [91, Chapt. III], [330, p. 89]). The Czech mathematician Karel Petr in 1905. also dealt with the proof of the implication of $\Rightarrow$. We will briefly describe the main idea of his proof, because it has an unusually beautiful geometric interpretation, see [297].

In his proof, Petr considers oriented piecewise linear curves passing through all vertices of the regular p-gon, where p is a prime number. Curves with opposite orientations are assumed to be different. There are altogether exactly $(p - 1)!$ curves. If we start from any vertex, then we can go through the remaining $p - 1$ vertices in just as many ways as there are permutations of $p - 1$ elements.

The *regular curves* will be those which do not change when rotated by an angle of $2\pi/p$ around the center of the regular p-gon. The other curves will be called *irregular* (see Fig. 3.1).

The number of regular curves is clearly $p - 1$. For instance, if $p = 5$, one of these curves is sketched in Fig. 3.1a and the next has an opposite orientation. The remaining two regular curves look like the regular pentagon with edges oriented clockwise and counter-clockwise. Petr shows that the number of irregular curves is pm for some suitable positive integer m. From this, the total number of curves is

$$(p - 1)! = p - 1 + pm,$$

which implies the congruence (3.5) to be proved.

Wilson's Theorem 3.6 is a necessary and sufficient condition for the primality of p, while Fermat's Little Theorem 2.17 represents only a necessary condition. Nevertheless, we do not know yet an efficient algorithm for calculating the factorial appearing in (3.5). That is why Wilson's Theorem is very impractical for testing the primality of p on computers. But it is a good theoretical tool in proof techniques. The factorial on the left-hand side of the congruence (3.5) can be substantially reduced, as we will see in the following theorems.

Fig. 3.1 An example of a regular curve (left) and an irregular curve (right) for $p = 5$

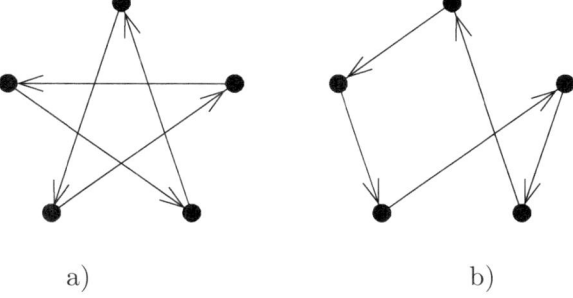

a)   b)

Theorem 3.8 *If p is a prime of the form $p = 4k + 1$, then*

$$\left(\left(\frac{p-1}{2}\right)!\right)^2 \equiv -1 \pmod{p}.$$

Proof Using Wilson's Theorem 3.6, we obtain

$$(p-1)! = 1 \cdot 2 \cdot 3 \cdots \frac{p-1}{2} \frac{p+1}{2} \frac{p+3}{2} \cdots (p-1) \equiv -1 \pmod{p}, \quad (3.7)$$

where the middle fractions are integers, since p is odd. Further, we see that modulo p we have

$$p - 1 \equiv -1, \quad p - 2 \equiv -2, \ldots, \quad \frac{p+1}{2} \equiv -\frac{p-1}{2}.$$

The relation (3.7) can be rearranged in pairs so that

$$(p-1)! \equiv 1(-1)2(-2) \cdots \frac{p-1}{2}\left(-\frac{p-1}{2}\right) = \left(\left(\frac{p-1}{2}\right)!\right)^2 (-1)^{(p-1)/2} \pmod{p}.$$

From this and Wilson's Theorem 3.6 it follows that

$$\left(\left(\frac{p-1}{2}\right)!\right)^2 \equiv (-1)(-1)^{(p-1)/2} = (-1)^{(p+1)/2} \equiv -1 \pmod{p}, \quad (3.8)$$

since $p = 4k + 1$.   □

From the previous theorem if follows that the congruence

$$x^2 \equiv -1 \pmod{p}$$

has a solution if p is of the form $4k + 1$.

Theorem 3.9 *If p is a prime of the form $p = 4k - 1$, then*

$$\left(\left(\frac{p-1}{2}\right)!\right)^2 \equiv 1 \pmod{p} \qquad\qquad (3.9)$$

and

$$\left(\frac{p-1}{2}\right)! \equiv \pm 1 \pmod{p}. \qquad\qquad (3.10)$$

Proof Let $p = 4k - 1$. The whole proof is the same as that of Theorem 3.8, only the last equivalence in (3.8) will be equal to 1, and thus (3.9) holds. However, since p is an odd prime, by Theorem 3.5 we can take the square root of both sides and get (3.10).   □

The article [279] by Mordell states what specific sign in (3.10) applies for primes of the form $p = 4k - 1$.

3.3 Dirichlet's Theorem

In 1837 Peter Gustav Lejeune Dirichlet (1805–1859) published an interesting theorem that uses very sophisticated analytical methods in number theory.

Theorem 3.10 (Dirichlet) *Let $a, d \in \mathbb{N}$ be coprime integers. Then there exist infinitely many primes in the arithmetic progression*

$$a, \; a + d, \; a + 2d, \; a + 3d, \; \ldots$$

A proof of this statement is in the seminal paper by Peter Gustav Lejeune Dirichlet [94]. Theorem 3.10 can be equivalently formulated so that the set

$$S = \{p \in \mathbb{P}; \;\; p \equiv a \pmod{d}\}$$

has infinitely many elements. Moreover, the density of S in the set of primes $\mathbb{P}$ is equal to $1/\phi(d)$, where ϕ is the Euler totient function, i.e.,

$$\lim_{x \to \infty} \frac{|\{p \in \mathbb{P}; \;\; p \equiv a \pmod{d} \text{ and } p \le x\}|}{|\{p \in \mathbb{P}; \;\; p \le x\}|} = \frac{1}{\phi(d)}.$$

A proof of this statement can be found e.g. in Ireland and Rosen [151, pp. 251–261]. However, one of the most beautiful and at the same time most surprising mathematical results from the beginning of the 21st century is the Green-Tao theorem published in Annals of Mathematics, see Ben Green and Terence Tao [126].

Theorem 3.11 (Green-Tao) *For any positive integer k there exists an arithmetic progression of length k consisting solely of primes.*

Its original very difficult and nearly seventy-page proof was recently reduced to about a third, see Conlon, Fox, and Zhao [76]. In particular, for $k = 3, \; 5, \; 6, \; 10$ the following arithmetic progressions

3, 5, 7

5, 11, 17, 23, 29

7, 37, 67, 97, 127, 157

199, 409, 619, 829, 1039, 1249, 1459, 1669, 1879, 2089

consist of primes (see Table 13.1 at the end of this book). The Green-Tao Theorem in fact states that for any arbitrarily large positive integer k there always exist prime numbers

$$p_1 < p_2 < \cdots < p_k$$

such that

$$p_2 - p_1 = p_3 - p_2 = \cdots = p_k - p_{k-1}.$$

We shall utilize this strong property in Sects. 9.2 and 9.4.

3.4 Fermat's Christmas Theorem

On December 25, 1640, Pierre de Fermat wrote a letter to Marin Mersenne in which he informed him about his new discovery. Fermat figured out that every prime number of the form $4k + 1$ for $k \in \mathbb{N}$ can be uniquely written as the sum of two squares of positive integers. Moreover, such a prime number is a hypotenuse of exactly one Pythagorean triangle with sides $a < b < c$. In the letter (in Old French) he literally writes:

Tout nombre premier, qui sourpasse de l'unité un multiple du quaternaire, est une seul fois la somme de deux quarrés, et une seule fois l'hypoténuse d'un triangle rectangle.

For example, the primes 5, 13, and 17 can be written as a sum two squares as follows

$$5 = 1^2 + 2^2, \quad 13 = 2^2 + 3^2 \quad \text{and} \quad 17 = 1^2 + 4^2.$$

This important discovery is called Fermat's Christmas Theorem, because Fermat wrote his letter to Mersenne just during Christmas. Further, we see that the numbers 5, 13, and 17 are the hypotenuses of Pythagorean triangles with sides $\langle 3, 4, 5 \rangle$, $\langle 5, 12, 13 \rangle$, and $\langle 8, 15, 17 \rangle$. We will return to this interesting feature later.

Any odd natural number greater than 1 is obviously of the form $4k \pm 1$ for some $k \in \mathbb{N}$. From Theorem 2.6, which is also due to Fermat, we already know that no number of the form $4k - 1$ for $k \in \mathbb{N}$ is a sum of two squares of integers. In Theorems 3.12 and 3.13 we will deal with numbers of the form $4k + 1$.

In proving the Christmas Theorem, Fermat used the following equalities due to Françoise Viète

$$(a^2 + b^2)(c^2 + d^2) = (ac + bd)^2 + (ad - bc)^2 = (ac - bd)^2 + (ad + bc)^2, \tag{3.11}$$

which are at present called the *Viète identities*. A detailed proof of Fermat's Christmas Theorem was given by Euler. A condensed "one-sentence" proof by Don Zaiger is mentioned in [6, p. 19].

Theorem 3.12 (Fermat's Christmas Theorem) *Every prime number of the form $p = 4k + 1$ can be uniquely written as the sum of two squares of positive integers.*

Proof First we prove the uniqueness of such a decomposition. So suppose that a given prime number p can be written in two ways as the sum of two squares

$$p = a^2 + b^2 = c^2 + d^2,$$

where a, b, c, d are positive integers. Then

$$a^2 c^2 - b^2 d^2 = a^2(a^2 + b^2 - d^2) - d^2(c^2 + d^2 - a^2) = a^2(a^2 + b^2) - d^2(c^2 + d^2)$$
$$= p(a^2 - d^2) \equiv 0 \pmod{p},$$

and by Euclid's Theorem 2.1 we have

$$ac \equiv bd \pmod{p} \quad \text{or} \quad ac \equiv -bd \pmod{p}.$$

Since all numbers a, b, c, d are less than $\sqrt{p}$, we find that

$$ac - bd = 0 \quad \text{or} \quad ac + bd = p.$$

If the second equality holds, then by the Viète identities (3.11) we get

$$p^2 = (a^2 + b^2)(c^2 + d^2) = (ac + bd)^2 + (ad - bc)^2 = p^2 + (ad - bc)^2,$$

i.e., $ad - bc = 0$. Hence, altogether

$$ac = bd \quad \text{or} \quad ad = bc.$$

Assume, for instance, that $ac = bd$. Then $a \mid d$, since $a \mid bd$ and $(a, b) = 1$. Therefore, there exists a positive integer m such that $d = ma$. From this it follows that $c = mb$, and thus
$$p = c^2 + d^2 = m^2(a^2 + b^2),$$

which yields $m = 1$, $a = d$, and $b = c$.

Assuming $ad = bc$, we would similarly get that $a = c$ and $b = d$. In this way we proved the uniqueness of the decomposition of the prime p into two squares (up to order).

Next, we will focus on proving the existence of this decomposition. Let

$$z \equiv \left(\frac{p-1}{2}\right)! \pmod{p},$$

where $0 \le z \le p - 1$. Since p is of the form $4k + 1$, by Theorem 3.8 we have

$$z^2 \equiv -1 \quad (\text{mod } p). \tag{3.12}$$

Consider an ordered pair $\langle x, y \rangle$, where $x, y \in \{0, 1, \ldots, n - 1\}, n = \lfloor \sqrt{p} \rfloor + 1$ and $\lfloor r \rfloor$ denotes the integer part of a real number r (i.e. the largest integer not exceeding r). There exist exactly n^2 such pairs, and therefore there exist n^2 integers of the form $x - zy$ for $x, y \in \{0, 1, \ldots, n - 1\}$, since z is fixed. Since there are exactly p different remainders modulo p and $n^2 > p$, it follows from Dirichlet's pigeonhole principle (see Theorem 1.5) that there exist two different ordered pairs $\langle x', y' \rangle$ and $\langle x'', y'' \rangle$ such that

$$x' - zy' \equiv x'' - zy'' \quad (\text{mod } p),$$

where $x', y', x'', y'' \in \{0, 1, \ldots, n - 1\}$. Hence,

$$x' - x'' \equiv z(y' - y'') \quad (\text{mod } p). \tag{3.13}$$

Setting

$$a = |x' - x''| \quad \text{and} \quad b = |y' - y''|,$$

we get $a, b \in \{0, 1, \ldots, n - 1\}$ and by (3.13), (1.4), and (3.12) it follows that

$$a^2 \equiv z^2 b^2 \equiv -b^2 \quad (\text{mod } p),$$

i.e.,

$$a^2 + b^2 \equiv 0 \quad (\text{mod } p). \tag{3.14}$$

However, both the numbers a and b are not zero at the same time, since $\langle x', y' \rangle$ and $\langle x'', y'' \rangle$ are different ordered pairs. If p is a prime, then $\sqrt{p}$ is not an integer. Consequently, by (3.14) we have $0 \le a < \sqrt{p}$ and $0 \le b < \sqrt{p}$. From this it follows that

$$0 < a^2 + b^2 < (\sqrt{p})^2 + (\sqrt{p})^2 = 2p.$$

But there is only one number between the numbers $1, 2, \ldots, 2p - 1$ divisible by the prime number p, and that is the number p. Thus from (3.14) we obtain

$$a^2 + b^2 = p$$

and the theorem is proved.   □

Remark Suppose that p is a prime number of a form $p = 4k + 1$. The monograph by Bressoud and Wagon [43, p. 281] presents Smith's algorithm for calculating explicitly the positive integers a and b from Theorem 3.12 such that $p = a^2 + b^2$. To do this, it is enough to choose z so that $z^2 \equiv -1 \quad (\text{mod } p)$, and use the Euclidean algorithm on the pair (p, z). The first two residues that are both less than $\sqrt{p}$ are the sought numbers a and b.

For example, choose $p = 61$, i.e. $7 < \sqrt{61} < 8$. We see that $z = 11$ satisfies the congruence

$$z^2 \equiv -1 \pmod{61}.$$

Then by the Euclidean algorithm from Sect. 1.5 we get

$$(61, 11) = (11, 6) = (6, 5),$$

and thus

$$61 = 6^2 + 5^2.$$

Remark The number z satisfying the congruence $z^2 \equiv -1 \pmod{p}$ can be found for "small" p using Theorem 3.8. For "larger" p's we can proceed as follows: Since $p = 4k + 1$, the number $(p - 1)/2$ is even. Using Euler's criterion (2.51), we find a quadratic nonresidue a, i.e., $a^{(p-1)/2} \equiv -1 \pmod{p}$. This can be done experimentally, because the number of quadratic residues is the same as the number of quadratic nonresidues. Now it is enough to choose the remainder $z, 0 \leq z < p$, such that $z \equiv a^{(p-1)/4} \pmod{p}$.

Theorem 3.13 *If n is a composite number that can written as the product of two or more primes of the form $4k + 1$, then n can be written as the sum of two squares in several ways.*

Proof First let $n = qr$ for some primes $q > r > 1$ of the form $4k + 1$. From Fermat's Christmas Theorem 3.12 and the Viète identities (3.11)

$$n = qr = (a^2 + b^2)(c^2 + d^2) = (ac + bd)^2 + (ad - bc)^2$$
$$= (ac - bd)^2 + (ad + bc)^2, \qquad (3.15)$$

it follows that every number, which is the product of two or more primes congruent to 1 modulo 4, can be written at the sum of two squares, where a, b, c, d are positive integers. By induction we can then make sure that every number that is the product of two or more primes congruent to 1 modulo 4, is the sum of two squares.

We still have to show that the squares in (3.15) whose sum gives n are different. Clearly,

$$ac + bd \neq \pm(ac - bd).$$

So assume that $ac + bd = ad + bc$ and $ac - bd = \pm(ad - bc)$. Then by summing we get $2ac = 2ad$, i.e. $c = d$, or $2ac = 2bc$, i.e. $a = b$. But this contradicts the fact that both r and q are odd. $\qquad \square$

Example For primes $q = 641$ and $r = 101$ we have $q \equiv 1 \pmod 4$ and $r \equiv 1 \pmod 4$. Then by Smith's algorithm we get $q = 25^2 + 4^2$, $r = 10^2 + 1$, and by (3.15),

$$64\,741 = qr = 254^2 + 15^2 = 246^2 + 65^2.$$

A general theorem determining in how many ways any positive integer can be written as the sum of two squares is given in Hardy and Wright [138, p. 241]. In Sects. 4.12 and 11.7, we will point out interesting applications of Fermat's Christmas Theorem 3.12 in algebra and bioinformatics. Fermat already knew about its further application in geometry.

Theorem 3.14 *For each prime number of the form $p = 4k + 1$ there is exactly one Pythagorean triple, whose largest number (corresponding to the hypotenuse) is p.*

Proof Let p be a prime number. According to Fermat's Christmas Theorem 3.12, we have $p = m^2 + n^2$. Since p is odd, the numbers m and n cannot both be odd or both even at the same time, i.e., they have different parities. Since p is prime, we have $(m, n) = 1$. Thus p is a primitive Pythagorean triple according to relations (2.5) and (2.6). □

3.5 Polynomials Generating Primes

In 1772, Leonhard Euler wrote to Johann III Bernoulli (1744–1807) that he had discovered a quadratic polynomial

$$p(x) = x^2 + x + 41, \tag{3.16}$$

which yields the following sequence of primes (cf. Table 13.1) for $x = 0, 1, \ldots, 39$:

$$41,\ 43,\ 47,\ 53,\ 61,\ 71,\ 83,\ 97,\ 113,\ 131,$$
$$151,\ 173,\ 197,\ 223,\ 251,\ 281,\ 313,\ 347,\ 383,\ 421,$$
$$461,\ 503,\ 547,\ 593,\ 641,\ 691,\ 743,\ 797,\ 853,\ 911,$$
$$971,\ 1033,\ 1097,\ 1163,\ 1231,\ 1301,\ 1373,\ 1447,\ 1523,\ 1601. \tag{3.17}$$

The number $p(40) = 40^2 + 40 + 41 = 40 \cdot 41 + 41 = 41^2$ is composite. It is remarkable that we get primes also for negative arguments for $x = -40, -39, \ldots, -1$ of the polynomial (3.16). So in total 80 consecutive integer arguments of x yield prime numbers with each of these primes appearing twice.

Another quadratic polynomial

$$q(x) = x^2 + x + 17$$

Fig. 3.2 Ułam's square spiral of positive integers. Primes are shown in circles to distinguish them from composite numbers

91	90	89	88	87	86	85	84	83	82
92	57	56	55	54	53	52	51	50	81
93	58	31	30	29	28	27	26	49	80
94	59	32	13	12	11	10	25	48	79
95	60	33	14	3	2	9	24	47	78
96	61	34	15	4	1	8	23	46	77
97	62	35	16	5	6	7	22	45	76
98	63	36	17	18	19	20	21	44	75
99	64	37	38	39	40	41	42	43	74
100	65	66	67	68	69	70	71	72	73

generates prime numbers for $x = -16, -15, \ldots, 15$, yielding 16 distinct prime values. Russell Ruby discovered the polynomial

$$r(x) = 36x^2 - 810x + 2753,$$

which gives 45 different primes for $x = 0, 1, \ldots, 44$. There are dozens of similar polynomials in the literature ($2x^2 + 29$, $x^2 - 79x + 1601$, $103x^2 - 3945x + 34381$, etc.). In 2002, Dress and Landreau found the fifth-degree polynomial

$$s(x) = \frac{1}{4}(x^5 - 133x^4 + 6729x^3 - 158379x^2 + 1720294x - 6823316)$$

which generates 57 distinct primes in absolute value for $x = 0, 1, \ldots, 56$. (Note that quintic polynomial equations are not solvable in radicals, see [214].)

Figure 3.2 shows Ułam's square spiral which was designed by Polish mathematician Stanislaw Marcin Ułam (1909–1984) in 1963. The next Fig. 3.3 shows the distribution of the first 16 000 primes in Ułam's spiral. Notice interesting structures containing lines with slopes ± 1 which correspond to quadratic polynomials of the form $4x^2 + bx + c$. For instance, the polynomial

$$u(x) = 4x^2 + 10x + 5$$

produces the following primes $u(0) = 5, u(1) = 19, u(2) = 41$, and $u(3) = 71$ lying on the line with slope -1 (see Fig. 3.2). Similarly, for the polynomial $v(x) = 4x^2 + 10x + 7$ we have $v(-4) = 31$, $v(-3) = 13$, $v(-2) = 3$, etc. Unwinding the spiral from the number 41 we would get in a similar way for the above Euler's polynomial $p = p(x)$ a line with slope -1 containing 40 consecutive prime numbers.

Fig. 3.3 Distribution of the first 16 000 primes on Ułam's spiral

Jakob Philipp Kulik (1793–1863), Professor of Mathematics at the University of Prague, came up with a similar idea much earlier. He studied the distribution of primes arranged in an isosceles triangle (see [216, p. 102]). He also compiled extensive tables of prime divisors of integers up to one hundred million. To this day, it remains a mystery how he could have realized this respectable work.

In 1976 J. P. Jones, D. Sato, H. Wada, and D. Wiens published a 25th degree polynomial of 26 variables $a, b, \ldots, z$ which are nonnegative integers. All positive values of this polynomial are prime numbers (see [416, p. 150] for details),

$$(k + 2)\{1 - [wz + h + j - q]^2 - [(gk + 2g + k + 1)(h + j) + h - z]^2$$
$$- [16(k + 1)^3(k + 2)(n + 1)^2 + 1 - f^2]^2 - [e^3(e + 2)(a + 1)^2 + 1 - o^2]$$
$$- [2n + p + q + z - e]^2 - [(a^2 - 1)y^2 + 1 - x^2]^2 - [16r^2y^4(a^2 - 1) + 1 - u^2]^2$$
$$- [((a + u^2(u^2 - a))^2 - 1)(n + 4dy^2) + 1 - (x + cu)^2]^2 - [n + l + v - y]^2$$
$$- [(a^2 - 1)l^2 + 1 - m^2]^2 - [ai + k + 1 - l - i]^2 - [p + l(a - n - 1)$$
$$+ b(2an + 2a - n^2 - 2n - 2) - m]^2 - [q + y(a - p - 1) + s(2ap + 2a$$
$$- p^2 - 2p - 2) - x]^2 - [z + pl(a - p) + t(2ap - p^2 - 1) - pm]^2\}.$$

However, counting primes in this way is very impractical. Notice that the polynomial in braces is positive only if the expressions in square brackets are zero. Then the value in braces is 1 and the desired prime number is equal to $k + 2$.

3.6 Riemann Hypothesis

Set

$$\zeta(z) = \sum_{n=1}^{\infty} \frac{1}{n^z}, \tag{3.18}$$

where the sum converges for an arbitrary complex number $z \in \mathbb{C}$ such that

$$\text{Re}(z) > 1.$$

The function ζ has complex conjugate roots and is called the *Riemann ζ-function*. At the same time it is true that $\zeta(1) = \infty$, which is the sum of the harmonic series, and $\zeta(2) = \pi^2/6$.

Now we prove the following surprising equality containing on the right-hand side the product over all primes p,

$$\zeta(z) = \prod_{p} \frac{1}{1 - p^{-z}}. \tag{3.19}$$

Note that already Leonhard Euler derived that (3.18) and (3.19) are equal for $z > 1$ real.

Let p_i be the ith prime. Then by the Fundamental Theorem of Arithmetic 2.2 each positive integer can be uniquely expressed by the product $p_1^{k_1} p_2^{k_2} p_3^{k_3} \cdots$ with nonnegative integer exponents $k_1, k_2, k_3, \ldots$ Hence, the set of positive integers $\mathbb{N}$ can be expressed by finite products as follows:

$$\mathbb{N} = \{p_1^{k_1} p_2^{k_2} p_3^{k_3} \cdots \mid k_1, k_2, k_3, \ldots \text{ are nonnegative integers}\}.$$

Then by the formula for the sum of a geometric progression we have

$$\sum_{j=0}^{\infty}\left(\frac{1}{p_i^z}\right)^j = \frac{1}{1 - p_i^{-z}},$$

for any $i = 1, 2, \ldots$ Multiplying together the first m of such equations and taking the limit, we get the desired equality

$$\prod_{i=1}^{\infty}\frac{1}{1 - p_i^{-z}}$$

$$=\left(1 + \frac{1}{2^z} + \frac{1}{2^{2z}} + \cdots\right)\left(1 + \frac{1}{3^z} + \frac{1}{3^{2z}} + \cdots\right)\cdots\left(1 + \frac{1}{p_i^z} + \frac{1}{p_i^{2z}} + \cdots\right)\cdots$$

$$=\sum_{k_1,k_2,k_3,\cdots\geq 0}\frac{1}{(p_1^{k_1} p_2^{k_2} p_3^{k_3}\cdots)^z} = \sum_{n=1}^{\infty}\frac{1}{n^z}.$$

From this it is clear how Euler came to his result. The product on the left-hand side of the above equality also shows why the number 1 is not considered to be a prime number.

The German mathematician Bernhard Riemann (1826–1866) derived another useful relation for the ζ-function

$$\pi^{-z/2}\Gamma\left(\frac{z}{2}\right)\zeta(z) = \pi^{-(1-z)/2}\Gamma\left(\frac{1-z}{2}\right)\zeta(1-z), \tag{3.20}$$

which describes the behavior of ζ-function when z is replaced by $1 - z$. Here the Γ-function is defined by

$$\Gamma(z) = \int_0^{\infty} e^{-t}t^{z-1}dt$$

and it is a natural extension of the factorial

$$\Gamma(n) = (n - 1)!$$

to the plane of complex numbers. The function $\zeta(1 - z)$ on the right-hand side of (3.20) has a pole at $z = 0$ and $\Gamma\left(\frac{1-z}{2}\right)$ has poles at $z = 1, 3, 5, 7, \ldots$. Hence, the right-hand side (3.20) has to be zero for $z = 3, 5, 7, \ldots$, and thus

$$0 = \zeta(-2) = \zeta(-4) = \zeta(-6) = \ldots.$$

These roots are called *trivial*. The function ζ can be extended continuously and smoothly to a meromorphic function defined on the whole complex plane except for the point 1, where it has a pole (i.e. $\zeta(1) = \infty$).

The Riemann hypothesis states that all non-trivial roots have a real part equal to $\frac{1}{2}$. Solving this hypothesis is now considered one of the most difficult and the most important problem in number theory, as it is closely related to the question of the distribution of primes (see (3.19)). To solve it a reward of 1 000 000 dollars was announced (see [429]).

3.7 Further Properties of Primes

The following theorem states that the sum of the reciprocals of all prime numbers is divergent. It was proved by Leonhard Euler. Now we will present its proof due to Paul Erdős, see Hardy and Wright [138].

Theorem 3.15 *Let $p_1 < p_2 < \ldots$ be all the primes. Then*

$$\sum_{i=1}^{\infty} \frac{1}{p_i} = \infty. \tag{3.21}$$

Proof Suppose to the contrary that the sum in (3.21) is convergent. Then we can choose a positive integer j so that the remainder after j terms is less than $\frac{1}{2}$, i.e.,

$$\frac{1}{p_{j+1}} + \frac{1}{p_{j+2}} + \cdots < \frac{1}{2}. \tag{3.22}$$

Let $N(x)$ denote the number of positive integers n not exceeding x that are not divisible by any prime $p > p_j$. Expressing such an n in the form

$$n = n_1^2 m, \tag{3.23}$$

where m is not divisible by any square of a prime number, then

$$m = 2^{b_1} 3^{b_2} \cdots p_j^{b_j},$$

where $b_i \in \{0, 1\}$. Thus there exist 2^j possible different values of m. From (3.23) we see that

$$n_1 \leq \sqrt{n} \leq \sqrt{x},$$

and thus there exist at most $\sqrt{x}$ different values of n_1. From this it follows that

$$N(x) \leq 2^j \sqrt{x}. \tag{3.24}$$

Let

$$x \geq 2^{2j+2} \tag{3.25}$$

be an arbitrary positive number and p a prime. Then the number of those positive integers $n \leq x$, that are divisible by p, is at most x/p. Hence, $x - N(x)$ which is the number of those $n \leq x$ divisible by at least one prime $p_{j+1}, p_{j+2}, \ldots$, is not greater than

$$\frac{x}{p_{j+1}} + \frac{x}{p_{j+2}} + \cdots < \frac{x}{2}. \tag{3.26}$$

However, from (3.24) and (3.26) we have

$$\frac{x}{2} < N(x) \leq 2^j \sqrt{x},$$

that is $x < 2^{2j+2}$, which contradicts (3.25). The series (3.21) is therefore divergent.   □

Twin primes are pairs of primes $\langle p, q \rangle$ for which $p < q$ and $q - p \leq 2$. They are for example, the following pairs

$$\langle 2, 3 \rangle, \quad \langle 3, 5 \rangle, \quad \langle 5, 7 \rangle, \quad \langle 11, 13 \rangle, \quad \langle 17, 19 \rangle, \quad \langle 10^9 + 7, 10^9 + 9 \rangle. \tag{3.27}$$

As of 2020, the largest known pair of twin primes is $2996863034895 \cdot 2^{1290000} \pm 1$.

From Theorem 3.15 we know that the sum of the reciprocals of all prime numbers grows beyond all bounds. On the other hand, in 1919 Viggo Brun showed that the sum of the reciprocals of twin primes (except for the first pair from (3.27))

$$\sum_q \frac{1}{q} = \left(\frac{1}{3} + \frac{1}{5}\right) + \left(\frac{1}{5} + \frac{1}{7}\right) + \left(\frac{1}{11} + \frac{1}{13}\right) + \cdots \tag{3.28}$$

converges to a constant which has since been calculated to many decimal places as the number $1.90216058\ldots$ and which is called *Brun's constant* [287]. This means that twin primes are very sparsely distributed in the set of all primes. However, for the time being we do not know if there are infinitely many twin primes.

In 1994, Thomas R. Nicely calculated Brun's constant on two different computers, the first one had the 486-processor and the second Intel's Pentium processor. When he evaluated reciprocal values of the twin primes $824\,633\,702\,441$ and $824\,633\,702\,443$ for Brun's sum (3.28), he found that the second computer incorrectly computed $1/824633702441$ beyond the eighth significant digit. This led him to the discovery of the famous "Pentium Bug" (see Cipra [72] and Koshy [176, p. 113]). This example nicely illustrates how number theory can help to solve real-life technical problems.

In 1949 P. Clement proved the following remarkable theorem (see [395, p. 120]):

Theorem 3.16 (Clement) *The number $p(p+2)$ divides $4((p-1)! + 1) + p$ if and only if $\langle p, p+2 \rangle$ are prime twins.*

The foursome $\langle p, q, r, s \rangle$ of primes is called a *prime quadruplet*, if $p < q < r < s$ and $s - p \leq 8$. For example $\langle 2, 3, 5, 7 \rangle$, $\langle 11, 13, 17, 19 \rangle$ or

$$10013950 + i \quad \text{for } i = 1, 3, 7, 9$$

are prime quadruplets. If there were only a finite number of prime twins, there would also obviously be a finite number of prime quadruplets.

Theorem 3.17 (Zhang) *Denote by p_n the nth prime. Then*

$$\liminf_{n \to \infty}(p_{n+1} - p_n) < 7 \cdot 10^7. \tag{3.29}$$

This result was proved by Yitang Zhang in Annals of Mathematics [422]. He also conjectures that $\liminf_{n \to \infty}(p_{n+1} - p_n) = 2$ without any proof. James Maynard [265] has lowered the upper bound given in (3.29) from $7 \cdot 10^7$ to 600, and Baker and Irving [18] have extended Maynard's results. The upper bound in (3.29) has since been improved to 246, see [304].

Finally, let us ask the following question:

Can every even number greater than 2 be written as the sum of two prime numbers?

This question arose during a mutual correspondence between Leonhard Euler and Christian Goldbach in 1742. We readily observe that $4 = 2 + 2, 6 = 3 + 3, 8 = 3 + 5, 10 = 5 + 5 = 3 + 7$, etc. The hypothesis that at least one such decomposition always exists is called *Goldbach's conjecture*. Up to the present time, no one solved it. According to some sources it was first stated by Euler inspired by Goldbach. In 1937, the Russian mathematician Ivan Matveyevich Vinogradov (1891–1983) proved that there exists a positive integer n_0 such that every odd number $n > n_0$ can be expressed as the sum of three prime numbers.

We close this chapter by one more remarkable criterion for primality testing from the article [258] based on the famous Pascal's triangle which was discovered by the French mathematician Blaise Pascal (1623–1662). Its rows contain binomial coefficients $\binom{n}{0}, \binom{n}{1}, \binom{n}{2}, \ldots, \binom{n}{n}$ for $n = 0, 1, 2, \ldots$ In Chinese literature, this triangle is named after the mathematician Yang Hui, who published it in his book as early as 1261 for $n = 6$, see [196], [260, p. 230].

Theorem 3.18 *A positive integer $k > 1$ is a prime if and only if*

$$n \mid \binom{n}{k - 2n} \quad \text{for all } n \text{ such that } \frac{k}{3} \leq n \leq \frac{k}{2}.$$

To get a closer look at what this theorem actually states, we shall create an illustration table. Its upper row contains successive numbers of columns $k = 0, 1, \mathbf{2}, \mathbf{3}, 4, \mathbf{5}, 6, \ldots$, where prime numbers are marked in bold. Let us write Pascal's triangle in further rows so that the next line is always shifted two places more to the right than the previous line. The left column of the table contains successive row numbers

$n = 0, 1, 2, \ldots$ If the binomial coefficient is divisible by the row number, we will also write it in boldface:

$n \backslash k$	0	1	**2**	3	4	**5**	6	**7**	8	9	10	**11**	12	**13**	14	15
0	1															
1			**1**	**1**												
2				1	**2**	1										
3					1	**3**	**3**	1								
4						1	4	6	**4**	1						
5								1	**5**	10	**10**	5	1			
6											1	**6**	15	**20**		$\ldots$

Theorem 3.18 claims that a column contains only bold binomial coefficients if and only if it corresponds to a prime number.

Chapter 4
Special Types of Primes

4.1 Mersenne Primes

The French mathematician Marin Mersenne (1588–1648, see Fig. 4.1) studied numbers of the form

$$M_p = 2^p - 1,$$

where p is a prime. They are named *Mersenne numbers* after him. If moreover $2^p - 1$ itself is a prime, then it is called a *Mersenne prime*.

Theorem 4.1 *If $2^p - 1$ is a prime, then p is also a prime.*

Proof Let $2^p - 1$ be a prime. Suppose to the contrary that p is composite, i.e., there exist integers $i > 1$ and $j > 1$ such that $p = ij$. However, the number $2^{ij} - 1$ can be factored as follows

$$2^{ij} - 1 = (2^i - 1)(2^{i(j-1)} + 2^{i(j-2)} + \cdots + 2^i + 1), \qquad (4.1)$$

where both the factors on the right-hand side of (4.1) are clearly greater than 1. This is a contradiction, since $2^p - 1$ is assumed to be prime. $\qquad\square$

Due to this theorem, we require that the exponent p in the definition of $M_p = 2^p - 1$ is a prime. We observe that

$$M_2 = 3, \quad M_3 = 7, \quad M_5 = 31, \text{ and } M_7 = 127$$

are primes (see also Tables 13.5 and 13.6). However, for $p = 11$ the number

$$2^{11} - 1 = 2047 = 23 \cdot 89$$

is composite. Similarly, $2^{23} - 1 = 47 \cdot 178481$ and $2^{29} - 1 = 233 \cdot 1103 \cdot 2089$. Hence, the converse of Theorem 4.1 does not hold. We will return to divisors of Mersenne numbers in Theorems 4.5 and 4.6.

© The Author(s), under exclusive license to Springer Nature Switzerland AG 2021
M. Křížek et al., *From Great Discoveries in Number Theory to Applications*,
https://doi.org/10.1007/978-3-030-83899-7_4

Fig. 4.1 Memorial plaque of Marin Mersenne at his birthplace in Oizé recalls that he can be considered as the founder of the Academy of Sciences in France

Although 51 Mersenne prime numbers were discovered up to 2020, little is known about their actual distribution (several empirical relations are given in [350, 407]). The number M_p is a prime, if

$p =$2, 3, 5, 7, 13, 17, 19, 31, 61, 89,

  107, 127, 521, 607, 1279, 2203, 2281, 3217, 4253, 4423,

  9689, 9941, 11213, 19937, 21701, 23209, 44497, 86243, 110503, 132049,

  216091, 756839, 859433, 1257787, 1398269, 2976221, 3021377, 6972593,

  13466917, 20996011, 24036583, 25964951, 30402457, 32582657, 37156667,

  42643801, 43112609, 57885161, 74207281, 77232917, 82589933, ...

The largest known Mersenne prime has 24 862 048 digits. This would occupy over 6000 pages with 40×100 digits per page.

Marin Mersenne in the preface to his work *Cogitata Physica-Mathematica* from 1644 incorrectly states that the numbers $2^p - 1$ are prime numbers for

$$p = 2,\ 3,\ 5,\ 7,\ 13,\ 17,\ 19,\ 31,\ 67,\ 127,\ 257$$

and they are composite for all other positive integers $p < 257$ (see Dickson [91, p. 12–13]).

There is a well-known story involving the case $p = 67$ (see e.g. Wells [416]). Frank Nelson Cole (1861–1926) factored the number M_{67} into two prime numbers, which took him three years of Sundays. In 1903 he presented a legendary lecture without words to a meeting of the American Mathematical Society, which Eric Temple Bell described as follows:

Cole, a quiet man anyway, approached the chalkboard and in complete silence proceeded to calculate the value of 2^{67}. Then he carefully subtracted 1 and as a result he gained a numerical monster

$$147\,573\,952\,589\,676\,412\,927.$$

Cole then moved to the other side of the board, wrote

$$193\,707\,721 \cdot 761\,838\,257\,287,$$

and worked through the tedious calculations by hand. Upon completing the multiplication and demonstrating that the result equaled M_{67}, Cole returned to his seat, not having uttered a word during the hour-long presentation. His audience greeted the presentation with a standing ovation. Mersenne's conjecture thus disappeared into the depths of mathematical heroic sagas.

According to Kraïtchik [177], Pierre de Fermat factored the numbers $2^p - 1$ into primes for $p = 11$, 23, 37. These results led him to discover Fermat's Little Theorem 2.17 for the base $a = 2$:

$$2^{q-1} \equiv 1 \quad (\text{mod } q) \tag{4.2}$$

for any prime $q > 2$, i.e., congruence (4.2) states that any prime q divides the difference $2^{q-1} - 1$.

Theorem 4.2 *A prime divides at most one Mersenne number.*

The proof immediately follows from the next theorem which claims that the Mersenne numbers are pairwise coprime.

Theorem 4.3 *Let $m, n \in \mathbb{N}$. Then $\left(2^m - 1, 2^n - 1\right) = 1$ if and only if $(m, n) = 1$.*

Proof It suffices to show that

$$\left(2^m - 1, 2^n - 1\right) = 2^{(m,n)} - 1. \tag{4.3}$$

Let $d = (m, n)$ and $d' = \left(2^m - 1, 2^n - 1\right)$. By (4.1)

$$2^d - 1 \mid 2^m - 1 \quad \text{and} \quad 2^d - 1 \mid 2^n - 1.$$

Thus, $2^d - 1 \mid d'$.

Equality (4.3) will now follow if we can show that $d' \mid 2^d - 1$. By Theorem 1.3 there exist integers a and b such that

$$am + bn = d. \tag{4.4}$$

Since $m, n, d > 0$ and $d \leq \min(m, n)$, we have that either $a \leq 0$ and $b \geq 0$ or it is the case that $a \geq 0$ and $b \leq 0$. Suppose that $a \leq 0$ and $b \geq 0$. Let $a = -k$, where $k \geq 0$. Then by (4.4),

$$bn = d + km.$$

From this we see that

$$2^{bn} - 1 = 2^d (2^{km} - 1) + 2^d - 1. \tag{4.5}$$

By (4.1), $2^m - 1 \mid 2^{km} - 1$ and $2^n - 1 \mid 2^{bn} - 1$. Since $d' \mid 2^m - 1$ and $d' \mid 2^n - 1$, we observe by (4.5) that

$$d' \mid (2^{bn} - 1) - 2^d (2^{km} - 1) = 2^d - 1.$$

Hence, the equivalence holds. $\qquad\qquad\qquad\qquad\qquad\qquad\qquad\qquad\qquad\qquad\square$

The following necessary and sufficient condition is proved in [225, Theorem 5.4] and [327, p. 126].

Theorem 4.4 (Lucas-Lehmer Test) *Let $y_1 = 4$ and $y_{k+1} = y_k^2 - 2$ for $k = 1, 2, \ldots$ Then, for $p > 2$, the Mersenne number $M_p = 2^p - 1$ is a prime if and only if M_p divides y_{p-1}.*

We see that the sequence y_i grows quite rapidly

$$4, 14, 194, 37634, 1416317954, \ldots.$$

The Lucas-Lehmer test, in fact, asserts that M_p is a prime for $p > 2$ if and only if $y_{p-1} \equiv 0 \pmod{M_p}$.

Denote by $M(n)$ the nth Mersenne prime, Consequently, $M(1) = 2^2 - 1 = 3$, $M(2) = 2^3 - 1 = 7$, $M(3) = 2^5 - 1 = 31, \ldots$ Fig. 4.2 shows a remarkable distribution of known Mersenne primes $M(n)$. On the basis of this distribution, Wagstaff predicted (see [407, p. 388]) that the expected number of Mersenne primes with exponent p between k and $2k$ is equal to $e^\gamma = 1.781\,072\,418\ldots$, where

$$e = \lim_{k \to \infty} \left(1 + \frac{1}{k}\right)^k = 2.718\,281\,828\ldots$$

is the *Euler number* and

$$\gamma = \lim_{k \to \infty} \left(1 + \frac{1}{2} + \frac{1}{3} + \cdots + \frac{1}{k} - \log k\right) = 0.577\,215\,665\ldots$$

is the *Euler constant*. This estimate greatly facilitates the search for further Mersenne primes.

In 1996, an internet project GIMPS (Great Internet Mersenne Prime Search) was created, in which thousands of mathematicians and computer experts from all over

Fig. 4.2 Values of $\log_2(\log_2 M(n))$ for $n = 1, 2, 3, \ldots$

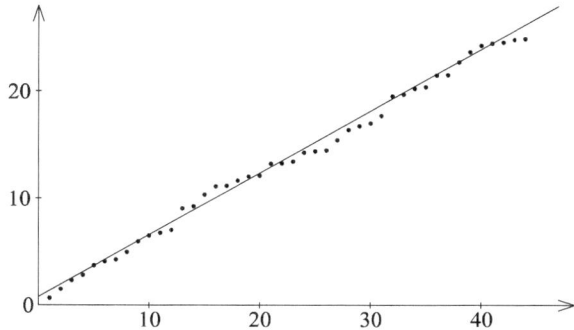

the world look for Mersenne primes with a large number of digits (see [430, 431]). The idea to create free software, which uses the Lucas-Lehmer test (see Theorem 4.4), came from George Woltman who founded this project. The corresponding programs can be downloaded from the network created by the company of Scott Kurowski. They utilize a quick multiplication (cf. [349]) which essentially reduces the number of arithmetic operations.

A complete list of prime factors of Mersenne numbers M_p for prime exponents $p \leq 257$ can be found in Riesel [327] (see also [45] for larger p). The general form of divisors of Mersenne numbers is given by the following result which was already known to Fermat (see Dickson [91, p. 12], Mahoney [252, p. 294]).

Theorem 4.5 *If $p > 2$ is a prime, then every prime factor of M_p has the form $2kp + 1$.*

Proof Let q be a prime factor of $2^p - 1$. Then $2^p \equiv 1 \pmod{q}$. Since p is a prime and $2^1 \not\equiv 1 \pmod{q}$, we see that p is the smallest exponent for which $2^p \equiv 1 \pmod{q}$. By a special case of Fermat's Little Theorem (4.2) we have $2^{q-1} \equiv 1 \pmod{q}$. Hence, p divides $q - 1$, i.e., there exists j such that $jp = q - 1$. Since p is odd and $q - 1$ is even, we find that $j = 2k$ for some natural number k. $\quad\square$

Theorem 4.6 *If n divides M_p for some prime $p > 2$, then $n \equiv \pm 1 \pmod 8$.*

For the proof see Ribenboim [322, p. 66].

There is a surprising connection between Mersenne primes and perfect numbers. A natural number is called *perfect*, if it is equal to the sum of all its divisors less than n. For instance, 6 and 28 are perfect, since

$$6 = 1 + 2 + 3 \quad \text{and} \quad 28 = 1 + 2 + 4 + 7 + 14.$$

A detailed overview of other such numbers and their discoverers is given in Bezuszka and Kenney [31].

Let n be an arbitrary natural number. Denote by $\sigma(n)$ the sum of all its positive divisors, i.e.,

$$\sigma(n) = \sum_{d \mid n} d.$$

For instance, $\sigma(12) = 1 + 2 + 3 + 4 + 6 + 12 = 28$. Further values of σ for $n \leq 32$ are listed in Table 13.3. Using the function σ, we can state an equivalent definition: n is *perfect*, if

$$\sigma(n) = 2n.$$

Moreover, we have

$$\sigma(mn) = \sigma(m)\sigma(n),$$

if m and n are coprime natural numbers.

A necessary and sufficient condition for an even number n to be perfect is that it be of the form $n = 2^{p-1}(2^p - 1)$, where $2^p - 1$ is a prime (i.e., $p > 1$ must be also prime by Theorem 4.1). Euclid (4th–3rd century BC) already knew that this condition is sufficient, but he did not know if it is also necessary. Two millennia later this question was positively answered by Leonhard Euler.

Theorem 4.7 (Euclid) *If $2^p - 1$ is a prime, then $n = 2^{p-1}(2^p - 1)$ is perfect.*

Proof Since the numbers 2^{p-1} and $2^p - 1$ are coprime, we have

$$\sigma(n) = \sigma(2^{p-1})\sigma(2^p - 1) = \frac{2^p - 1}{2 - 1}(1 + 2^p - 1) = (2^p - 1)2^p = 2n.$$

Therefore, n is perfect.   □

Theorem 4.8 (Euler) *All even perfect numbers have the form*

$$n = 2^{p-1}(2^p - 1),$$

where $p > 1$ and $2^p - 1$ is a prime.

Proof If n is even, then we can write $n = 2^{p-1}u$, where $p > 1$ and u is odd. Since 2^{p-1} and u are coprime, the sum of divisors of n equals

$$\sigma(n) = \sigma(2^{p-1})\sigma(u) = (2^p - 1)\sigma(u).$$

If n is perfect, we have

$$\sigma(n) = 2n = 2^p u,$$

and thus

$$(2^p - 1)\sigma(u) = 2^p u.$$

Since $2^p - 1$ and 2^p are coprime, we find that $\sigma(u) = 2^p t$ and $u = (2^p - 1)t$, where t is a natural number. However, since u has at least the following divisors $1, t, 2^p - 1$, and $t(2^p - 1)$ for $t > 1$, the sum of divisors of u fulfills the inequality

$$\sigma(u) \geq 1 + t + 2^p - 1 + t(2^p - 1) = 2^p(1 + t),$$

which contradicts the equality $\sigma(u) = 2^p t$. Therefore, $t = 1$. But then $\sigma(u) \geq 1 + 2^p - 1 = 2^p$ and the required equality $n = 2^{p-1}(2^p - 1)$ holds only if $u = 2^p - 1$ is a prime. $\square$

Setting $q = 2^p - 1$ and $m = p - 1$, we can formulate Euclid's and Euler's theorems together as follows:

The number $q = 1 + 2 + 2^2 + \cdots + 2^m$ is prime if and only if $2^m q$ is perfect.

From Theorems 4.7 and 4.8 we get another interesting relation between even perfect numbers n and the Mersenne primes M_p, namely,

$$n = 2^{p-1}(2^p - 1) = \frac{2^p}{2}(2^p - 1) = 1 + 2 + \cdots + (2^p - 1) = \sum_{i=1}^{M_p} i.$$

Theorem 4.9 *If we sum the digits of any even perfect number greater than 6, then sum the resulting number, and repeat this process until we get a single digit, then that digit will be one.*

Proof First we show that for any perfect number $n > 6$ the congruence $n \equiv 1$ (mod 9) is valid. Let p be an odd prime. Since $2 \equiv -1$ (mod 3) and the number $p - 1$ is even, by (1.4) we get $2^{p-1} \equiv 1$ (mod 3), i.e.,

$$2^{p-1} = 3k + 1$$

for some natural number k. Thus, by Euler's Theorem 4.8 we obtain

$$n = 2^{p-1}(2^p - 1) = (3k + 1)\big(2(3k + 1) - 1\big) = (3k + 1)(6k + 1)$$
$$= 18k^2 + 9k + 1 \equiv 1 \quad (\text{mod } 9).$$

The number n can be written in the form

$$n = c_k 10^k + c_{k-1} 10^{k-1} + \cdots + 10c_1 + c_0,$$

where the digit $c_k \neq 0$. Since $10 \equiv 1$ (mod 9), by (1.4) we have

$$10^j \equiv 1^j = 1 \quad (\text{mod } 9)$$

for any $j \geq 0$. Hence, for the sum of all the digits we obtain

$$n \equiv c_k + \cdots c_1 + c_0 \equiv 1 \quad (\text{mod } 9).$$

Now we will repeat this procedure until we get a single digit that will obviously be equal to 1. $\square$

The following theorem can be found in Kraïtchik [177].

Theorem 4.10 (Heath) *Every even perfect number* $2^{p-1}(2^p - 1)$ *for* $p > 2$ *is the sum of* $2^{(p-1)/2}$ *consecutive odd cubes.*

Proof First of all notice that, by Theorems 4.1 and 4.9, p is a prime. Let $p > 2$. Putting $k = (p-1)/2$, $m = 2^k$, and $s = 1^3 + 3^3 + 5^3 + \cdots + (2m-1)^3$, then by induction we find that

$$
\begin{aligned}
s &= \sum_{k=1}^{m}(2k-1)^3 = \sum_{k=1}^{m}(8k^3 - 12k^2 + 6k - 1) \\
&= 8\frac{m^2(m+1)^2}{4} - 12\frac{m(m+1)(2m+1)}{6} + 6\frac{m(m+1)}{2} - m = m^2(2m^2 - 1).
\end{aligned}
$$

Here the following relation

$$
1^3 + 2^3 + \cdots + m^3 = (1 + 2 + \cdots + m)^2 = \frac{1}{4}m^2(m+1)^2
$$

was used. Now we see that $s = 2^{2k}(2^{2k+1} - 1) = 2^{p-1}(2^p - 1)$.   □

We will return to perfect numbers in Sect. 8.2.

There is a number of unanswered questions concerning Mersenne numbers M_p. For example, one of the conjectures states that there are infinitely many Mersenne primes, and thus also infinitely many even perfect numbers. However, up to now we do not know if there exists an odd perfect number. There are only necessary conditions for such a number to exist. Let N denote an odd perfect number. Euler showed that

$$
N = p^\alpha M^2,
$$

where p is a prime, $p \equiv \alpha \equiv 1 \pmod 4$, and M is odd. A proof of this result is given in [408, pp. 84–85]. Touchard [401] proved that N is of the form $N = 12m + 1$ or $N = 36m + 9$. A simple proof of Touchard's Theorem is given in [146]. Moreover, Ochem and Rao [292] proved that $N > 10^{1500}$, N has at least 101 not necessarily distinct prime factors, and the largest prime power q^a dividing N satisfies $q^a > 10^{62}$. It has been further demonstrated that N has at least ten distinct prime factors (one of them is greater than 10^8), see [124, 289]. For earlier results on odd perfect numbers see [155, 288].

There is another unsolved conjecture: whether the sequence $m_{k+1} = 2^{m_k} - 1$ starting with $m_1 = 2$ contains only primes. Indeed, the first five terms $m_1 = 2$, $m_2 = 2^2 - 1 = 3$, $m_3 = 2^3 - 1 = 7$, $m_4 = 2^7 - 1 = 127$, and

$$
m_5 = 2^{127} - 1 = 170141183460469231731687303715884105727
$$

are the following primes 2, M_2, M_3, M_7, and M_{127}. For the time being, the character of m_6 is not known. However, if m_k were to be composite for some k, then by (4.1) the number m_{k+1} would also be composite.

We also do not know if there are infinitely many composite Mersenne numbers or if every Mersenne prime is square-free (cf. Rotkiewicz [337] and also later Warren and Bray [411]). We know only that if a prime p divides a Mersenne number M_q then (cf. Sect. 4.3)

$$p^2 \mid M_q \iff 2^{p-1} \equiv 1 \pmod{p^2} \quad \text{(Wieferich congruence)}.$$

Finally, note that Mersenne numbers are used in digital signal processing—see [80, 93], [102, p. 425], [123]. The Mersenne number transform is defined similarly as the Fermat number transform [199]. For more information about Mersenne primes and their generalizations see [91, 233, 323].

4.2 Fermat Primes

The French mathematician Pierre de Fermat (1601–1665) is considered to be the founder of modern number theory (see e.g. Theorems 2.6, 2.8–2.12, 2.16, 2.17, 3.12). However, one of his erroneous claims literally caused a revolution in number theory. Fermat believed that all numbers of the form

$$F_m = 2^{2^m} + 1 \quad \text{for } m = 0, 1, 2, \ldots \tag{4.6}$$

are primes. The first five members of this sequence are really primes,

$$F_0 = 3, \quad F_1 = 5, \quad F_2 = 17, \quad F_3 = 257, \quad F_4 = 65537. \tag{4.7}$$

But in 1732 Leonhard Euler (see [106, p. 104]) found that

$$F_5 = 641 \cdot 6700417,$$

and thus refuted Fermat's claim. We can verify that $641 \mid F_5$ without calculator. Since $641 = 5^4 + 2^4 = 5 \cdot 2^7 + 1$ divides the following two numbers

$$5^4 2^{28} + 2^{32} = (5^4 + 2^4) 2^{28},$$
$$5^4 2^{28} - 1 = (5 \cdot 2^7 + 1)(5^3 2^{21} - 5^2 2^{14} + 5 \cdot 2^7 - 1),$$

cf. (2.57), 641 must also divide their difference which is F_5. Thus, a natural question arises, whether there are infinitely many prime numbers of the form (4.6).

The numbers F_m are called *Fermat numbers* (see Table 13.2). If F_m is a prime, then we say that it is a *Fermat prime*.

Let us first formulate a necessary condition for $n \in \mathbb{N}$ in order to make it possible for the number $2^n + 1$ to be a prime. This condition tells us why Fermat chose the exponent in (4.6) in exponential form.

Theorem 4.11 *Let n be a natural number. If $2^n + 1$ is a prime, then $n = 2^m$ for some $m \in \{0, 1, 2, \dots\}$.*

Proof If k is a natural number and $\ell \geq 3$ is odd, then

$$2^{k\ell} + 1 = (2^k + 1)(2^{k(\ell-1)} - 2^{k(\ell-2)} + \cdots - 2^k + 1). \tag{4.8}$$

From this it follows that the number $2^n + 1$ is composite if the exponent n is divisible by an odd natural number $\ell \geq 3$. However, this does not happen in the sequence (4.6). Therefore, n must be a power of two.   □

Note also that for $n = 4r + 2$, *Lucas' formula* holds (see Kraïtchik [177])

$$2^n + 1 = (2^{2r+1} - 2^{r+1} + 1)(2^{2r+1} + 2^{r+1} + 1).$$

From (4.6) we obtain the following recurrence formula

$$F_{m+1} = (F_m - 1)^2 + 1, \tag{4.9}$$

and thus,

$$F_{m+1} - 2 = F_m(F_m - 2).$$

From this we get by induction an interesting property

$$F_{m+1} - 2 = F_m F_{m-1} \cdots F_1 F_0,$$

which states that $F_{m+1} - 2$ is divisible by all smaller Fermat numbers, i.e.,

$$F_k \mid (F_m - 2) \quad \text{for } k = 0, 1, \dots, m - 1. \tag{4.10}$$

From this it easily follows as in Theorem 4.2 that a prime divides at most one Fermat number, see [199, p. 33] for Goldbach's Theorem.

Until 1796, Fermat numbers were more or less a mathematical curiosity. Interest in them grew (see [183, 187, 202, 204]) when Carl Friedrich Gauss discovered an incredible connection between Fermat primes and the Euclidean construction of regular polygons (i.e., using a compass and straightedge).

Theorem 4.12 (Gauss) *There exists a Euclidean construction of a regular n-gon if and only if the number of its sides is equal to*

$$n = 2^i F_{m_1} F_{m_2} \cdots F_{m_j},$$

where $i \geq 0$, $j \geq 0$, $n \geq 3$ are integers and F_{m_1}, F_{m_2}, ..., F_{m_j} are distinct Fermat primes.

For a proof based on Galois theory see [199, Chapt. 16]. Investigation of primality of Fermat numbers thus started to be an important task.

The regular n-gon with an odd number of sides can therefore be constructed for (see [182, 186, 189])

$$n = 3, \ 5, \ 15, \ 17, \ 51, \ 85, \ 255, \ 257, \ldots, \tag{4.11}$$

where $15 = 3 \cdot 5, 51 = 3 \cdot 17, 85 = 5 \cdot 17, 255 = 3 \cdot 5 \cdot 17, \ldots$ are products of Fermat primes. Thanks to modern mathematical methods and powerful computer technologies it was proved that

$$F_m \text{ is composite for } 5 \leq m \leq 32,$$

even though we do not know any prime factor of F_{20} and F_{24} as of 2020. The number F_{24} has over 5 million digits. A proof that it was a composite number required 10^{17} arithmetic operations (see Crandall, Mayer, and Papadopoulos [81]). As of 2003, it has been the most extensive calculation that resulted in a single bit YES/NO type information. So far, we do not know if the current list of Euclidean constructible regular polygons is already complete. The factoring status of Fermat numbers is given in [436]. As of 2020 the largest number m for which the Fermat number F_m has been shown to be composite is $m = 18\,233\,954$ and over 300 Fermat numbers have been found to be composite.

For testing the primality of Fermat numbers, the following theorem is mostly applied (see Pépin [296]). It is a special case of Proth's Theorem 3.2.

Theorem 4.13 (Pépin's test) *For $m \geq 1$ the Fermat number F_m is a prime if and only if*

$$3^{(F_m-1)/2} \equiv -1 \pmod{F_m}. \tag{4.12}$$

The French mathematician Jean François Théofile Pépin (1826–1904) proved this theorem in 1877, where the base 3 is replaced by 5.

Further we introduce another interesting connection between Fermat primes and directed graphs (i.e. *digraphs*). Let

$$H = \{0, 1, \ldots, n - 1\},$$

where $n > 1$. Assume that f is a mapping from H to itself. Starting with an arbitrary element x_0 from H, we may define a sequence of successive elements of H by

$$x_{i+1} = f(x_i), \quad i = 0, 1, \ldots$$

Since H is finite, the sequence (x_i) has to be cyclic starting from some element x_k. If $x_k, x_{k+1}, \ldots, x_t$ are distinct and

$$x_{k+1} = f(x_k),$$

$$\vdots$$

$$x_t = f(x_{t-1}),$$
$$x_k = f(x_t),$$

then the elements $x_k, x_{k+1}, \ldots, x_t$ constitute a *cycle*. A cycle of length 1 is called a *fixed point*.

The *iteration graph* of the mapping f is a digraph all of whose vertices are elements of H such that there exists an oriented edge from x to $f(x)$ for all $x \in H$ and there are no other edges.

The iteration graph is called a *binary graph* if its symmetrization has exactly two components (i.e., the corresponding nondirected graph consists of two disjoint connected subgraphs) and the following three conditions hold:

1. The vertex 0 is an isolated fixed point.
2. The vertex 1 is a fixed point and there exists a directed edge from the vertex $n-1$ to 1.
3. For any vertex for the set $\{1, 2, \ldots, n-1\}$ there exist two edges or no edge directed towards this vertex, and there are no other edges.

Further, we will consider a special discrete iteration. For each $x \in H$ let $f(x)$ be the remainder of x^2 modulo n, i.e.,

$$f(x) \in H \quad \text{and} \quad f(x) \equiv x^2 \pmod{n}. \tag{4.13}$$

This leads to the following iteration scheme $x_{i+1} \equiv x_i^2 \pmod{n}$.

In Szalay [389] the following theorem is proved (cf. Figs. 4.3 and 4.4).

Theorem 4.14 (Szalay) *The iteration graph of the mapping f defined by (4.13) is a binary graph if and only if the modulus n is a Fermat prime.*

Figure 4.4 nicely illustrates why these graphs are called binary graphs. The larger component is a binary tree if we remove vertex 1 and both edges directed towards 1.

Fig. 4.3 The iteration graph of the function f defined by $f(x) \equiv x^2 \pmod{14}$

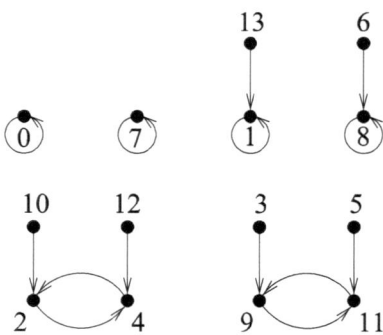

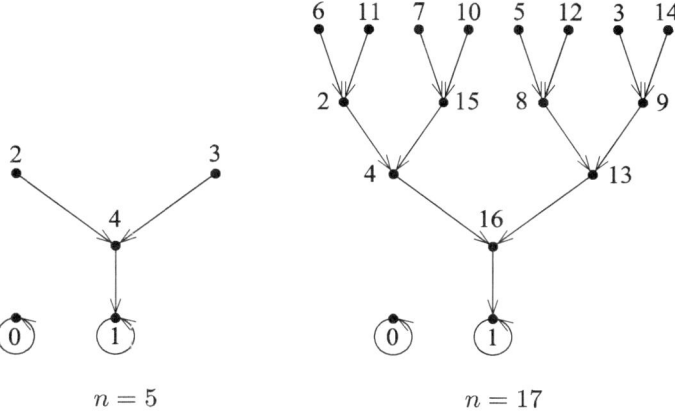

Fig. 4.4 Binary graphs corresponding to the Fermat numbers F_1 and F_2

In [207], we prove that among primes only Fermat primes have the property that the set of primitive roots is equal to the set of quadratic non-residues modulo F_m (cf. Theorem 2.29).

Figure 2.6 shows all primitive roots modulo 17. Now notice that the primitive roots modulo n lie on the "top" of each digraph from Fig. 4.4. If F_m is a prime, then by Pépin's test (4.12) the number 3 is always a primitive root modulo F_m for $m \geq 1$. Figure 4.4 thus provides a graphical illustration of Pépin's test. In this test the number 3 can be replaced by any primitive root modulo n.

Now we present two theorems on primitive roots which are not presented in the monograph [199].

Theorem 4.15 *The number F_m is prime if and only if it has a primitive root.*

Proof First we show that a Fermat number is never a perfect power. Since $F_0 = 3$ and 3 is not a perfect power, we may assume that $m \geq 1$. By (4.9) we easily find that F_m cannot be a square, because the only consecutive squares are 0 and 1. So, assume that $F_m = x^n$ for some odd number n. Then

$$2^{2^m} = F_m - 1 = x^n - 1 = (x - 1)(x^{n-1} + x^{n-2} + \cdots + 1), \qquad (4.14)$$

where $x^{n-1} + x^{n-2} + \cdots + 1 = (x^n - 1)/(x - 1)$ is an odd integer. Since the left-hand side of (4.14) is a power of 2, this forces $(x^n - 1)/(x - 1)$ to be equal to 1. Hence, $n = 1$ and the rest of the proof now follows from Theorem 2.22. □

Theorem 4.16 *Let b an arbitrary primitive root modulo a Fermat prime F_m for $m \geq 0$. Then b^{2k-1} for $k = 1, \ldots, (F_m - 1)/2$ are all incongruent primitive roots modulo F_m.*

Proof We easily find that this theorem holds for $m \leq 1$. So let $m > 1$. By [199, p. 42] and [360], the base 3 in Pépin's test (4.12) can be replaced by any natural number b

such that for the Jacobi symbol we have $\left(\dfrac{b}{F_m}\right) = -1$ for all $m > 1$. In other words, F_m is a prime if and only if

$$b^{(F_m-1)/2} \equiv -1 \pmod{F_m}.$$

From this we get

$$\left(b^{2k-1}\right)^{(F_m-1)/2} = \left(b^{(F_m-1)/2}\right)^{2k-1} \equiv (-1)^{2k-1} = -1 \pmod{F_m},$$

i.e., b^{2k-1} for $k = 1, \ldots, (F_m - 1)/2$ are primitive roots modulo F_m. Further we see that these roots are mutually incongruent, since for $1 \le \ell < k \le (F_m - 1)/2$ we have

$$b^{2k-1} - b^{2\ell-1} = b^{2\ell-1}\left(b^{2(k-\ell)} - 1\right) = b^{2\ell-1}\left(b^{k-\ell} + 1\right)\left(b^{k-\ell} - 1\right) \not\equiv 0 \pmod{F_m},$$

where the last incongruence follows directly from the definition of primitive roots.
$\square$

In [209] we present several further necessary and sufficient conditions for the primality of F_m. A surprising connection between the five known Fermat primes and the five Platonic solids (i.e. regular polyhedra, see Fig. 12.3) is described in [192].

In 1878 the French mathematician François Édouard Anatole Lucas (1842–1891) proved the following theorem which has become a powerful tool for finding prime factors of Fermat numbers (see [246]).

Theorem 4.17 (Lucas) *If p divides F_m for some $m > 1$, then there exists a natural number k such that*

$$p = k\, 2^{m+2} + 1.$$

Its proof is given in [199, p. 59]. The usefulness of this theorem can be illustrated by an example, which was treated by A. E. Western in 1903. He wanted to find whether F_{18} is composite. The number of its digits is respectable—almost 80 000, since

$$\log_{10}(2^{2^{18}} + 1) + 1 \approx \log_{10} 2^{2^{18}} + 1 = 2^{18}\log_{10} 2 + 1 \approx 78\,914.$$

For comparison, the observable universe has only about 10^{80} elementary particles, such as protons, electrons, neutrons.

By Lucas' Theorem 4.17, we need to find a natural number k such that $k\,2^{20} + 1$ divides F_{18} and $k\,2^{20} + 1$ is prime. In this way, Western relatively easily found that the sought number k is 13, because for smaller values of k (except for $k = 7$) all the numbers $k\,2^{20} + 1$ are composite.

However, how can we verify that $p = 13 \cdot 2^{20} + 1 = 13631489$ really divides the huge Fermat number F_{18} without computers? This can be easily done by the following string of congruences:

$$2^{2^5} = \ \ 65536^2 \equiv 1048261 \quad (\mathrm{mod}\ p),$$
$$2^{2^6} \equiv 1048261^2 \equiv 3164342 \quad (\mathrm{mod}\ p),$$
$$2^{2^7} \equiv 3164342^2 \equiv 9153547 \quad (\mathrm{mod}\ p),$$

$$\vdots \qquad \vdots$$

$$2^{2^{17}} \equiv 1598622^2 \equiv 1635631 \quad (\mathrm{mod}\ p),$$
$$2^{2^{18}} \equiv 1635631^2 \equiv 13631488 \quad (\mathrm{mod}\ p),$$

where on the right-hand sides there are the remainders of division by the prime number p. Hence,

$$2^{2^{18}} + 1 \equiv 0 \quad (\mathrm{mod}\ 13631489).$$

In 2002, Payam Samidoost proved the following theorem concerning the special case $k = 9$ in the Lucas Theorem 4.17.

Theorem 4.18 (Samidoost) *No prime of the form $9 \cdot 2^n + 1$ with even n can divide a Fermat number.*

Proof Let q be a prime and $k \geq 2$ be an integer. We say that the integer a is a kth-power residue modulo q if there exists an integer r such that $r^k \equiv a$ (mod q). Suppose that $q \equiv 1$ (mod k). By a well-known result of Euler, if $a \not\equiv 0$ (mod q) and $q \equiv 1$ (mod k), then a is a kth-power residue modulo q if and only if $a^{(q-1)/k} \equiv 1$ (mod q) (see Lemma 6.14 in [199, p. 64]).

Suppose by way of contradiction that $p = 9 \cdot 2^n + 1$ is a prime such that $p \mid F_m$, where n is even. We can assume that $m \geq 2$, since $F_m < 9$ for $m = 0$ or 1. First suppose that $n = 4j + 2$, where $j \geq 0$. We claim that 2 is a cubic residue modulo p. By Theorem 4.17, $n \geq m + 2$. Since $F_m = 2^{2^m} + 1$, we see that

$$2^{2^m} \equiv -1 \quad (\mathrm{mod}\ p).$$

Hence,

$$2^{(p-1)/3} = 2^{3 \cdot 2^n} = \left(2^{2^m}\right)^{3 \cdot 2^{n-m}} \equiv (-1)^{3 \cdot 2^{n-m}} \equiv 1 \quad (\mathrm{mod}\ p),$$

and 2 is a cubic residue modulo p. But in order for 2 to be a cubic residue modulo p, it is necessary that in the unique representation of p in the form $r^2 + 3s^2$, the number s be a multiple of 3 (see for example Ireland and Rosen [151, pp. 118–119]). However, we observe that

$$p = 9 \cdot 2^{4j+2} + 1 = (3 \cdot 2^{2j+1} - 1)^2 + 3\left(2^{j+1}\right)^2$$

has a representation of the form $r^2 + 3s^2$, where s is not a multiple of 3. Therefore, 2 is not a cubic residue modulo p, which is a contradiction.

Now suppose that $p = 9 \cdot 2^{4j} + 1$, where $n = 4j \geq m + 2 \geq 4$. The last digit of 2^{4j} is 6 for any $j \geq 1$. Therefore, the sum $9 \cdot 2^{4j} + 1$ is always divisible by 5 and p is not a prime. The result now follows.   □

Theorem 4.19 *Let F_m be composite and let $k2^n + 1 \neq F_m$ be an arbitrary factor. Then there exists an odd $\ell \geq 3$ such that $F_m = (k2^n + 1)(\ell 2^n + 1)$ and*

$$\max(k, \ell) \geq F_{m-2}.$$

For the proof see [190] (cf. also [191]).

Remark In Matiyasevich [261], it is proved that every recursively enumerable subset Y of the set of positive integers can be expressed in the form:
There exist $N \geq 0$ and a polynomial $p(y, x_1, \ldots, x_N)$ with integer coefficients such that

$$y \in Y \quad \Longleftrightarrow \quad \exists x_1, \ldots, x_N \geq 0 : \quad p(y, x_1, \ldots, x_N) = 0.$$

Thus, the set Y equals the set of parameters for which the equation $p = 0$ has a solution. Setting

$$q(y, x_1, \ldots, x_N) = y(1 - p^2(y, x_1, \ldots, x_N)),$$

we find that Y is equal to the set of positive values of the polynomial q, whose variables are nonnegative integers. For the set of Fermat primes such a polynomial q has 14 variables and is of the form (see Sect. 3.5 and Jones [157])

$$
\begin{aligned}
q(a, b, c, \ldots, m, n) &= (6g + 5)(1 - (bh + (a - 12)c + n(24a - 145) - d)^2 \\
&\quad - (16b^3 h^3 (bh + 1)(a + 1)^2 + 1 - m^2)^2 - (3g + 2 - b)^2 - (2be + e - bh - 1)^2 \\
&\quad - (k + b - c)^2 - ((a^2 - 1)c^2 + 1 - d^2)^2 - (4(a^2 - 1)i^2 c^4 + 1 - f^2)^2 \\
&\quad - ((d + \ell f)^2 - ((a + f^2(f^2 - a))^2 - 1)(b + 2jc)^2 - 1)^2).
\end{aligned}
$$

All positive values of this polynomial are Fermat primes (except for 3), where the variables go through all nonnegative integers. The construction of this polynomial is based on Pépin's test 4.11, so it is difficult to use it for finding new Fermat primes, since the complexity of Pépin's test grows exponentially with the index of the Fermat number.

4.3 Wieferich Primes

If $p^2 \mid F_m$, where p is prime, then we can say much more about the form of p than in Lucas' Theorem 4.17. The following theorem is proved in Ribenboim [319, p. 88].

Theorem 4.20 *Let $m \geq 2$ and $p^2 \mid F_m$, where p is a prime. Then there exists k odd and $r \geq m + 2$ such that*

$$p = k2^r + 1 \quad \text{and} \quad \frac{k^{p-1} - 1}{p} \equiv 1 \pmod{p}.$$

For the time being no Fermat numbers are known which are divisible by the square of a prime number [199]. The following theorem is usually used to test whether F_m is divisible by the square of a prime number (see Ribenboim [319]).

Theorem 4.21 *If a prime p divides F_m, then*

$$p^2 \mid F_m \iff 2^{p-1} \equiv 1 \pmod{p^2} \quad \text{(Wieferich congruence)}. \tag{4.15}$$

Proof If $p^2 \mid F_m$, then

$$2^{2^m} \equiv -1 \pmod{p^2}.$$

From this we see by (1.4) that $2^{2^{m+2}} \equiv 1 \pmod{p^2}$, and thus $2^{k2^{m+2}} \equiv 1 \pmod{p^2}$ for all $k \in \mathbb{N}$. Since by Lucas' Theorem 4.17, it follows that $p = k2^{m+2} + 1$ for some k, we obtain $2^{p-1} \equiv 1 \pmod{p^2}$.

Assume conversely that $2^{p-1} \equiv 1 \pmod{p^2}$. From the Binomial Theorem one can easily prove (see LeVeque [231]) that if $a \geq 2$ is an integer and t is the greatest integer such that $\operatorname{ord}_{p^t}(a) = \operatorname{ord}_p(a)$, then $\operatorname{ord}_{p^r}(a) = \operatorname{ord}_p(a)$ for $r \in \{1, \ldots, t\}$, and $\operatorname{ord}_{p^t}(a) = p^{r-t}\operatorname{ord}_p(a)$ for $r > t$. Since $p \nmid (p-1)$, we have $\operatorname{ord}_{p^2}(2) = \operatorname{ord}_p(2)$. From this it follows that $2^k \equiv 1 \pmod{p^2}$ if and only if $2^k \equiv 1 \pmod{p}$. Since $p \mid F_m$, we obtain $2^{2^m} \equiv -1 \pmod{p}$. Hence, $2^{2^{m+1}} \equiv 1 \pmod{p}$ which implies that $2^{2^{m+1}} \equiv 1 \pmod{p^2}$. In other words,

$$p^2 \mid \left(2^{2^{m+1}} - 1\right) = \left(2^{2^m} + 1\right)\left(2^{2^m} - 1\right).$$

Since $p \mid F_m = 2^{2^m} + 1$ and $\left(2^{2^m} + 1\right) - \left(2^{2^m} - 1\right) = 2$, we see that $p \nmid \left(2^{2^m} - 1\right)$. Therefore, $\left(p^2, 2^{2^m} - 1\right) = 1$, and thus $p^2 \mid \left(2^{2^m} + 1\right) = F_m$. $\qquad\square$

Remark Although an extensive computer search was performed (see [95], cf. also [79, 169] up to $6.7 \cdot 10^{15}$, so far we know only two prime numbers that satisfy the Wieferich congruence on the right-hand side of equivalence (4.15),

$$p = 1093 \quad \text{and} \quad p = 3511.$$

An elementary proof that these numbers satisfy the Wieferich congruence is given in [130] (see also the table in [278]). The primes 1093 and 3511 are called *Wieferich primes*. The first one was discovered by Waldemar Meissner in 1913 and the second one by Nicolaas G. W. H. Beeger in 1922, that is, long before the era of electronic computers began (see Ribenboim [323, p. 334]). However, these two primes do not divide any Fermat prime F_m, since none of them has the form $k2^{m+2} + 1$ for $m > 1$, as required by Lucas' Theorem 4.17.

Before Fermat's Last Theorem 2.11 was proved by Andrew Wiles and Richard Taylor [396, 418], mathematicians investigated the equation

$$x^p + y^p = z^p, \tag{4.16}$$

where p is an odd prime, $xyz \neq 0$, and $(x, y, z) = 1$. Fermat's Last Theorem is traditionally divided into two cases. In the *first case of Fermat's Last Theorem* one assumes that $p \nmid xyz$. In the second case which is much harder, it is required that $p \mid xyz$. Equivalence (4.15) expresses a close connection between the first case of Fermat's Last Theorem and Fermat numbers:

Theorem 4.22   (Wieferich) *If the first case of Fermat's Last Theorem holds for an odd prime exponent p, then p satisfies the Wieferich congruence*

$$2^{p-1} \equiv 1 \pmod{p^2}.$$

This theorem is proved in Wieferich [417] (see also Ribenboim [320]). It states that if equation (4.16) had a solution for some odd prime $p \nmid xyz$, then the Wieferich congruence would hold. However, since Fermat's Last Theorem has already been proved, Wieferich's Theorem has only a historical significance. In monographs [199, p. 69] and [323, p. 333–346] other interesting connections between Fermat numbers and Wieferich primes are given.

4.4   Elite Primes

A prime p is called *elite,* if only a finite number of Fermat numbers are quadratic residues modulo p.

In other words, a prime p is elite, if there exists an index k such that for all $m > k$ the quadratic congruence

$$x^2 \equiv F_m \pmod{p}$$

has no solution, i.e., for the Jacobi symbol we have (see [199, p. 42])

$$\left(\frac{p}{F_m}\right) = -1 \quad \text{for all } m > k.$$

Elite primes are thus closely connected with Fermat numbers.

Example   We show that 3, 5, and 7 are elite primes. From relations (4.7) and (4.10) we see that $F_m \equiv 2 \pmod 3$ for $m \geq 1$. Since $F_m \equiv 1 \pmod 4$, then by Theorem 2.26 and the Law of Quadratic Reciprocity (see Theorem 2.27) we obtain

$$\left(\frac{3}{F_m}\right) = \left(\frac{F_m}{3}\right) = \left(\frac{2}{3}\right) = -1.$$

This fact lies at the heart of Pépin's test.

By (4.10) we have $F_m \equiv 2 \pmod 5$ for $m \geq 2$, since $F_1 = 5$. Then by Theorems 2.26 and 2.27 we get

$$\left(\frac{5}{F_m}\right) = \left(\frac{F_m}{5}\right) = \left(\frac{2}{5}\right) = -1.$$

As already mentioned, Pater Pépin in his original article (see [296]) of 1877 used the base 5 instead of base 3, which is in relation (4.12). According to Aigner [5], bases 3 and 5 in Pépin's test can be replaced by the number 7. To do that, it is enough to show that the Jacobi symbol $\left(\frac{7}{F_m}\right)$ equals -1 for all m greater than some k. We show that $\left(\frac{7}{F_m}\right) = -1$ for all $m \geq 1$. Since

$$F_m = 16^{2^{m-2}} + 1 \equiv F_{m-2} \pmod 7,$$

we have (cf. van Maanen [250, p. 349])

$$F_m \equiv F_0 \equiv 3 \pmod 7 \text{ for } m \text{ even,}$$
$$F_m \equiv F_1 \equiv 5 \pmod 7 \text{ for } m \text{ odd.}$$

Since $F_m \equiv 1 \pmod 4$ for $m \geq 1$, we observe by the Law of Quadratic Reciprocity for the Jacobi symbol (see property (viii) of Theorem 2.28) that

$$\left(\frac{7}{F_m}\right) = \left(\frac{F_m}{7}\right) = \left(\frac{3}{7}\right) = -1 \quad \text{for every even } m \geq 2$$

and

$$\left(\frac{7}{F_m}\right) = \left(\frac{F_m}{7}\right) = \left(\frac{5}{7}\right) = -1 \quad \text{for every odd } m \geq 1.$$

We will briefly indicate how it can be verified that 41 is an elite prime number. To do that, we just show that $\left(\frac{41}{F_m}\right) = -1$ for all F_m for which $m \geq 2$. First one has to prove that

$$\left(\frac{41}{F_{m+4}}\right) = \left(\frac{41}{F_m}\right) \quad \text{for } m \geq 2.$$

By a direct calculation we have to check that $\left(\frac{41}{F_m}\right) = -1$ for $2 \leq m \leq 5$.

Elite primes can thus be substituted for the number 3 or 5 in the classical Pépin's test.

Due to recurrence relation (4.9) one can prove (see Müller [281]) that for any natural number n the sequence $(F_m \pmod n))_{m=0}^{\infty}$ is periodic from some index m on. By Aigner [5], any prime number written in the form $p = 2^r h + 1$, where $h \geq 1$

is odd and $r \in \mathbb{N}$, this period starts at F_r or earlier. The length ℓ will be called the *Fermat period*, if ℓ is the smallest natural number satisfying the congruence

$$F_{r+\ell} \equiv F_r \pmod{p}.$$

The terms F_{r+s} for $s = 0, \ldots, \ell - 1$ are called the *Fermat residues modulo p*.

Elite primes were first defined and studied by the Austrian mathematician Alexander Aigner in the 1986 article [5]. He discovered 14 of them less than 35 million. Later Tom Müller using computers found another two elite primes less than one billion (see Müller [281]). The first 16 elite prime numbers are:

3, 5, 7, 41, 15361, 23041, 26881, 61441, 87041, 163841, 544001, 604801, 6684673, 14172161, 159318017, 446960641.

Müller discovered another 26 elite prime numbers greater than one billion. He used the following statement.

Theorem 4.23 (Müller) *Let $p = 2^r h + 1$ be a prime number, where h is odd. Then p is elite if and only if the multiplicative order of each Fermat residue modulo p is a multiple of 2^r.*

Alain Chaumont and Tom Müller [68] found another 5 elite prime numbers less than 250 billion:

1 151 139 841, 3 208 642 561, 38 126 223 361, 108 905 103 361, 171 727 482 881.

There is a number of open problems and conjectures around elite prime numbers. Let us list only the most important ones.

Conjecture 1. The number of elite primes is infinite.

Nevertheless, in [200] we prove that

$$\sum_{p \in \mathbb{E}} \frac{1}{p} \tag{4.17}$$

converges, where $\mathbb{E}$ is the set of elite primes. This means that the set $\mathbb{E}$ is not "dense" enough to cause divergence of the sum (4.17), as in the case of all primes (see Theorem 3.15). Denote by $E(x)$ the number of elite primes not exceeding x. Then from the inequalities

$$\prod_{i=0}^{t} F_{t+i} < \prod_{i=0}^{2t} F_i < 2^{2^{2t+1}}$$

one can derive that

$$E(x) = \mathcal{O}\left(\frac{x}{(\log x)^2}\right) \quad \text{for } x \to \infty,$$

see [200, 283]. Here $\mathcal{O}$ denotes the *Landau symbol*, which in this case means that there exists a constant $C > 0$ such that

$$E(x) \leq C \frac{x}{(\log x)^2} \quad \text{for } x \to \infty.$$

This upper bound is probably quite rough. Based on numerical evidence, Müller [281] proposed three more conjectures.

Conjecture 2. $E(x) = \mathcal{O}(\log x)$ for $x \to \infty$.

Conjecture 3. The number of elite primes of the form $2^r \cdot 15 + 1$ is infinite.

Conjecture 4. The length of the Fermat period ℓ is not bounded, i.e., there exist elite primes with arbitrarily large period.

In contrast with the elite primes, Müller in [282] defines an *anti-elite prime* as a prime p for which F_m is a quadratic residue modulo p for all but finitely many m. For example, 13 and 97 are anti-elite primes and so are all Fermat primes greater than 5 (see Table 13.5 and [198] for more details).

4.5 Regular and Irregular Primes

Regular primes, which we define below, were introduced by the German mathematician Ernst Eduard Kummer (1810–1893) during his attempts to prove Fermat's Last Theorem 2.11. Recall that this theorem claims that the equation

$$x^n + y^n = z^n \tag{4.18}$$

has no solution in positive integers when $n > 2$. Fermat himself proved by his own method of infinite descent proved that the Diophantine equation (4.18) has no solution for $n = 4$ (see Theorem 2.12).

Around 1850, Kummer made great progress in his effort to prove Fermat's Last Theorem if p is a regular prime number. To define regular primes, we first introduce how the Bernoulli numbers B_k are defined. They appear in coefficients of the power series of the function

$$\frac{x}{e^x - 1} = \sum_{k=0}^{\infty} B_k \frac{x^k}{k!}. \tag{4.19}$$

We have

$$B_0 = 1, \quad B_1 = -\frac{1}{2}, \quad B_2 = \frac{1}{6}, \quad B_3 = 0, \quad B_4 = -\frac{1}{30}, \ldots, \quad B_{10} = \frac{5}{66}, \ldots,$$

where the *Bernoulli numbers* B_k are defined recurrently by

$$\binom{k+1}{k} B_k + \binom{k+1}{k-1} B_{k-1} + \cdots + \binom{k+1}{1} B_1 + B_0 = 0.$$

They are named after Jacob Bernoulli. It is easy to see that all B_k are rational numbers. It is also possible to deduce that $B_{2k+1} = 0$ for any $k \in \mathbb{N}$.

We say that an odd prime is *regular*, if it does not divide any numerator of the Bernoulli numbers $B_0, B_2, B_4, \ldots, B_{p-3}$. A prime which is not regular is called *irregular*.

For instance, the primes

$$3, 5, 7, 11, 13, 17, 19, 23, 29, 31, 41, \ldots$$

are regular while the primes

$$37, 59, 67, 101, 103, 131, 149, \ldots$$

are irregular.

If p is irregular, its *irregularity index* is equal to the number of natural numbers k such that p divides the numerator of B_{2k} for $2 \leq 2k \leq p - 3$, see [53]. The largest irregularity index known until 2020 is 9. It is attained for the irregular prime 1 767 218 027. Nevertheless, it is generally believed that the index of irregularity can be arbitrarily large.

Using somewhat heuristic arguments, it was derived that the ratio of the number of all regular prime numbers smaller than n to the number of all primes less than n converges to $e^{-1/2} \approx 0.60653$.

This conjecture is supported by a fact derived in the article [52]. It is claimed here that of 283 145 primes less than $4 \cdot 10^6$ there are 171 548 regular (thus 60.587 %). Nevertheless, it is still not known whether there exists an infinite number of regular prime numbers. Jensen [156] in 1915 proved that there are infinitely many irregular primes. The largest irregular prime number known until 2020 has 13 862 digits. It divides the numerator of the Bernoulli number B_{2370} and was discovered in 2003. For his dedicated effort to prove Fermat's Last Theorem Kummer received a prize of 3000 francs from the French Academy of Sciences without having entered the competition.

4.6 Sophie Germain Primes

In 1819, the French mathematician Sophie Germain (1776–1831) became famous for a partial proof of the so-called first case of Fermat's Last Theorem (cf. Theorem 4.22), i.e., the equation $x^p + y^p = z^p$ has no solution in natural numbers for a prime number exponent $p > 2$ such that p does not divide the product xyz. She proved that if p and $2p + 1$ are primes, then the first case of Fermat's Last Theorem holds for the exponent p.

An odd prime number p for which $2p + 1$ is also a prime is thus called a *Sophie Germain prime*. For instance, 2, 3, 5, 11, and 23 are Sophie German primes. The number $2p + 1$ associated with a Sophie Germain prime is called a *safe prime*.

These prime numbers have a lot of interesting properties. For instance, if p is a Sophie Germain prime, then in [207] we prove that all quadratic nonresidues are primitive roots modulo $2p + 1$ except for $2p$ which is a quadratic nonresidue, but it is not a primitive root. We will return to this property in Theorem 4.30.

Another theorem which was already known to Fermat shows a close connection between Mersenne numbers and Sophie Germain primes. It was proved later by Leonhard Euler and also independently by Joseph Louis Lagrange.

Theorem 4.24 *Let p be a prime such that $p \equiv 3 \pmod 4$. Then $2p + 1$ divides the Mersenne number M_p if and only if $2p + 1$ is a prime.*

A proof is given e.g. in Ribenboim [323, pp. 90–91] and Robbins [330, p. 149]. Hence, if $p = 11, 23, 83, \ldots$, then M_p has prime factors $23, 47, 167, \ldots$, respectively. By Euler's criterion given in Theorem 2.25 we get $2^{((2p+1)-1)/2} = 2^p \equiv -1 \pmod{2p+1}$ or $1 \pmod{2p+1}$ for a Sophie Germain prime p. Hence, $2p + 1$ divides $2^p + 1$ or $2^p - 1$. Another connection between the Mersenne primes and Sophie Germain primes is given in Theorem 4.29.

Up to now, we do not know if there exist infinitely many Sophie Germain primes. If there were infinitely many Sophie Germain primes p for which $p \equiv 3 \pmod 4$, then by Theorem 4.24 there would exist infinitely many composite Mersenne numbers, since $2p + 1$ divides $2^p - 1$.

The largest known Sophie Germain prime number until 2020 was

$$p = 2618163402417 \cdot 2^{1290000} - 1,$$

i.e., $2p + 1$ is also a prime.

For any $n \in \mathbb{N}$ consider a digraph $G(n)$ with vertices $0, 1, 2, \ldots, n - 1$. Its definition is the same as that in (4.13). We show that if p is a Sophie Germain prime, then its digraph $G(2p + 1)$ will have a surprising structure. Its nontrivial components will resemble little suns or a steering wheel for controlling a ship (see Rogers [332, p. 323] and Figs. 4.5, 4.6, and 4.7).

Rogers in [332] describes the structure of each component of $G(n)$, if n is prime. In [210, 370–372] we describe the structure of $G(n)$ also for composite n.

Let $\omega(n)$ denote the number of different primes that divide n. By Szalay [389], the number of fixed points of the digraph $G(n)$ is equal to $2^{\omega(n)}$ (we will prove it in Theorem 5.1). This leads to the following statement (cf. Figs. 4.6 and 4.7).

Theorem 4.25 *If n is a prime, then there exist exactly two fixed points of the digraph $G(n)$, namely 0 and 1.*

The following statements are taken from the article [210]. They use properties of the Carmichael lambda function λ from Sect. 2.9.

Theorem 4.26 *In a digraph $G(n)$ there exists a cycle of length t if and only if $t = \mathrm{ord}_d 2$ for some odd positive divisor d of the number $\lambda(n)$.*

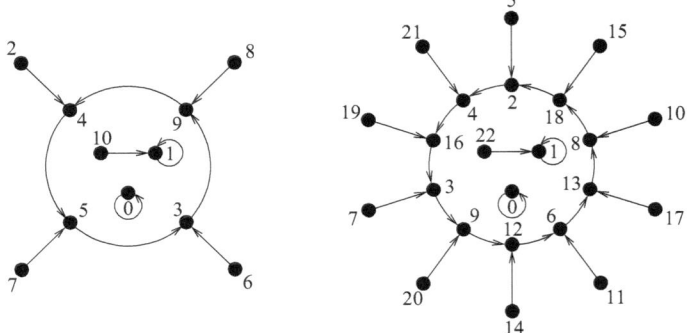

Fig. 4.5 Iteration digraphs associated to $n = 11$ and $n = 23$

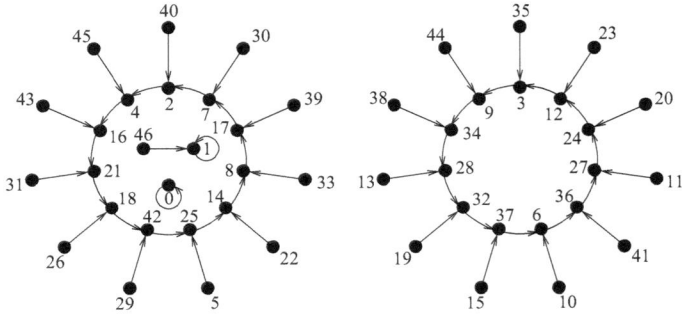

Fig. 4.6 Iteration digraph for $n = 47$

Proof Assume that a is a vertex in a t-cycle of $G(n)$. Then t is the smallest natural number for which

$$a^{2^t} \equiv a \pmod{n}.$$

From this we obtain that t is the smallest natural number for which

$$a^{2^t} - a \equiv a\left(a^{2^t-1} - 1\right) \equiv 0 \pmod{n}. \tag{4.20}$$

Since $(a, a^{2^t-1} - 1) = 1$, from (4.20) it follows that if $n_1 = (a, n)$ and $n_2 = n/n_1$, then t is the smallest natural number such that

$$a \equiv 0 \pmod{n_1},$$
$$a^{2^t-1} \equiv 1 \pmod{n_2}. \tag{4.21}$$

Hence, $(n_1, n_2) = 1$ and by the Chinese Remainder Theorem 1.4 we obtain the existence of an integer b such that

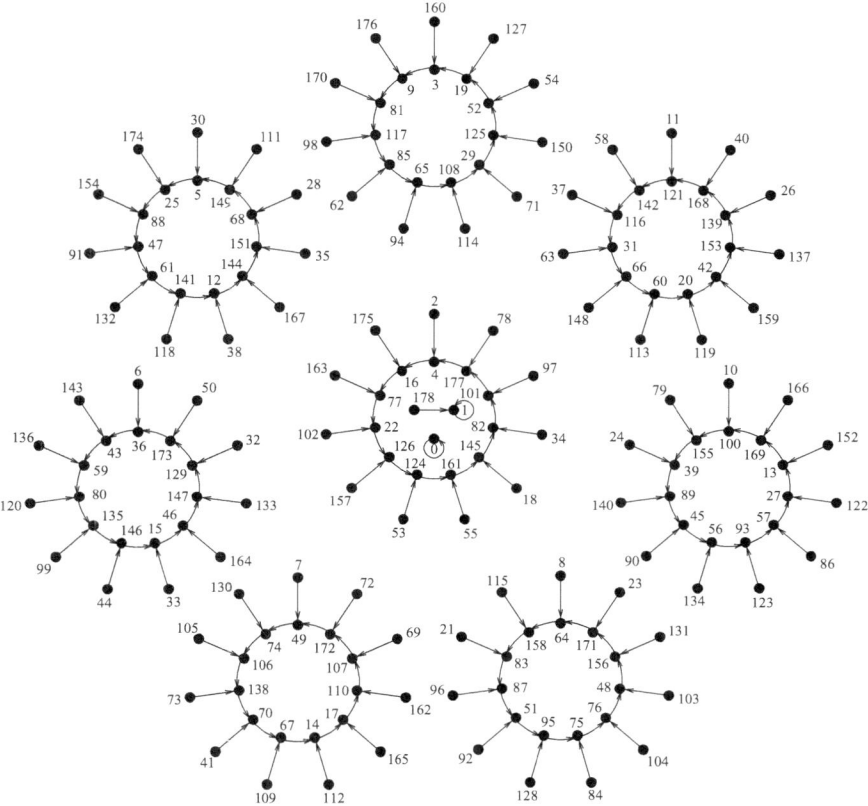

Fig. 4.7   The iteration digraph for $n = 179$

$$b \equiv 1 \pmod{n_1},$$
$$b \equiv a \pmod{n_2}. \tag{4.22}$$

From (4.21) and (4.22) it further follows that t is the smallest natural number such that

$$b^{2^t-1} \equiv 1 \pmod{n}. \tag{4.23}$$

Let $d = \mathrm{ord}_n b$. Then $d \mid (2^t - 1)$. Since, by (4.23), t is the smallest natural number for which $d \mid (2^t - 1)$, we see that $t = \mathrm{ord}_2 d$. Clearly, d is odd, because $d \mid (2^t - 1)$. Moreover, $d \mid \lambda(n)$ due to Carmichael's Theorem 2.24, since by (4.23) it follows that $(b, n) = 1$.

Suppose conversely that d is an odd positive divisor of $\lambda(n)$ and let $t = \mathrm{ord}_2 d$. According to Carmichael's Theorem 2.24, there exists a remainder g modulo n such that $\mathrm{ord}_n g = \lambda(n)$. Let $h = g^{\lambda(n)/d}$. Then $\mathrm{ord}_n h = d$. Since $d \mid (2^t - 1)$ (but

$d \nmid (2^k - 1)$ whenever $1 \le k < t$), we see that t is the smallest natural number for which

$$h^{2^t - 1} \equiv 1 \pmod{n}. \tag{4.24}$$

Then

$$h \cdot h^{2^t - 1} = h^{2^t} \equiv h \pmod{n},$$

and thus h is a vertex in the t-cycle of $G(n)$. □

Theorem 4.27 *Let p be a Sophie Germain prime. Then $G(2p + 1)$ has two trivial components: the isolated point 0 and the component $\{1, 2p\}$ whose fixed point is 1. Any other component has $2t$ vertices and contains a t-cycle, where $t = \mathrm{ord}_p 2$. The number of oriented edges coming to each vertex of a t-cycle is exactly 2.*

Proof Since $n = 2p + 1$ is a prime, by the definition of the Carmichael lambda function (see Sect. 2.9) we get

$$\lambda(2p + 1) = 2p.$$

The number $2p$ has exactly two odd divisors 1 and p. Setting $d = 1$ in the previous Theorem 4.26, we get by Theorem 4.25 that there exist exactly two fixed points 0 and 1. The number 0 is the only solution of the congruence $x^2 \equiv 0 \pmod{n}$, and thus 0 is an isolated fixed point. Moreover, $x = 1$ and $x = 2p$ are the only solutions of the congruence $x^2 \equiv 1 \pmod{n}$, since n is a prime. We have to show that the corresponding component containing $\{1, 2p\}$ does not have other vertices. Since p and n are odd numbers, we see that $n \equiv 3 \pmod{4}$. Therefore, $2p$ is not a quadratic residue modulo n which means that the congruence $x^2 \equiv 2p \pmod{n}$ has no solution.

Now let us put $d = p$ in Theorem 4.26. Every other component $G(2p + 1)$ thus contains a cycle of length $t = \mathrm{ord}_p 2$ for $t > 1$. If a vertex a belongs to this t-cycle, then the congruence $x^2 \equiv a \pmod{n}$ has a solution and therefore, a is a quadratic residue modulo n. Since n is an odd prime, this congruence has exactly two solutions c and $-c$. One of them lies on the t-cycle and the second one outside. Since $n = 2p + 1 \equiv 3 \pmod{4}$, one of the residues c or $-c$ must be a quadratic residue and the second one a quadratic nonresidue modulo n.

Assume that c is not a quadratic residue modulo n. Then c lies outside the t-cycle and the there is an oriented edge from c to a. Since c is not a quadratic residue modulo n, there is no edge entering to c. Therefore, the corresponding component has exactly $2t$ vertices. □

In [210] we also deal with a converse statement to Theorem 4.27.

If p is a Sophie Germain prime, then all components of the digraph $G(2p + 1)$ that do not contain vertices 0 and 1 will be called *Sophie Germain little suns*.

Theorem 4.28 *Let p be a Sophie Germain prime. Then the number of Sophie Germain little suns in the digraph $G(2p + 1)$ is equal to*

$$\frac{p-1}{\text{ord}_p 2}. \tag{4.25}$$

Proof According to Theorem 4.27, the number of vertices of $G(2p+1)$ which lie outside the trivial components is $2p - 2$. From Theorem 4.27 we also know that every Sophie Germain little sun has $2\,\text{ord}_p 2$ vertices which proves the theorem. □

It is of interest to note that the expression in (4.25) also denotes the number of irreducible factors of the polynomial $1 + x + x^2 + \ldots + x^{p-1}$ over $\mathbb{F}_2[x]$, where $\mathbb{F}_2$ denotes the finite field with two elements (see Theorem 2.47 (ii) of [232] by Lidl and Niederreiter).

Remark If $2p + 1$ is a prime and $p > 1$, then $2p - 2$ is not divisible by 3. Hence, by (4.25) the number of Sophie Germain little suns is never divisible by 3 and the length of all associated t-cycles is also never divisible by 3.

Now we prove a somewhat more general statement, namely that $G(2p + 1)$ never contains a q-cycle for $q = 3, 5, 7, 13, 17, 19, \ldots$, where q ranges over all the odd exponents of Mersenne primes $M_q = 2^q - 1$ with $q > 2$. (Note that $G(7)$ contains a 2-cycle.)

Theorem 4.29 *Let M_q be a Mersenne prime for $q > 2$. Then there is no Sophie Germain prime p such that $G(2p + 1)$ contains a q-cycle.*

Proof Assume to the contrary that there exists a Sophie Germain prime p and a Mersenne prime M_q for $q > 2$ such that $G(2p + 1)$ contains a q-cycle. Then by Theorem 4.27 $q = \text{ord}_p 2$, and thus $p = 2^q - 1$. However, the number

$$2p + 1 = 2^{q+1} - 1 = (2^{(q+1)/2} + 1)(2^{(q+1)/2} - 1)$$

is composite for the prime $q > 2$ which is a contradiction. □

Consequently, this statement again connects Mersenne primes with Sophie Germain primes.

Example Let $p = 89$. Since $2^{11} \equiv 1 \pmod{89}$, we see that $\text{ord}_{89} 2 = 11$. Hence, by Theorem 4.28 the number of Sophie Germain little suns in the digraph $G(179)$ is $88/11=8$ (see Fig. 4.7).

Let us still note that digraphs $G(n)$ corresponding to Mersenne numbers n are investigated in Szalay [389]. In the articles [371, 372], we investigate the structure of digraphs corresponding to the congruence $f(x) \equiv x^k \pmod{N}$ for $k \geq 2$, which is more general than (4.14).

From Theorem 4.16 we know that the number of primitive roots of Fermat primes is equal to the number of quadratic nonresidues. The other prime numbers do not have this property. Assume now that p is a Sophie Germain prime. The following theorem shows that the number of primitive roots of the prime $2p + 1$ is only one less than the number of quadratic nonresidues which is p.

Theorem 4.30 *If p is a Sophie Germain prime, then the prime $q = 2p + 1$ has $p - 1$ primitive roots modulo q.*

Proof A Sophie Germain p is by definition odd. Hence, $(2, p) = 1$. By Theorem 2.22, (2.39), and using relation (2.37) twice, we find that the number of primitive roots of the prime number q is

$$\phi(\phi(q)) = \phi(2p) = \phi(2)\phi(p) = \phi(p) = p - 1.$$

The above completes the proof.   □

According to Theorem 2.29, each primitive root is a quadratic nonresidue modulo q. The following theorem shows that the quadratic nonresidue $q - 1$ is not a primitive root (see Figs. 4.5–4.7).

Theorem 4.31 *If p is a Sophie Germain prime, then the number $2p$ is the only quadratic nonresidue which is not a primitive root modulo $q = 2p + 1$.*

Proof We see that
$$(2p)^2 = (q - 1)^2 \equiv 1 \pmod{q}.$$

From this and the inequality $q \geq 7$ it follows that $2p$ is not a primitive root modulo q, since $\phi(q) = q - 1 \geq 6$.

The prime q is of the form $4k - 1$, since p is an odd prime. By Theorem 3.3, the congruence
$$x^2 \equiv 2p \equiv -1 \pmod{q}$$

does not have a solution, and thus $2p$ is a quadratic nonresidue.   □

4.7 Euclidean Primes

Euclid's Theorem 2.3 guarantees that there are infinitely many prime numbers. Its proof was by contradiction. We assumed that there exist only a finite number of primes $p_1, p_2, \ldots, p_n$ and then we investigated the number (see (2.2))

$$m = p_1 p_2 \cdots p_n + 1. \tag{4.26}$$

For this reason, prime numbers of the form (4.26) are called *Euclidean prime numbers*. For example,

$$2 + 1 = 3, \ 2 \cdot 3 + 1 = 7, \ 2 \cdot 3 \cdot 5 + 1 = 31, \ 2 \cdot 3 \cdot 5 \cdot 7 + 1 = 211, \ 2 \cdot 3 \cdot 5 \cdot 7 \cdot 11 + 1 = 2311$$

are Euclidean primes. Notice that the first three Euclidean primes are the same as the first three Mersenne primes.

However, according to (2.3), not every number of the form (4.26) is a prime. Denote by $p\#$ the product of all primes less than or equal to a prime p. Then $p\# + 1$ is a Euclidean prime only for

$$p = 2, 3, 5, 7, 11, 31, 379, 1019, 1021, 2657, 3229, 4547, 4787, 11549, \ldots$$

See e.g. [38, 60]. Similarly one can investigate the *primorial numbers* of the form $p\# - 1$, which are primes for

$$p = 3, 5, 11, 13, 41, 89, 317, 337, 991, 1873, 2053, 2377, 4093, 4297, 4583, \ldots$$

Let $A(p)$ denote the number of primitive roots modulo the prime p. From Theorem 4.18 we know that the number of incongruent primitive roots of Fermat primes is almost equal to $\frac{1}{2}p$, more precisely,

$$\frac{A(p)}{p-1} = \frac{1}{2}.$$

All other primes have the relative number of primitive roots smaller. By Theorem 4.30, the number of primitive roots of prime numbers of the form $2p + 1$, where p is a prime, is also almost equal to 50%,

$$\frac{A(2p+1)}{2p-2} = \frac{1}{2},$$

cf. Table 13.4. On the other hand, from the proof of the following theorem and Theorem 4.33 it follows that the Euclidean primes have a very small number of primitive roots relative to the size of p (see [211]).

Theorem 4.32 *For any $\varepsilon > 0$ there exists a prime p such that*

$$\frac{A(p)}{p} < \varepsilon.$$

Proof From Theorem 2.22 and relation (2.37) we have $A(p) = \phi(p-1)$. Denote by p_i the ith prime. According to Dirichlet's Theorem 3.10, for any $n \in \mathbb{N}$ there exists $r \in \mathbb{N}$ such that $rp_1p_2 \cdots p_n + 1$ is equal to some prime p. Then $p - 1$ can be factored as follows

$$p - 1 = sp_1^{k_1} p_2^{k_2} \cdots p_n^{k_n}, \tag{4.27}$$

where $(s, p_1, p_2, \ldots, p_n) = 1$ and $k_i \geq 1$ for $i = 1, \ldots, n$. From this, (4.27), (2.38), and (2.39) we get

$$\frac{\phi(p-1)}{p} < \frac{\phi(p-1)}{p-1} = \frac{\phi(s)\phi(p_1^{k_1})\cdots\phi(p_n^{k_n})}{p-1}$$

$$= \frac{\phi(s)}{s}\frac{p_1^{k_1-1}(p_1-1)}{p_1^{k_1}}\cdots\frac{p_n^{k_n-1}(p_n-1)}{p_n^{k_n}} = \frac{\phi(s)}{s}\frac{p_1-1}{p_1}\cdots\frac{p_n-1}{p_n}$$

$$\leq \left(1-\frac{1}{p_1}\right)\cdots\left(1-\frac{1}{p_n}\right),$$

since $\phi(s) \leq s$. It remains to prove that

$$\lim_{n\to\infty} a_n = 0, \tag{4.28}$$

where

$$a_n = \prod_{i=1}^{n}\left(1-\frac{1}{p_i}\right).$$

From the formula for the sum of the geometric sequence it follows that

$$\sum_{j=0}^{\infty}\left(\frac{1}{p_i}\right)^j = \frac{1}{1-p_i^{-1}},$$

and thus

$$\frac{1}{a_n} = \prod_{i=1}^{n}\frac{1}{1-p_i^{-1}} = \prod_{i=1}^{n}\sum_{j=0}^{\infty}\left(\frac{1}{p_i}\right)^j$$

$$= \prod_{i=1}^{n}(1+p_1^{-1}+p_1^{-2}+\cdots)\cdots(1+p_n^{-1}+p_n^{-2}+\cdots) = \sum_m\frac{1}{m},$$

where in the last sum we make summation over all natural numbers m which are divisible only by the primes $p_1, p_2, \ldots, p_n$. However, we see by the Fundamental Theorem of Arithmetic 2.2 that as $n \to \infty$, the right-hand side approaches the harmonic series $\sum_{m=1}^{\infty}\frac{1}{m}$, which is divergent. Hence, the left-hand side diverges to infinity as $n \to \infty$, and thus (4.28) holds.   □

From the proof of the foregoing theorem we get for the Euclidean primes $p = p_1 p_2 \cdots p_n + 1$ the following upper bound

$$\frac{A(p)}{p} < \left(1-\frac{1}{p_1}\right)\cdots\left(1-\frac{1}{p_n}\right).$$

Note that the product of the right-hand side converges very slowly and monotonically to zero as $n \to \infty$. The ratio on the left-hand side can be easily calculated for small p. The prime 3 has a unique primitive root 2 and the prime 7 has two primitive roots

3 and 5. The number of primitive roots of further Euclidean primes is $A(31) = 8$, $A(211) = 48$, $A(2311) = 480$, etc. We observe that the ratios $A(p)/p$ corresponding to the first five Euclidean primes $p = 3, 7, 31, 211, 2311$, i.e.,

$$\frac{1}{3} \approx 0.333, \quad \frac{2}{7} \approx 0.285, \quad \frac{8}{31} \approx 0.258, \quad \frac{48}{211} \approx 0.227, \quad \frac{480}{2311} \approx 0.207,$$

form a slowly decreasing sequence (see [211]).

Remark We note by Mertens' Third Theorem (see [270]) that

$$\lim_{x \to \infty} \log x \prod_{\text{primes } p \leq x} \left(1 - \frac{1}{p}\right) = e^{-\gamma} = 0.5614594836\ldots,$$

where $\gamma = 0.5772156649\ldots$ is the Euler constant, sometimes also called the *Euler-Mascheroni constant*.

Theorem 4.33 *If p is a Euclidean prime, then for all primes $q < p$ we have*

$$\frac{A(q)}{q} > \frac{A(p)}{p}.$$

Proof Let $p = p_1 p_2 \cdots p_n + 1$ be a Euclidean prime and let $q < p$ be an arbitrary prime. For $q = 2$ the theorem holds, since

$$\frac{A(p)}{p} = \frac{\phi(p-1)}{p} < \frac{\phi(p-1)}{p-1} = \frac{\phi(p_1)\cdots\phi(p_n)}{p_1 \cdots p_n}$$
$$= \frac{p_1 - 1}{p_1} \cdots \frac{p_n - 1}{p_n} \leq \frac{1}{2} = \frac{A(q)}{q}. \tag{4.29}$$

Hence, we may assume that $q > 2$. Then $q - 1$ has less than n different primes in its prime number factorization

$$q - 1 = q_1^{\ell_1} q_2^{\ell_2} \cdots q_m^{\ell_m},$$

where $q_1 < q_2 < \cdots < q_m$ are primes and $\ell_i \geq 1$. Since $m < n$, we find that

$$p_i \leq q_i \quad \text{for } i = 1, \ldots, m.$$

From the properties of the Euler totient function (2.38) and (2.39) we obtain

$$\frac{\phi(q_i^{\ell_i})}{q_i^{\ell_i}} = \frac{\phi(q_i)}{q_i} = \frac{q_i - 1}{q_i} \quad \text{for } i = 1, \ldots, m.$$

From this, (4.29), and the inequality

$$\frac{p_i - 1}{p_i} \leq \frac{q_i - 1}{q_i} \quad \text{for } i = 1, \ldots, m$$

we get

$$\frac{A(p)}{p} < \frac{p_1 - 1}{p_1} \cdots \frac{p_n - 1}{p_n} \leq \frac{q_1 - 1}{q_1} \cdots \frac{q_n - 1}{q_n} \frac{q - 1}{q}$$
$$= \frac{\phi(q_1^{\ell_1})}{q_1^{\ell_1}} \cdots \frac{\phi(q_m^{\ell_m})}{q_m^{\ell_m}} \frac{q - 1}{q} = \frac{\phi(q_1 - 1)}{q - 1} \frac{q - 1}{q} = \frac{A(q)}{q},$$

where the second inequality follows from the fact that the p_i are consecutive primes, $m < n$, and $q_m < q$.   □

4.8 Factorial Primes

In the proof of Euclid's Theorem 2.3, relation (2.2) (i.e. also (4.26)) can be replaced by

$$m = p! + 1, \tag{4.30}$$

where p is the largest prime, if there is only a finite number of primes. Since for every natural number $q \leq p$ the ratio m/q always produces the remainder 1, no prime less than or equal to p divides m. Hence, by the Fundamental Theorem of Arithmetic 2.2 the number m is the next prime or m is divisible by a prime greater than p, which is a contradiction.

Primes of the form (4.30) are called *factorial primes* (see [38, 60]), even though we do not require p to be a prime number.

For instance, if

$$n = 1, 2, 3, 11, 27, 37, 41, 73, 77, 116, 154, 320, 340, 399, 427, 872, 1477, \ldots$$

then $n! + 1$ is a prime. Primes of the form $n! - 1$ are also called *factorial primes*. We get them if

$$n = 3, 4, 6, 7, 12, 14, 30, 32, 33, 38, 94, 166, 324, 379, 469, 546, 974, \ldots$$

Similar types of primes are investigated, as well. For example multifactorial primes (see Caldwell and Dubner [59]).

A very interesting conjecture is that if p is the smallest prime larger than $n!$, then $p - n!$ is also a prime or 1. The verification of this conjecture for any $n \leq 2000$

follows from [437, A033932]. A similar conjecture states that if q is the largest prime smaller than $n!$, then $n! - q$ is also a prime or 1 for $n > 2$.

4.9 Palindromic Primes

The word palindrome comes from Greek and means "run back again". A *palindrome* is a string of characters which reads the same backward as forward, such as madam, racecar. There are also numeric palindromes. The most famous palindrome in the Czech lands is perhaps the number 135797531, which is associated with laying the foundation stone for the Charles Bridge in 1357 of the Julian calendar on July 9 at 5 o'clock and 31 min, see Horský [147].

A *palindromic prime* is a prime which is at the same time a palindrome, for instance,

3, 5, 7, 11,

101, 131, 151, 181, 191, 313, 353, 373, 383, 727, 757, 787, 787, 797, 919

10301, 10501, ... ,

Notice that there are no numbers with four digits in this list. In fact, even a more general statement holds:

Theorem 4.34 *The only palindromic prime with an even number of digits is* 11.

Proof Let the number $m = 9090\ldots90$ have $2k$ digits. We see that

$$m \cdot 11 = 99999\ldots990,$$

and thus we have

$$11(m + 1) = 11m + 11 = 10^{2k+1} + 1. \tag{4.31}$$

Consider now a palindromic number p of the form

$$p = a_1 10^{2n-1} + a_2 10^{2n-2} + \cdots + a_2 10 + a_1$$

with an even number $2n$ of digits, where $a_1 \neq 0, a_2, \ldots, a_n$ are decimal digits. From this it follows that

$$\begin{aligned} p &= a_1(10^{2n-1} + 1) + a_2(10^{2n-2} + 10) + \cdots + a_n(10^n + 10^{n-1}) \\ &= a_1(10^{2n-1} + 1) + 10a_2(10^{2n-3} + 1) + \cdots + 10^{n-1}a_n(10 + 1). \end{aligned}$$

However, by (4.31) all its terms are divisible by 11, and therefore any palindrome $p > 11$ with an even number of digits is always composite. □

The largest known palindromic prime in 2020 was the number

$$10^{474500} + 999 \cdot 10^{237249} + 1,$$

which has 474501 digits. Special cases of these numbers are primes composed only of ones. They are called *repunit* from the word "repeating unit". According to Theorem 4.34, they must have an odd number of digits (except for 11). For instance,

$$11, \ 1111111111111111111, \ 11111111111111111111111, \qquad (4.32)$$

are palindromic primes with 2, 19, and 23 digits. Other such known prime numbers are

$$R_n = \frac{10^n - 1}{9} \qquad (4.33)$$

for $n = 317$, 1031 (cf. Theorem 4.34). They have n digits. The repunit R_n is a probable prime for $n = 49081$, 86453, 109297, 270343, 5794777, 8177207. We also refer to [100] for the so-called antipalindromic primes to the base 3.

4.10 Cyclic and Permutation Primes

A prime number is called a *cyclic prime,* if any cyclic permutation of its digits yields again a prime. For example, 3779 is a cyclic prime, since

$$3779, \quad 7793, \quad 7937 \quad \text{and} \quad 9377.$$

are also primes. Further nontrivial cyclic primes are

$$11, 13, 17, 37, 79, 113, 197, 199, 337, 1193, 11939, 19937, 193939, 199933,$$

but also the numbers given in (4.32) and (4.33).

A prime is called a *permutation prime,* if an arbitrary permutation of its digits yields again a prime. For the time being we know only the following permutation primes

$$2, 3, 5, 7, 11, 13, 17, 37, 79, 113, 199, 337$$

and all primes composed of ones. If $d \in \{2, 3, \ldots, 9\}$, then each n-digit number of the form $dd \ldots d$ is obviously divisible by d, i.e., it is composite for $n > 1$.

4.11 Further Types of Primes

If we replace the modulus p by p^2 in the congruence of Wilson's Theorem 3.6, we get the following definition. A prime p is called a *Wilson prime*, if

$$(p - 1)! \equiv -1 \pmod{p^2}.$$

Up to now only the following three Wilson primes 5, 13, and 563 were discovered.

In the article [26] *Siamese primes* were introduced as a pair of primes of the form $n^2 - 2$ and $n^2 + 2$. The sequence of these pairs starts as follows

$$\langle 7, 11 \rangle, \quad \langle 79, 83 \rangle, \quad \langle 223, 227 \rangle, \quad \langle 439, 443 \rangle, \quad \langle 1087, 1091 \rangle, \quad \langle 13687, 13691 \rangle.$$

Reciprocal values of primes are rational numbers, i.e., they have a periodic decimal expansion. The length of this period can be determined a priori using Theorem 4.35 below. For instance, the fractions

$$\frac{1}{7} = 0.142857142857\ldots, \quad \frac{1}{13} = 0.076923076923\ldots$$

both have a period of length 6.

A prime $p \notin \{2, 5\}$ is called *unique*, if there does not exist any other prime q whose reciprocal value $1/q$ has the minimal period of the same length as $1/p$ (in decimal system). The unique primes are, e.g.,

$$3, \ 11, \ 37, \ 101, \ 9091, \ 9901, \ 333667, \ 909091, \ 99990001, \ldots,$$

whose minimal periods have lengths:

$$1, \ 2, \ 3, \ 4, \ 10, \ 12, \ 9, \ 14, \ 24, \ldots$$

Hence, 1/37 has a minimal of length 3 and no other prime has this property. All palindromic primes of the form (4.32) are also unique primes.

Theorem 4.35 *Let $n = 2^\alpha 5^\beta q$, where $\alpha \geq 0$, $\beta \geq 0$, $q > 1$, $(q, 10) = 1$, and $\mu = \max(\alpha, \beta)$. Let k be the smallest natural number such that $q \mid (10^k - 1)$. If $m < n$ are coprime natural numbers, then the decimal expansion of the fraction $\frac{m}{n}$ has μ nonrepeating digits and k repeating digits.*

The proof can be found in Hardy [138, p. 111].

There is a large number of other types of primes (see e.g. Wells [416]). One can look for prime numbers such that when "trimming" from the right or from the left, they yield again only prime numbers, e.g., 73939133, 7393913, 739391, 73939, 7393, 739, 73, 7 or 632647, 32647, 2547, 647, 47, 7. The largest known prime which produces only primes when trimming from the left is 357686312646216567629137.

Theorem 4.36 *Let k be a fixed positive integer. Then there are infinitely many primes whose last k digits are all 9s.*

Proof Note that $a = 10^k - 1$ and $d = 10^k$ are coprime numbers for any $k \in \mathbb{N}$ and that each term in the arithmetic progression $(10^k - 1 + 10^k i)_{i=0}^{\infty}$ ends with k 9s. By Dirichlet's Theorem 3.10, this sequence contains infinitely many primes. □

Primes of the form $c999\ldots999$ with $c \in \{1, 2, 4, 5, 7, 8\}$ are called the *Bat'a primes*. In Sects. 7.6, 8.4, 8.6, and 8.7 we will investigate Fibonacci, Lucas, Thabit, Cullen, and Woodall primes.

4.12 Gaussian Primes

Prime numbers can also be defined in other algebraic structures than just in the set of natural numbers. In the complex plane first define *Gaussian numbers* as sums $a + ib$, where $a, b \in \mathbb{Z}$ and i is the imaginary unit (see Fig. 4.8). The size of $g = a + ib$ is given by $|g| = \sqrt{a^2 + b^2}$.

We see that the sum and the product of two Gaussian numbers $a + ib$ and $c + id$ is again a Gaussian number

$$(a + ib) + (c + id) = (a + c) + i(b + d),$$
$$(a + ib)(c + id) = (ac - bd) + i(ad + bc).$$

There exist exactly four Gaussian numbers ± 1 and $\pm i$ whose size is 1 (see Fig. 4.8).

Denote by $\mathbb{Z}[i]$ the set of all Gaussian numbers. In 2008, Elias Lampakis [222] proved by means of the elliptic curve $y^2 + x^3 = 432$ that the equation (cf. (2.25))

$$x^3 + y^3 = z^3$$

has no solution in $\mathbb{Z}[i]$ for $xyz \neq 0$.

Now we will show how Gaussian primes are defined on the set $\mathbb{Z}[i]$. By Fermat's Christmas Theorem 3.12, any prime of the form $4k + 1$ can be uniquely written as the

Fig. 4.8 Gaussian numbers in the complex plane have integer coordinates

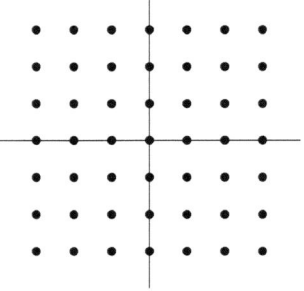

sum of two squares. In complex arithmetic, this prime can be factored as the product of two complex conjugate Gaussian numbers whose absolute values are $\sqrt{4k+1}$, i.e., they lie between 1 and $4k+1$. For instance,

$$13 = 3^2 + 2^2 = (3 - 2i)(3 + 2i), \quad 17 = 4^2 + 1^2 = (4 - i)(4 + i).$$

Also the prime 2 can be factored into two conjugate Gaussian numbers whose absolute values are $\sqrt{2}$,

$$2 = (1 - i)(1 + i).$$

These are examples of composite Gaussian numbers. However, for primes of the form $4k - 1$ a similar factorization is impossible (see Theorem 2.6). This enables us to introduce the following definition.

A Gaussian number $g = a + ib$ is called a *Gaussian prime*, if $|g| > 1$ and g cannot be written as a product of two Gaussian numbers whose absolute value is greater than 1 and smaller than $|g|$.

The following theorem is given in Wells [416, p. 112].

Theorem 4.37 *The number $g = a + ib$, $a, b \in \mathbb{Z}$, is a Gaussian prime if and only if*
(1) $a^2 + b^2$ is a prime for $a \neq 0 \neq b$, or
(2) $|g|$ is a prime of the form $4k - 1$ for $a = 0$ or $b = 0$.

In Fig. 4.9 we see the distribution of Gaussian primes whose absolute values are less than $\sqrt{1000}$. All Gaussian primes whose absolute value does not exceed 7 and $0 < \operatorname{Im} g \leq \operatorname{Re} g$, are of the form

$$1 + i, \quad 2 + i, \quad 3, \quad 3 + 2i, \quad 4 + i, \quad 5 + 2i, \quad 5 + 4i, \quad 6 + i, \quad 7.$$

Gaussian prime quadruplets are four Gaussian primes of the form

$$\langle a - 1 + ib, \ a + i(b - 1), \ a + i(b + 1), \ a + 1 + ib \rangle, \quad a, b \in \mathbb{Z},$$

for instance,

$$4 + 5i, \quad 5 + 4i, \quad 5 + 6i, \quad 6 + 5i,$$

see Fig. 4.9, where you can also find other Gaussian prime quadruplets.

Every Mersenne prime is at the same time also a Gaussian prime. At present, the largest known Gaussian prime is the largest known Mersenne prime. A number of other properties of Gaussian primes is given in Bressoud and Wagon [43].

Fig. 4.9 Schematic illustration of Gaussian primes in the complex plane. Sides of particular squares have lengths equal to 1. The center of this figure corresponds to the zero Gaussian number $0 + i0$

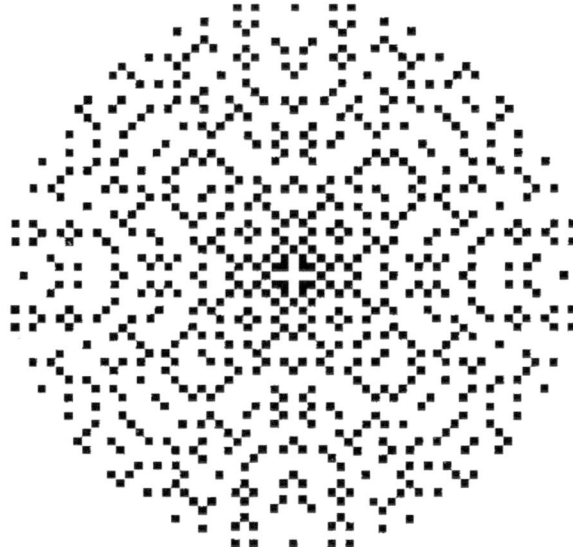

4.13   Eisenstein Primes

Ferdinand Gotthold Max Eisenstein (1823–1852) defined a similar type of prime numbers in the complex plane. Let us put

$$\omega = \frac{-1 + i\sqrt{3}}{2}.$$

We see that ω is the third root of one, i.e., it is a solution of the equation $z^3 = 1$ in the complex plane and moreover, we have

$$1 + \omega + \omega^2 = 0 \quad \text{and} \quad \omega^2 = \overline{\omega}, \tag{4.34}$$

where $\overline{\omega}$ denotes the complex conjugate number to ω.

The numbers of the form $a + \omega b$, where $a, b \in \mathbb{Z}$, are called *Eisenstein numbers* (see Fig. 4.10).

By (4.34) one can show that the sum and the product of two such numbers is again an Eisenstein number. The following six Eisenstein numbers $\pm 1, \pm \omega, \pm \overline{\omega}$ have the absolute value 1 (see Fig. 4.10).

According to (4.34), the number (cf. Fig. 4.10)

$$3 = 1 - \omega - \omega^2 + \omega^3 = (1 - \omega)(1 - \omega^2) = (1 - \omega)(1 - \overline{\omega})$$

Fig. 4.10 Eisenstein
numbers in the complex
plane

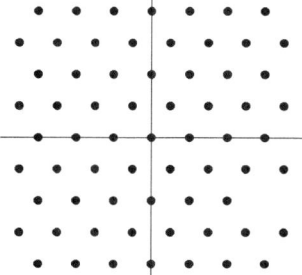

Fig. 4.11 Schematic
illustration of Eisenstein
primes in the complex plane

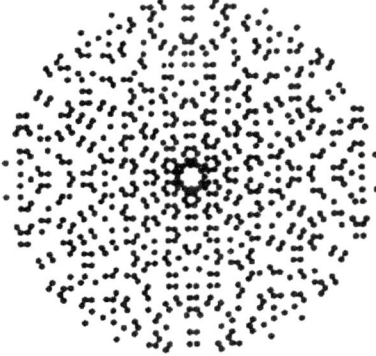

can be written as the product of two Eisenstein numbers whose absolute value is $\sqrt{3}$
which lies between 1 and 3.

From the Eratosthenes sieve (2.4) it follows that each prime greater than 3 is only of
the form $6k \pm 1$ for some suitable $k \in \mathbb{N}$. However, primes of the form $p = 6k + 1$
can be written as the product of two Eisenstein numbers whose absolute value is
greater than 1 and less than $6k + 1$. For instance,

$$7 = (2 - \omega)(2 - \overline{\omega}), \quad 13 = (3 - \omega)(3 - \overline{\omega}) \quad \text{and} \quad 19 = (3 - 2\omega)(3 - 2\overline{\omega}).$$

On the other hand a similar factorization cannot be done for the prime number 2 and
all primes of the form $6k - 1$.

An Eisenstein number e is called an *Eisenstein prime,* if $|e| > 1$ and e cannot be
expressed as the product of two Eisenstein numbers whose absolute value is greater
than 1 and less than $|e|$.

The distribution of the Eisenstein primes in the Gaussian complex plane (see
Fig. 4.11) has an interesting hexagonal symmetry. Notice the six small hexagons that
surround the central hexagon. These are Eisenstein prime sextuplets, for instance,

$$3 + 2\omega, \quad 3 + \omega, \quad 4 + \omega, \quad 5 + 2\omega, \quad 5 + 3\omega, \quad 4 + 3\omega.$$

At present no other such examples are known. As of 2020, the largest known (real) Eisenstein prime is $10223 \cdot 2^{31172165} + 1$, discovered by Péter Szabolcs. Real Eisenstein primes are congruent to 2 (mod 3), and all Mersenne primes greater than 3 are congruent to 1 (mod 3). Hence, no Mersenne prime is an Eisenstein prime.

Chapter 5
On a Connection of Number Theory with Graph Theory

5.1 Definitions and Notations

This chapter extends results given in the works [48] by Bryant, [66, 67] by Chassé, [389] by Szalay (motivated by [166, 331]), and [332] by Rogers, which provide an interesting connection between number theory, graph theory, geometry, and also group theory (cf. [115]). In the paper [389], Szalay investigated properties of the iteration digraph representing a dynamical system occurring in number theory. Each natural number has a specific iteration digraph corresponding to it (see Figs. 4.3, 4.4, 4.5, 4.6 and 4.7). We will classify sets of positive integers whose iteration digraphs have particular structural characteristics.

For $n \geq 1$ we again define

$$H = \{0, 1, \ldots, n-1\}$$

and let f be a map of H into itself as in Sect. 4.2. Thus, there exists exactly one directed edge from x to $f(x)$ for all $x \in H$ (cf. Fig. 5.1).

Note that f need not be a polynomial modulo n when n is not a prime. To see this, set $n = 4$, $f(0) = 0$, and $f(i) = 1$ for $i \neq 0$. Then $f(2) \not\equiv f(0) \pmod{2}$ which implies that f is not a polynomial modulo 4. For standard references concerning graphs and digraphs see [65, 136].

From here on, until Sect. 5.4, we shall deal only with a special function f, namely, for each $x \in H$ let $f(x)$ be the remainder of x^2 modulo n, i.e.,

$$f(x) \in H \quad \text{and} \quad f(x) \equiv x^2 \pmod{n}.$$

This corresponds to the iteration scheme $x_{j+1} \equiv x_j^2 \pmod{n}$. In Sect. 5.5, we introduce its generalization.

Cycles are assumed to be oriented counterclockwise, see Fig. 5.2 for $n = 11$ (and also Figs. 4.3, 4.4, 4.5, 4.6 and 4.7). The cycles of length t are said to be t-cycles. As in Sect. 4.2, cycles of length 1 are called *fixed points*.

© The Author(s), under exclusive license to Springer Nature Switzerland AG 2021
M. Křížek et al., *From Great Discoveries in Number Theory to Applications*,
https://doi.org/10.1007/978-3-030-83899-7_5

Fig. 5.1 The iteration digraph corresponding to $n = 8$

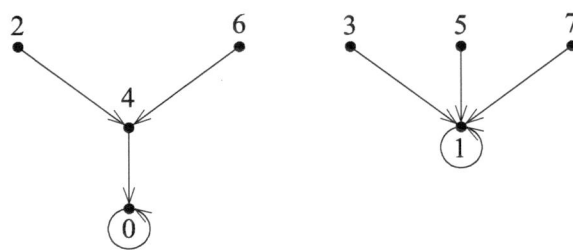

Fig. 5.2 The iteration digraph corresponding to $n = 11$

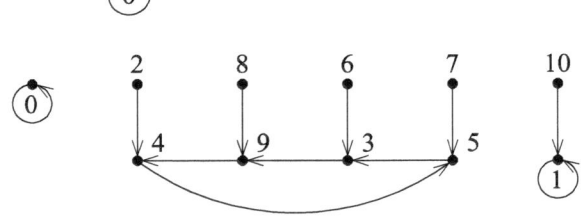

We identify the vertex a of H with residues modulo n. For notational convenience we will make statements such as $(a, n) = 1$, treating the vertex a as a number. Moreover, when we refer, for instance, to the vertex a^2, we identify it with the remainder d such that $d \equiv a^2 \pmod{n}$ and $0 \le d < n$.

For a particular value of n, we denote the iteration digraph of f by $G(n)$. It is obvious that $G(n)$ with n vertices also has exactly n directed edges.

Let $\omega(n)$ denote the number of distinct primes dividing n and let the prime power factorization of n be given by

$$n = \prod_{i=1}^{s} p_i^{k_i},$$

where $p_1 < p_2 < \cdots < p_s$ are primes and $k_i > 0$, i.e. (see Table 13.3)

$$s = \omega(n) = |\{p \in \mathbb{P};\ p \mid n\}|.$$

The next two theorems are proved in Szalay [389].

Theorem 5.1 (Szalay) *The number of fixed points of $G(n)$ is equal to $2^{\omega(n)}$.*

Proof Let n have the prime power factorization given above. Suppose that a is a fixed point of $G(n)$. Then $a^2 \equiv a \pmod{n}$ which implies that

$$a^2 - a = a(a - 1) \equiv 0 \pmod{p_i^{k_i}}$$

for $i \in \{1, 2, \ldots, s\}$. Since $(a, a - 1) = 1$, we have that $a \equiv 0$ or $a \equiv 1 \pmod{p_i^{k_i}}$ for each $i \in \{1, 2 \ldots, \omega(n)\}$. It now follows from the Chinese Remainder Theorem 1.4 that the number of fixed points of $G(n)$ equals $2^{\omega(n)}$. $\qquad\square$

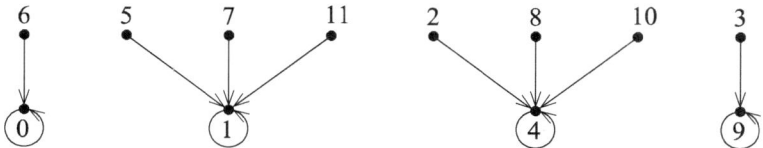

Fig. 5.3 The iteration digraph corresponding to $n = 12$

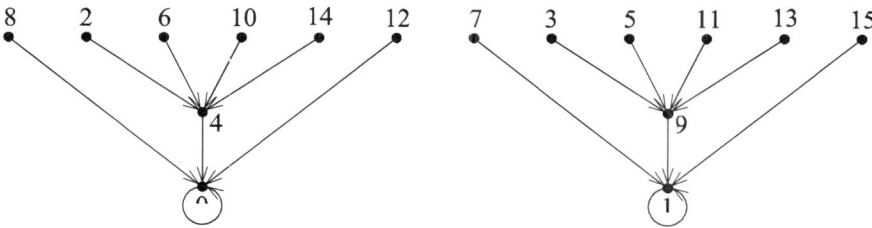

Fig. 5.4 The iteration digraph corresponding to $n = 16$

A *component* of the iteration digraph is a subdigraph which is a maximal connected subgraph of the symmetrization of this digraph (i.e., the associated nondirected graph). When n is a prime, Rogers [332] describes completely the structure of each component of $G(n)$.

The digraph $G(n)$ is called *symmetric* if its set of components can be split into two sets in such a way that there exists a bijection between these two sets such that the corresponding digraphs are isomorphic (compare with Fig. 5.3).

Theorem 5.2 (Szalay) *The iteration digraph $G(n)$ is symmetric if $n \equiv 2$ (mod 4) or $n \equiv 4$ (mod 8).*

Theorem 5.2 gives only a sufficient condition for the symmetry of $G(n)$ which is not necessary (compare with $G(16)$ in Fig. 5.4).

5.2   Structure of Iteration Digraphs

The *indegree* of a vertex $a \in H$ of $G(n)$, denoted by $\mathrm{indeg}_n(a)$, is the number of directed edges coming into a, and the *outdegree* of a is the number of directed edges leaving the vertex a. For simplicity, the subscript n will be omitted from now on. By the definition of f, the outdegree of each vertex of $G(n)$ is equal to 1. For an isolated fixed point, the indegree and outdegree are both equal to 1.

Theorem 5.3 *The number 0 is an isolated fixed point of $G(n)$ if and only if n is square free.*

Proof If $p^2 \mid n$ for some prime p, then for the Legendre symbol we get

$$\left(\frac{n}{p}\right)^2 = n \cdot \frac{n}{p^2} \equiv 0 \pmod{n},$$

and 0 is not an isolated fixed point, since n/p is mapped into 0.

Conversely, if n is square free, then it is evident that $x \equiv 0 \pmod{n}$ is the only solution to the congruence $x^2 \equiv 0 \pmod{n}$, and hence 0 is an isolated fixed point of $G(n)$.   □

The isolated and nonisolated fixed points 0 of $G(11)$ and $G(12)$ are illustrated in Figs. 5.2 and 5.3, respectively.

Let $N_n(a)$ denote the number of incongruent solutions of the congruence

$$x^2 \equiv a \pmod{n}.$$

Then obviously

$$N_n(a) = \text{indeg}(a). \tag{5.1}$$

Theorem 5.4 *There are no isolated cycles of $G(n)$ of length greater than 1. The digraph $G(n)$ has an isolated fixed point $a \neq 0$ if and only if $2 \mid n$ and n is square free. In this case, $a = n/2$.*

Proof Assume that $a \neq 0$ is part of an isolated cycle of $G(n)$. We first show that n is an even square-free integer and next demonstrate that $a = n/2$ and that a is a fixed point.

Let $b^2 \equiv a \pmod{n}$. Since $(-b)^2 \equiv b^2 \pmod{n}$ and $\text{indeg}(a) = N_n(a) = 1$, we get that $-b \equiv b \pmod{n}$. This implies that $2b \equiv 0 \pmod{n}$. Since $a \not\equiv 0 \pmod{n}$, we see that $2 \mid n$ and $b \equiv n/2 \pmod{n}$.

Now suppose that $p^2 \mid n$ for some prime p. If $p = 2$, then $a \equiv (n/2)^2 \equiv 0 \pmod{n}$, which is a contradiction. Next assume that p is odd and $2 \parallel n$. Note that if m is an odd integer, then

$$\frac{n}{2}m \equiv \frac{n}{2} \pmod{n}. \tag{5.2}$$

Since $n/2$ is odd, it now follows that

$$a \equiv \frac{n}{2}\frac{n}{2} \equiv \frac{n}{2} \equiv \frac{n}{2}\frac{n}{2p^2} \equiv \frac{n^2}{(2p)^2} \pmod{n},$$

which contradicts the assumption that $N_n(a) = 1$. Hence, n is square free. We now observe by (5.2) that

$$a \equiv b^2 \equiv \frac{n}{2}\frac{n}{2} \equiv \frac{n}{2} \pmod{n}. \tag{5.3}$$

Consequently, $a \equiv n/2 \pmod{n}$ and a is a fixed point of $G(n)$.

We assume conversely that $2 \mid n$ and that n is square free. Then $n/2$ is odd and $n/2 \not\equiv 0 \pmod{n}$. By (5.2) and (5.3), we find that

$$\frac{n}{2}\frac{n}{2} \equiv \frac{n}{2} \pmod{n}, \tag{5.4}$$

and $n/2$ is a fixed point of $G(n)$. Suppose that $b^2 \equiv n/2 \pmod{n}$. Since $n/2$ is odd and n is even, we obtain $b \equiv 1 \pmod{2}$. Since $n/2$ is square free and $(n/2) \mid n$, it is easily seen that $b \equiv 0 \pmod{n/2}$. Noting that $(2, n/2) = 1$, we get from the Chinese Remainder Theorem 1.4 that b is uniquely determined modulo n, and hence by (5.4), $b \equiv n/2 \pmod{n}$. Therefore, $n/2$ is an isolated fixed point of $G(n)$, and the result follows.   □

The next statement follows immediately from Theorems 5.3 and 5.4.

Theorem 5.5 *Each digraph $G(n)$ has at most two isolated fixed points, and $G(n)$ has exactly two isolated fixed points if and only if $2 \mid n$ and n is square free. Then the only isolated fixed points are 0 and $n/2$.*

Recall that a graph is regular if all its vertices have the same degree. The digraph $G(n)$ is said to be *semiregular* if there exists a positive integer d such that each vertex of $G(n)$ either has indegree d or 0 (see Fig. 5.4 for $d = 4$ and $n = 16$).

For a natural integer n we set (see Table 13.3)

$$\varepsilon(n) = \begin{cases} -1 & \text{if } 2 \parallel n, \\ 1 & \text{if } 8 \mid n, \\ 0 & \text{otherwise.} \end{cases} \tag{5.5}$$

Further, we specify two particular subdigraphs of $G(n)$. Let $G_1(n)$ be the induced subdigraph of $G(n)$ on the set of vertices which are coprime to n and $G_2(n)$ be the induced subdigraph on the remaining vertices, not coprime with n. We observe that $G_1(n)$ and $G_2(n)$ are disjoint and that

$$G(n) = G_1(n) \cup G_2(n),$$

that is, no edge goes between $G_1(n)$ and $G_2(n)$. For example, the second component of Fig. 5.3 is $G_1(12)$ whereas the remaining three components make up $G_2(12)$. It is clear that 0 is always a vertex of $G_2(n)$. If $n > 1$ then 1 and $n - 1$ are always vertices of $G_1(n)$.

Theorem 5.6 *The digraph $G_1(n)$ is semiregular for all positive integers n. Moreover, if a is a vertex of $G_1(n)$ then*

$$\text{indeg}(a) = 0 \quad or \quad \text{indeg}(a) = 2^{\omega(n)+\varepsilon(n)}.$$

If $n \geq 2$, then $G_2(n)$ is semiregular if and only if $n = p^k$, where p is an odd prime for $k \in \{1, 2\}$, or $n = 2^k$ for $k \in \{1, 2, 3, 4, 6\}$.

The digraph $G(n)$ is semiregular if and only if $n = 2^k$ for $k = 0, 1, 2, 4$.

The proof of Theorem 5.6 is given in [370]. It is based on the following statement which asserts that there are no vertices with indegree 1 except for isolated fixed points.

Theorem 5.7 *If $(a, n) = 1$ and $N_n(a) > 0$, then $N_n(a) = 2^{\omega(n)+\varepsilon(n)}$.*

Proof The result clearly holds when $n = 1$, so assume that $n > 1$. Since the residues coprime to n form a group under multiplication modulo n, it is easily seen that $N_n(a) = N_n(1)$ if $N_n(a) > 0$ and $(a, n) = 1$. Thus it suffices to determine only $N_n(1)$.

We first find $N_{p^k}(1)$, where p is a prime and $k \geq 1$. Notice that

$$a^2 \equiv 1 \pmod{p^k} \tag{5.6}$$

if and only if

$$a^2 - 1 = (a + 1)(a - 1) \equiv 0 \pmod{p^k}. \tag{5.7}$$

Suppose that p is an odd prime. Since $(a + 1, a - 1) \mid 2$, it follows that (5.6) holds if and only if $a \equiv \pm 1 \pmod{p^k}$. Thus $N_{p^k}(1) = 2$.

Now assume that $p = 2$. Note that if (5.7) is satisfied, then 4 divides precisely one of the terms $a + 1$ and $a - 1$, and 2 exactly divides the other term. Hence, (5.6) holds if and only if $a \equiv 1 \pmod 2$ when $1 \leq k \leq 3$ and $a \equiv \pm 1 \pmod{2^{k-1}}$ when $k \geq 4$. Thus, $N_{2^k}(1) = 2^{1+\varepsilon(2^k)}$. The result now follows from (5.5) and the Chinese Remainder Theorem 1.4. □

Theorem 5.8 *Let d be any positive integer. Then there exist positive integers n and a such that a is a vertex of $G_2(n)$ and $\mathrm{indeg}(a) = d$.*

Proof If $d = 1$, we let n be a prime and $a = 0$. Now suppose that $d > 1$. Let k_1 be such that $d = 2^{k_1} d_1$, where d_1 is odd. We choose a positive integer d_2 such that $\omega(d_2) = k_1$ and $(2d_1, d_2) = 1$ (if $k_1 = 0$ then we set $d_2 = 1$). We now let

$$n = d_1^2 d_2 \quad \text{and} \quad a = d_1^2.$$

It now follows from Theorem 5.7 and the Chinese Remainder Theorem 1.4 that

$$\mathrm{indeg}(a) = N_n(a) = N_{d_1^2}(d_1^2) N_{d_2}(d_1^2) = d_1 2^{k_1} = d,$$

which completes the proof. □

A vertex a of $G(n)$ is said to be at *level i*, $i \geq 1$, if there exists a directed path of maximum length i which terminates at a and contains no directed edge belonging to a cycle. If such a path does not exist, the vertex is said to be at *level 0*. We say that a component of $G(n)$ has ℓ *levels* if the highest level of a vertex in the component is

$\ell - 1$. For instance, if $n = 16$ (see Fig. 5.4) then 2 and 8 are at level 0, the vertex 4 is at level 1, the vertex 0 is at level 2, and both the components have 3 levels.

Let C be a component of $G(n)$ which contains exactly ℓ levels. It is clear that each vertex at level ℓ is part of a cycle, since each vertex has outdegree 1. Moreover, if a and b are two distinct vertices at level i of C, where $0 \le i < \ell$, then there does not exist a directed path from a to b or from b to a. We thus have the following statement.

Theorem 5.9 *Each component has exactly one cycle, i.e., the number of components of $G(n)$ is equal to the number of its cycles.*

Note that Theorem 5.9 is a general property of the iteration digraph of any mapping $f : H \to H$.

5.3 Application of the Carmichael Lambda Function

Assume now that $\lambda(n)$ introduced in Sect. 2.9 has the following prime power factorization:

$$\lambda(n) = \prod_{j=1}^{r} q_j^{\ell_j}, \tag{5.8}$$

where $q_1 < q_2 < \cdots < q_r$ are primes and $\ell_i > 0$. It is evident from the definition of λ that $q_1 = 2$ if $n > 2$.

Let $A_t(G(n))$ denote the number of cycles of $G(n)$ of length t. In [389], Szalay gave a recursive formula for $A_t(G(n))$, which is in closed form when $t = 1$ or when $2^t - 1$ is a Mersenne prime, without using subdigraphs $G_1(n)$ or $G_2(n)$. In [332], Rogers completely determines $A_t(G(n))$ when $t \ge 1$ and n is a prime. He proves that the subgraphs attached to each of the vertices of all the cycles, except for the fixed point 0, are binary trees of the same structure. See also Chassé [66] for related results when n is prime.

When n is any positive integer, the number of square roots of any quadratic residue in $G_1(n)$ is equal to the number of square roots of 1 modulo n. It thus follows that for all $n \ge 2$, the subgraphs attached to each vertex of any cycle of $G_1(n)$ are all the same (see Figs. 5.2 and 5.5). The proof of this property is an immediate consequence of Theorems 5.6 and 5.15, which will be proved in the next section.

We have the following theorem concerning $A_t(G_1(n))$ and $A_t(G_2(n))$.

Theorem 5.10 *If $n \in \mathbb{N}$, then $A_1(G_1(n)) = 1$ and $A_1(G_2(n)) = 2^{\omega(n)} - 1$. Moreover, if $t > 1$ and $A_t(G_2(n)) > 0$, then $A_t(G_1(n)) > 0$.*

Proof First assume that $t = 1$. By Szalay's Theorem 5.1,

$$A_1(G(n)) = 2^{\omega(n)}. \tag{5.9}$$

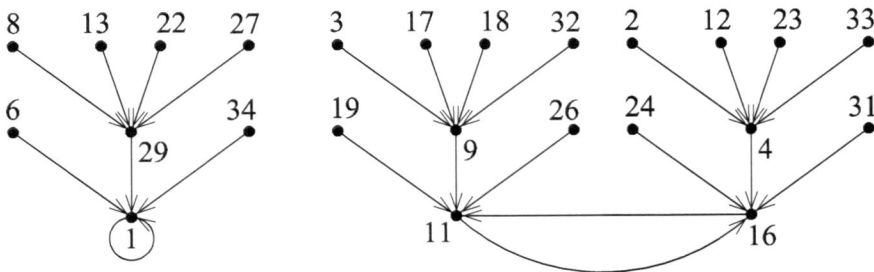

Fig. 5.5 The subdigraph $G_1(35)$

Suppose that a is a fixed point in $G_1(n)$. Then

$$a^2 - a = a(a-1) \equiv 0 \quad (\text{mod } n). \tag{5.10}$$

Since $(a, n) = 1$, we see from (5.10) that $a \equiv 1$ (mod n). Hence, $A_1(G(n)) = 1$ and $A_1(G_2(n)) = 2^{\omega(n)} - 1$.

Now suppose that $t > 1$ and that the vertex $a \in G_2(n)$ is part of a t-cycle. By (4.21), there exist integers n_1, n_2 such that $n_1 > 1, n_2 > 1, (n_1, n_2) = 1$, and $n_1 n_2 = n$ for which t is the least positive integer such that

$$a \equiv 0 \quad (\text{mod } n_1),$$
$$a^{2^t-1} \equiv 1 \quad (\text{mod } n_2). \tag{5.11}$$

By the Chinese Remainder Theorem 1.4 and the proof of Theorem 4.26 there exists a unique vertex $b \in G_1(n)$ such that b is part of a t-cycle and

$$b \equiv 1 \quad (\text{mod } n_1),$$
$$b \equiv a \quad (\text{mod } n_2). \tag{5.12}$$

Since for each vertex $a \in G_2(n)$ satisfying (5.11), there exists a distinct vertex $b \in G_1(n)$ satisfying (5.12), we see that if $t \geq 2$ and $A_t(G_2(n)) > 0$, then $A_t(G_1(n)) > 0$ for all $n \in \mathbb{N}$. $\qquad \square$

Example We note that if $t > 1$, then it can occur that either $A_t(G_1(n)) > A_t(G_2(n)) > 0$ or $A_t(G_2(n)) > A_t(G_1(n)) > 0$. For instance, if $n = 29 \cdot 43 = 1247$, then $A_3(G_1(n)) = 16$, whereas $A_3(G_2(n)) = 4$. On the other hand, if $n = 7 \cdot 13 \cdot 19 \cdot 31 = 53\,599$, then $A_2(G_1(n)) = 40$, while $A_2(G_2(n)) = 80$.

Remark It can be further determined when $A_t(G_1(n)) > A_t(G_2(n)) = 0$. By the proofs of Theorems 4.26 and 5.10, this can occur only if there exists an odd divisor $d > 1$ of $\lambda(n)$ such that $t = \text{ord}_d 2$. Let $A_t(G(n)) > 0$ for some $t > 1$. Then, by

(4.21), and the proofs of Theorems 2.24 and 4.24, $A_t(G_2(n)) = 0$ if and only if there do not exist an integer $n_1 > 1$ and an odd integer $d_1 > 1$ such that $n_1 \mid n$, $n/n_1 > 1$, $(n_1, n/n_1) = 1$, $t = \mathrm{ord}_{d_1} 2$, and $d_1 \mid \lambda(n_1)$.

For example, let $n = 13 \cdot 29 = 377$ and let us examine the 6-cycles of $G(n)$. Notice that $\lambda(n) = 84$ and that the only odd divisor d of $\lambda(n)$ for which $\mathrm{ord}_d 2 = 6$ is $d = 21$. Observe that $21 \nmid \lambda(13) = 12$ and $21 \nmid \lambda(29) = 28$. Thus $A_6(G_2(377)) = 0$. We find by inspection that $A_6(G_1(377)) = 2$.

Remark Let $n \geq 3$. Let $R(n)$ denote the number of quadratic residues in $G_1(n)$ and $Q(n)$ denote the number of quadratic nonresidues in $G_1(n)$. Noting that $G_1(n)$ has $\phi(n)$ vertices and the total number of directed edges in $G_1(n)$ is also equal to $\phi(n)$, it follows from Theorem 5.6 that

$$R(n) = \frac{\phi(n)}{2^{\omega(n)+\varepsilon(n)}}.$$

Since $\omega(n) + \varepsilon(n) \geq 1$ by the definition of $\omega(n)$ and $\varepsilon(n)$, we see that

$$Q(n) = \phi(n) - \frac{\phi(n)}{2^{\omega(n)+\varepsilon(n)}} \geq \frac{\phi(n)}{2^{\omega(n)+\varepsilon(n)}} = R(n),$$

i.e., the number of quadratic nonresidues of $G_1(n)$ for $n \geq 3$ is greater than or equal to the number of quadratic residues. It thus follows that the number of vertices in $G_1(n)$ which have indegree 0 is greater than or equal to the number of vertices in $G_1(n)$ having a positive indegree. In particular, the number of vertices in $G_1(n)$ not lying on a cycle is greater than or equal to the number of vertices in $G_1(n)$ which lie on a cycle (cf. Figs. 5.1, 5.2, 5.3, 5.4 and 5.5).

Example Let $n = 120 = 2^3 \cdot 3 \cdot 5$. Then (cf. Table 13.3)

$$\phi(120) = \phi(8)\phi(3)\phi(5) = 4 \cdot 2 \cdot 4 = 32$$

and

$$2^{\omega(120)+\varepsilon(120)} = 2^{3+1} = 16.$$

We now observe that

$$R(120) = \frac{\phi(120)}{2^{\omega(120)+\varepsilon(120)}} = 2$$

and

$$Q(120) = \phi(120) - R(120) = 32 - 2 = 30 \geq R(120).$$

By inspection, one sees that the only quadratic residues in $G_1(120)$ are 1 and 49, each with indegree 16.

5.4 Application of the Euler Totient Function

Set

$$S = \{n \geq 1 \mid \phi(n) \text{ is a power of } 2\},$$

where $\phi(n)$ is the Euler totient function.

By (2.37)–(2.39), a positive integer n belongs to S if and only if $n = 2^\alpha F_{m_1} \cdots F_{m_j}$ for some $\alpha \geq 0$ and $j \geq 0$, where $F_{m_i} = 2^{2^{m_i}} + 1$ are distinct Fermat primes . Comparison with the celebrated Theorem 4.12 due to Gauss shows that $n \in S$ for $n \geq 3$ if and only if the regular polygon with n sides has a Euclidean construction with ruler and compass.

Theorem 5.11 *There are no cycles in $G_1(n)$ of length greater than 1 if and only if $n \in S$. Moreover, there are no cycles in $G_2(n)$ of length greater than 1 if and only if $n \in S$ or n is a prime power.*

Proof According to the definition of the Carmichael lambda-function, $\lambda(n) = 2^i$ for some integer $i \geq 0$ if and only if $\phi(n) = 2^j$ for some integer $j \geq i$. Let d be a positive odd integer. Then we also observe that $\text{ord}_d 2 = 1$ if and only if $d = 1$. By Theorem 5.10, $A_t(G(n)) = 0$ for all $t \geq 2$ if and only if $A_t(G_1(n)) = 0$ for all $t \geq 2$.

First suppose that $n \in S$. Then 1 is the only positive odd divisor of $\lambda(n)$ and it follows from Theorem 4.26 that

$$A_t(G(n)) = A_t(G_1(n)) = A_t(G_2(n)) = 0$$

for all $t \geq 2$.

If $n \notin S$, then $\lambda(n)$ has an odd divisor $d > 1$. We see by Theorems 4.26 and 5.10 that $A_t(G_1(n)) \geq 1$ for $t = \text{ord}_d 2 > 1$.

Observe that if n is a prime power, then the only cycle in $G_2(n)$ is the fixed point 0. Assume now that n is not a prime power and $n \notin S$. We show that $A_t(G_2(n)) \geq 1$ for some $t \geq 2$. Notice that there exists a prime power $p^k \notin S$ such that $p^k \parallel n$ for some $k \geq 1$. Furthermore, $\lambda(p^k)$ has an odd divisor $d > 1$. Moreover, $d \mid \lambda(n)$ since $\lambda(p^k) \mid \lambda(n)$. Let $t = \text{ord}_d 2$. Then $t > 1$. By (4.24), there exists a residue a such that t is the least positive integer for which

$$a^{2^t - 1} \equiv 1 \pmod{p^k}. \tag{5.13}$$

Let $n = p^k n_1$, where $n_1 > 1$. Using the Chinese Remainder Theorem 1.4, we can now find a vertex $b \in G_2(n)$ such that

$$b \equiv a \pmod{p^k},$$
$$b \equiv 0 \pmod{n_1}. \tag{5.14}$$

Note that $(b, p^k) = (a, p^k) = 1$ by (5.13) and (5.14). It follows again from (5.13) and (5.14) that t is the least positive integer such that

$$b^{2^t} - b = b\left(b^{2^t-1} - 1\right) \equiv 0 \pmod{n}.$$

Hence, t is the least positive integer such that

$$b^{2^t} \equiv b \pmod{n},$$

and b is part of a t-cycle in $G_2(n)$. $\square$

As immediate consequences we get the following two theorems.

Theorem 5.12 *There are no cycles in $G(n)$ of length greater than 1 if and only if $n \in S$.*

Theorem 5.13 *The Fermat number F_m is composite if and only if there exists a cycle in $G(F_m)$ of length greater than 1.*

The next theorem generalizes a result of Rogers [332] from primes to natural numbers.

Theorem 5.14 *Each component of $G_1(n)$ has exactly $v_2(\lambda(n)) + 1$ levels, where $v_2(c)$ denotes the exponent in the highest power of 2 dividing c.*

Proof The result clearly holds for $n \in \{1, 2\}$. So assume that $n > 2$. Let a be a vertex in a component C of $G_1(n)$ for which $d = \mathrm{ord}_n a$. Then $d \mid \lambda(n)$ by Carmichael's Theorem 2.24. By the proof of Theorem 4.26, a is part of a t-cycle in C, and thus at the highest level of C, if and only if d is odd.

Suppose that $b^2 \equiv a \pmod{n}$. Since

$$d = \mathrm{ord}_n a = \frac{\mathrm{ord}_n b}{(2, \mathrm{ord}_n b)}, \tag{5.15}$$

it follows that

$$\mathrm{ord}_n a \mid \mathrm{ord}_n b \mid 2\,\mathrm{ord}_n a,$$

and $\mathrm{ord}_n b = 2\,\mathrm{ord}_n a$ if $2 \mid \mathrm{ord}_n a$. We also see from (5.15) that all vertices in the same cycle of $G_1(n)$ have the same order modulo n, i.e., there exists an odd integer $d > 1$ such that

$$\mathrm{ord}_n a = d \tag{5.16}$$

for all vertices a in the t-cycle of $G_1(n)$.

Let $\ell = v_2(\lambda(n))$ and assume that $2^\ell \parallel \mathrm{ord}_n a$. Noting that $2 \mid \lambda(n)$ for $n > 2$, we see that if there exists a vertex $b \in G_1(n)$ such that $b^2 \equiv a \pmod{n}$, then $2^{\ell+1} \parallel \mathrm{ord}_n b$, which contradicts the fact that $\mathrm{ord}_n b \mid \lambda(n)$. Hence, a is at level 0 in this instance. Consequently, any component C of $G_1(n)$ has at most $\ell + 1$ levels.

Let a be a vertex of $G_1(n)$ for which $d = \mathrm{ord}_n a$ is odd. We will find a vertex b of $G_1(n)$ such that

$$b^{2^\ell} \equiv a \pmod{n}, \quad \mathrm{ord}_n b^{2^{\ell-1}} = 2\,\mathrm{ord}_n a. \tag{5.17}$$

Then by the above discussion, $\mathrm{ord}_n b$ will be equal to $2^\ell d$ and b^{2^i} will be at level i for $i \in \{0, 1, \ldots, \ell\}$. Hence, C will have exactly $\ell + 1$ levels.

Let $n = p_1^{k_1} \cdots p_s^{k_s}$, where $p_1 < p_2 < \cdots < p_s$ are primes and $k_i > 0$. By the definition of $\lambda(n)$, there exists a prime power factor $p_j^{k_j}$ of n such that $v_2(\lambda(p_j^{k_j})) = \ell$ for some $j \in \{1, 2, \ldots, s\}$.

Let d_i be the order of a modulo $p_i^{k_i}$ for $i = 1, 2, \ldots, s$. Then $d_i \mid d$, d_i is odd, and a is part of a cycle in $G_1(p_i^{k_i})$ of length $t_i = \mathrm{ord}_{d_i}(p_i^{k_i})$. First suppose that $v_2(\lambda(p_i^{k_i})) < \ell$. Choose the vertex $b_i \in G_1(p_i^{k_i})$ in the same t_i-cycle as the vertex a modulo $p_i^{k_i}$ which is ℓ vertices from a in the clockwise direction (possibly going around the cycle more than once). Then $b_i^{2^\ell} \equiv a \pmod{p_i^{k_i}}$.

Now suppose that $2^\ell \parallel \lambda(p_i^{k_i})$ and either $p_i^{k_i} = 2$ or 4 or p_i is an odd prime. Then the vertices in $G_1(p_i^{k_i})$ form a cyclic group modulo $p_i^{k_i}$ with $\lambda(p_i^{k_i})$ elements. Consequently, there exists a primitive root $g_i \pmod{p_i^{k_i}}$ for which

$$g_i^{\lambda(p_i^{k_i})/d_i} \equiv a \pmod{p_i^{k_i}}.$$

Let $c_i = \lambda(p_i^{k_i})/(2^\ell d_i)$ and let $b_i \equiv g_i^{c_i} \pmod{p_i^{k_i}}$. Then $b_i^{2^\ell} \equiv a \pmod{p_i^{k_i}}$ and the order of $b_i^{2^{\ell-1}}$ modulo $p_i^{k_i}$ is equal to $2d_i$.

Finally, assume that $p_1 = 2$, $\ell \geq 3$, and $k_1 = \ell$, where $2^{k_1} \parallel \lambda(n)$. Then $a \equiv 1 \pmod{2^{k_1}}$, since the order of a modulo 2^{k_1} is odd. Moreover,

$$5^{2^\ell} \equiv 1 \equiv a \pmod{2^{k_1}}$$

and the order of $5^{2^\ell - 1}$ modulo 2^{k_1} is equal to 2. Let $b_1 \equiv 5 \pmod{2^{k_1}}$.

Applying the Chinese Remainder Theorem 1.4, we obtain a vertex $b \in G_1(n)$ such that $b \equiv b_i \pmod{p_i^{k_i}}$ for $i = 1, \ldots, s$. We note that if c_i is the order of b_i modulo $p_i^{k_i}$, then

$$\mathrm{ord}_n b = [c_1, c_2, \ldots, c_s].$$

It now follows that b indeed satisfies the required properties given in (5.17).   □

Remark We see from Theorems 5.6, 5.8, 5.11, and 5.14 that $G_1(n)$ exhibits a more regular behavior than $G_2(n)$.

The next theorem enables us to separate the Fermat primes from the other odd primes. It generalizes an interesting result presented independently by Szalay and Rogers on Fermat primes (see [332, 389], and also [199]).

Theorem 5.15 *The digraph $G(n)$ has exactly 2 components if and only if n is a Fermat prime or n is a power of 2.*

Proof This follows directly from Theorems 5.1, 5.9, and 5.12.   □

In Fig. 5.1, we see the structure of $G(2^3)$. Figure 4.9 shows the iteration digraph for the Fermat prime $F_2 = 17$. In [207], we show that only for a Fermat prime or twice a Fermat prime or 4, is the set of primitive roots equal to the set of quadratic nonresidues.

As a consequence of Theorems 5.1 and 5.15 we get:

Theorem 5.16 *The Fermat number F_m is composite if and only if there exists $x \in \{2, 3, \ldots, F_m - 1\}$ such that $x^2 \equiv x \pmod{F_m}$.*

Remark According to Theorem 5.3, the Fermat number F_m is square free if and only if $x^2 \not\equiv 0 \pmod{F_m}$ for all $x \in \{1, \ldots, F_m - 1\}$.

For Theorem 5.17 given below we assume that $\lambda(n)$ has the prime power factorization as given in (5.8). This theorem generalizes a result by Rogers [332] from a prime n to a natural number n.

Theorem 5.17 *The digraph $G(n)$ has exactly 3 components if and only if $n = 9$ or $n = 25$ or n is a prime and $q = (n - 1)/2^{\ell_1}$ is also a prime such that 2 is a primitive root modulo q.*

Proof Suppose $G(n)$ has exactly 3 components. By Theorems 5.9, 5.10, 5.11, and 5.15, this occurs if and only if $n = p^k$, where p is an odd prime and $k \geq 1$, and $G_1(n)$ has a unique cycle of length greater than 1. Let t be the length of this unique cycle in $G_1(n)$. By (5.16), there exists a fixed odd integer $d > 1$ for which $\mathrm{ord}_n a = d$ if a is a vertex in the t-cycle of $G_1(n)$. Moreover, by Theorem 4.26, $t = \mathrm{ord}_d 2$ and $d \mid \lambda(n)$. Hence, $\lambda(n)$ has a unique odd divisor $d > 1$.

Since the vertices in $G_1(n)$ form a cyclic group under multiplication modulo n, the number of vertices a in $G_1(n)$ for which $\mathrm{ord}_n a = d$ is equal to $\phi(d)$. Thus, there exists a unique t-cycle in $G_1(n)$ if and only if

$$\frac{\phi(d)}{t} = \frac{\phi(d)}{\mathrm{ord}_d 2} = 1.$$

Therefore, we require that 2 be a primitive root modulo d.

Now suppose that $k \geq 2$. Then $\lambda(n) = p^{k-1}(p - 1)$. Hence, $\lambda(n)$ has a unique odd divisor $d > 1$ if and only if $k = 2$ and $p - 1$ is a power of 2. Then $d = p$ and d is a Fermat prime. However, 2 is a primitive root modulo the Fermat prime $F_m = 2^{2^m} + 1$ if and only if $m \in \{0, 1\}$. Thus $n = 3^2 = 9$ or $n = 5^2 = 25$.

Further, assume that $k = 1$. Then $\lambda(n) = p - 1$, and $p - 1$ has a unique odd prime divisor $q > 1$ if and only if

$$n - 1 = \lambda(n) = 2^{\ell_1} q,$$

where q is an odd prime. The t-cycle in $G_1(n)$ is then unique if and only if 2 is a primitive root modulo q. The result now follows. $\qquad\square$

Remark For example, $G(n)$ has exactly three components if (compare with Fig. 5.2)

$$n = 7, \ 9, \ 11, \ 13, \ 23, \ 25, \ 41, \ 53.$$

The following theorem given without proof can be proved in a manner similar to Theorems 5.15 and 5.17.

Theorem 5.18 *The digraph $G(n)$ has exactly 4 components if and only if one of the following four conditions hold:*

(i) $n = 27$ or $n = 125$ or $n = 289$.
(ii) n is a prime and $p = (n - 1)/2^{\ell_1}$ is also a prime for which $\text{ord}_p 2 = (p - 1)/2$.
(iii) n is a prime such that $(n - 1)/2^{\ell_1}$ is the square of a prime q for which 2 is a primitive root modulo q^2.
(iv) n is a product of two coprime integers $q_1 > 1$ and $q_2 > 1$, each of which is equal to a Fermat prime or to a power of 2.

Remark For example, the digraph $G(n)$ has exactly four components if (compare with Fig. 5.3)

$$n = 6, \ 10, \ 12, \ 15, \ 19, \ 20, \ 24, \ 27, \ 29, \ 34, \ 37, \ 40, \ 47, \ 48, \ 51.$$

Similarly, the digraph has exactly five components for

$$n = 81, \ 109, \ 173, \ 251, \ 433, \ 625, \ 1109, \ 1229, \ 1997, \ldots$$

The vertices of $G_1(n)$ correspond to those residues which are relatively prime to n. The structure of $G_1(n)$ is well understood. We investigate in detail the structure of $G_2(n)$ for composite moduli in [371]. The following Theorems 5.19 and 5.20 are proved there. They are also consequences of Theorems 4.26 and 5.10.

Theorem 5.19 *The subdigraph $G_1(n)$ contains a t-cycle if and only if there exists a positive odd integer d such that $t = \text{ord}_d 2$ and $d \mid \lambda(n)$.*

Theorem 5.20 *If there exists a t-cycle in $G(n)$, then there exists a t-cycle in $G_1(n)$.*

Theorem 5.21 *Suppose that n is a prime or a prime power. Suppose further that for each positive integer t, $G_1(n)$ has a t-cycle if and only if $G_2(n)$ has a t-cycle. Then $n = 2^k$ for $k \geq 1$ or n is a Fermat prime.*

Proof Since 0 and 1 are fixed points of $G(n)$, both $G_1(n)$ and $G_2(n)$ have a cycle of length 1. By the definition of $G_2(n)$, the only cycle in $G_2(n)$ is the fixed point 0.

Now suppose that n is not a Fermat prime or a number of the form 2^k for $k \geq 1$. If $n = p^i$, where p is an odd prime and $i \geq 2$, then $p \mid \lambda(n)$. Let $t = \text{ord}_p 2$. Then $t > 1$ and $G_1(n)$ has a t-cycle by Theorem 5.20. If $n = p$, where the odd prime p is not a Fermat prime, then $\lambda(p) = p - 1$ and $p - 1$ has an odd prime divisor q. If $t = \text{ord}_q 2$, then $t > 1$ and $G_1(n)$ again has a t-cycle which is not in $G_2(n)$.

We finally suppose that n is a Fermat prime or $n = 2^k$, where $k \geq 1$. Then $\lambda(n) = 2^i$, where $i \geq 0$. By Theorem 5.19, the only cycles in $G_1(n)$ are of length 1. The result now follows.   □

Figure 5.2 (with $n = 11$) gives an example in which $\omega(n) = 1$ and $G_1(n)$ has a t-cycle, but $G_2(n)$ does not have a t-cycle for $t = 4$. Figure 4.4 (with $n = 5$ and $n = 17$) and Fig. 5.1 (with $n = 8$) provide examples in which $\omega(n) = 1$ and both $G_1(n)$ and $G_2(n)$ only have t-cycles for $t = 1$.

Remark By inspection, we find that the least positive integer n for which $\omega(n) \geq 2$ and there exists a positive integer t for which $G_1(n)$ has a t-cycle, but $G_2(n)$ does not have a t-cycle, is $n = 203 = 7 \cdot 29$. In this case $G_1(203)$ has a 6-cycle, whereas $G_2(203)$ does not have a 6-cycle. Note that $\lambda(203) = 2^2 \cdot 3 \cdot 7$ and $21 \mid \lambda(203)$. However, $21 \nmid \lambda(7) = 2 \cdot 3$ and $21 \nmid \lambda(29) = 2^2 \cdot 7$. Moreover, $\mathrm{ord}_{21} 2 = 6$, whereas $\mathrm{ord}_3 2 = 2$ and $\mathrm{ord}_7 2 = 3$.

5.5 Generalized Power Digraphs

In this section we extend some results from the previous sections to the following class of digraphs. For fixed integers $n \geq 1$ and $k \geq 2$ and for any $x \in H = \{0, 1, \ldots, n - 1\}$ let $f(x)$ be the remainder of x^k modulo n, i.e.,

$$f(x) \in H \quad \text{and} \quad f(x) \equiv x^k \pmod{n}. \tag{5.18}$$

Each pair of natural numbers n and $k \geq 2$ has a specific iteration digraph corresponding to it (see e.g. Figs. 5.6, 5.8, and 5.9). For particular values of n and k, we denote the iteration digraph of f by $G(n, k)$.

For $k \geq 3$, we define regular and semiregular digraphs $G(n, k)$ similarly as for $G(n, 2)$. Note that the set of semiregular digraphs $G(n, k)$ includes the subset of regular digraphs.

Clearly, $G(n, k)$ is regular only if $G(n, k)$ has no vertices of indegree 0. Since each component of $G(n, k)$ has a unique cycle, we see that $G(n, k)$ is regular if and only if each component of $G(n, k)$ is a cycle and each vertex of $G(n, k)$ has indegree 1. Since any vertex of indegree 0 is a noncycle vertex and there is a path from any noncycle vertex to the cycle in its component, we see that $G(n, k)$ is regular if and only if each vertex of positive indegree has indegree equal to 1. Noting that each vertex of $G(n, k)$ has outdegree 1, we observe that $G(n, k)$ is regular as a digraph if and only if $G(n, k)$ is regular as an undirected graph. Figure 5.6 provides an example of a regular digraph, while Fig. 5.4 gives an example of a semiregular digraph which is not regular. In [372], we characterize all semiregular and regular digraphs $G(n, k)$.

Further, we again specify two particular subdigraphs of $G(n, k)$. Let $G_1(n, k)$ be the induced subdigraph of $G(n, k)$ on the set of vertices which are coprime to n and

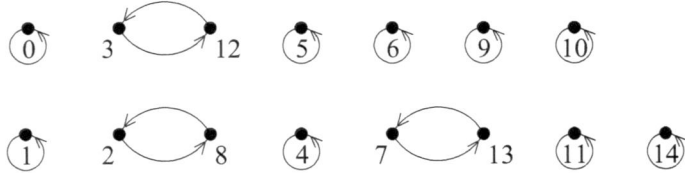

Fig. 5.6 The iteration digraph corresponding to $n = 15$ and $k = 3$

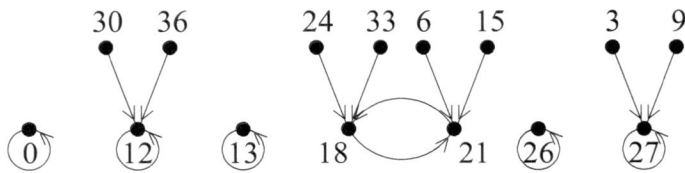

Fig. 5.7 The iteration subdigraph $G_2(39, 3)$

$G_2(n, k)$ be the induced subdigraph on the remaining vertices not coprime with n. It is clear that 0 is always a fixed point of $G_2(n, k)$.

By the next two theorems, $G_1(n, k)$ is always semiregular. Observe that in Fig. 5.1, the subdigraph $G_2(8, 2)$ is semiregular but $G(8, 2)$ is not semiregular. Note further that in Figs. 5.3 and 5.7, $G_2(n, k)$ is not semiregular, but each of its components is semiregular.

Theorem 5.22 *Let $n \geq 1$ and $k \geq 2$ be integers. Then*

  (i) $G_1(n, k)$ *is regular if and only if $(\lambda(n), k) = 1$;*
 (ii) $G_2(n, k)$ *is regular if and only if either n is square free and $(\lambda(n), k) = 1$, or $n = p$, where p is a prime;*
(iii) $G(n, k)$ *is regular if and only if n is square free and $(\lambda(n), k) = 1$.*

Theorem 5.23 *Let $n \geq 2$ and $k \geq 2$ be integers. If $(\lambda(n), k) > 1$ then $G_1(n, k)$ is semiregular but not regular.*

These theorems are proved in [372, pp. 48–49].

Theorem 4.4 of [372] gives necessary and sufficient conditions for $G_2(n, k)$ to be semiregular. Let $M \geq 2$ be an integer. The digraph $G(n, k)$ is said to be *symmetric of order M* if its set of components can be partitioned into subsets of size M, each containing M isomorphic components.

Figure 5.8 shows a symmetric digraph $G(39, 3)$ of order 3, while Fig. 5.9 exhibits a symmetric digraph of order 5. In Szalay [389], it was shown that $G(n, 2)$ is symmetric of order 2 if $2 \parallel n$ or $2^2 \parallel n$. In Carlip and Mincheva [61], it was also proved that $G(n, 2)$ is symmetric of order 2 if $n = 16p$, where p is a Fermat prime.

The digraph in Fig. 5.1 is not symmetric of order M for any $M \geq 2$ while the digraphs in Figs. 4.3, 5.3, and 5.4 are each symmetric of order 2.

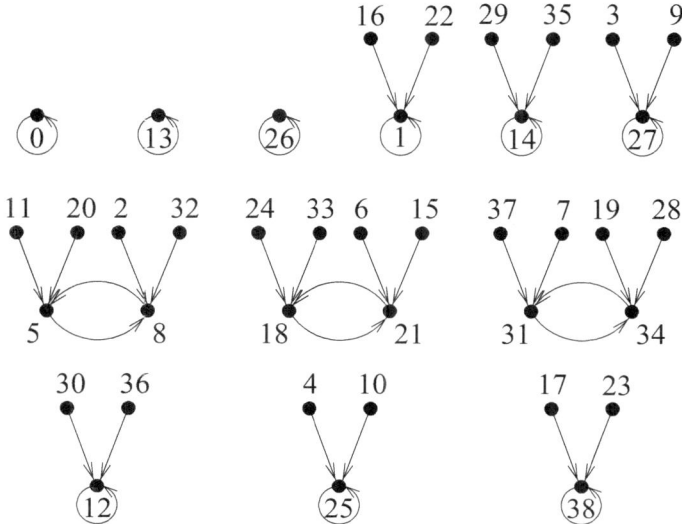

Fig. 5.8 The symmetric iteration digraph $G(39, 3)$ of order 3

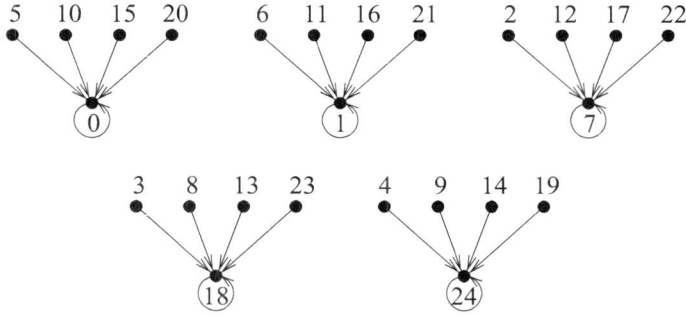

Fig. 5.9 The symmetric iteration digraph $G(25, 5)$ of order 5

Theorem 5.24 *Suppose that $p \parallel n$, where p is a prime. Then $G(n, p)$ is symmetric of order p.*

For a proof see [373, p. 2004].

Theorem 5.24 was partially extended by the papers [150, 178] in which necessary and sufficient conditions were given for the digraph $G(n, k)$ to be symmetric of order p, where $n = p^{\alpha} n_1$, p is an odd prime, $p \nmid n$, $\alpha \geq 1$, and n_1 is square free. The paper [178] treated the case in which $\alpha = 1$, while [150] dealt with the case $\alpha \geq 1$ and n_1 is odd. In [373], sufficient conditions were given for $G(n, k)$ to be symmetric of order M, where $M \geq 2$ is an arbitrary square-free number.

In [88, 89], Deng and Yuan gave necessary and sufficient conditions for a large set of digraphs $G(n, k)$ to be symmetric. They completely characterized all symmetric digraphs of order M when $M = 2$ or M is divisible by an odd prime, in which case they showed that M is square free.

In [375], we demonstrate that their classification is, in fact, complete by showing that there are no symmetric digraphs $G(n, k)$ of order 2^s for $s \geq 2$. Thus, we have the following statement.

Theorem 5.25 *There exists a symmetric digraph $G(n, k)$ of order $M \geq 2$ if and only if M is square free.*

Consider now a digraph $G(n, k)$ and factor $\lambda(n)$ as

$$\lambda(n) = \ell v, \tag{5.19}$$

where ℓ is the largest divisor of $\lambda(n)$ relatively prime to k.

Theorem 5.26 *There exists a t-cycle in $G_1(n, k)$ if and only if*

$$t = \mathrm{ord}_d k$$

for some factor d of ℓ. Moreover, $\mathrm{ord}_\ell k$ is the length of the longest cycle in $G_1(n, k)$.

This was proved in B. Wilson [420, pp. 232–233]. As an immediate consequence we get

Theorem 5.27 *Let ℓ be defined by (5.19). Then every cycle in $G_1(n, k)$ is a fixed point if and only if $k \equiv 1 \pmod{\ell}$.*

The next statement follows from [420, p. 234].

Theorem 5.28 *Let $t \geq 1$ be a fixed integer. Then any two components in $G_1(n, k)$ containing t-cycles are isomorphic.*

Theorem 5.29 *If there exists a t-cycle in $G_2(n, k)$, then there exists a t-cycle in $G_1(n, k)$.*

This theorem is proved in [374]. The following two theorems are also proved in [374] and present results about fixed points and the longest cycle in $G(n, k)$. As in the case in which $k = 2$, we let $A_t(G(n, k))$ denote the number of t-cycles in $G(n, k)$, where $k \geq 2$. We also let $L(G(n, k))$ denote the length of the longest cycle in $G(n, k)$.

Theorem 5.30 *Let $n > 1$ be a fixed integer. Then*

(i) *If k is even, then $A_1(G(n, k)) \geq 2^{\omega(n)}$ and $A_1(G_1(n, k)) \geq 1$. In particular, if $k = 2$, then $A_1(G(n, k)) = 2^{\omega(n)}$ and $A_1(G_1(n, k)) = 1$.*

(ii) *If $k \geq 3$ is odd and $2 \parallel n$, then $A_1(G(n, k)) \geq 2 \cdot 3^{\omega(n)-1}$ and $A_1(G_1(n, k)) \geq 2^{\omega(n)-1}$. In particular, if $k = 3$, then $A_1(G(n, k)) = 2 \cdot 3^{\omega(n)-1}$ and $A_1(G_1(n, k)) = 2^{\omega(n)-1}$.*

(iii) *If $k \geq 3$ is odd and either n is odd or $4 \parallel n$, then $A_1(G(n, k)) \geq 3^{\omega(n)}$ and $A_1(G_1(n, k)) \geq 2^{\omega(n)}$. In particular, if $k = 3$, then $A_1(G(n, k)) = 3^{\omega(n)}$ and $A_1(G_1(n, k)) = 2^{\omega(n)}$.*

(iv) *If $k \geq 3$ is odd and $8 \mid n$, then $A_1(G(n, k)) \geq 5 \cdot 3^{\omega(n)-1}$ and $A_1(G_1(n, k)) \geq 4 \cdot 2^{\omega(n)-1}$. In particular, if $k = 3$, then $A_1(G(n, k)) = 5 \cdot 3^{\omega(n)-1}$ and $A_1(G_1(n, k)) = 4 \cdot 2^{\omega(n)-1}$.*

(v) *The vertex 0 is an isolated fixed point of $G(n, k)$ if and only if n is square free.*

Theorem 5.31 *Let $n \geq 1$ be a fixed integer. Then*

(i) *$\max_{k \geq 2} L(G(n, k)) = \lambda(\lambda(n))$.*

(ii) *If $k \geq 2$ is an arbitrary integer and $A_t(G(n, k)) > 0$, then $t \mid \lambda(\lambda(n))$.*

(iii) *The digraph $G(n, k)$ contains only cycles of length 1 (fixed points) for all $k \geq 2$ if and only if n is one of the 8 positive divisors of 24.*

(iv) *$\max_{k \geq 2} L(G(n, k)) = 2$ if and only if n is one of the 136 positive divisors of $2^5 \cdot 3^2 \cdot 5 \cdot 7 \cdot 13 = 131\,040$ not dividing 24.*

(v) *If n is not a divisor of 24, then $\max_{k \geq 2} L(G(n, k))$ is an even integer.*

(vi) *$\max\limits_{k \geq 2 \text{ even}} L(G(n, k)) = 1$ if and only if $\phi(n)$ is a power of 2.*

(vii) *Suppose that $n > 5$. If it is not the case that n is a prime of the form $n = 2p^i + 1$, where p is an odd prime and $i \geq 1$, then*

$$\max_{k \geq 2} L(G(n, k)) < \frac{n}{4}.$$

If n is a prime of the form $2p^i + 1$, then

$$\max_{k \geq 2} L(G(n, k)) = p^{i-1}(p - 1) = \frac{n-1}{2} - \frac{n-1}{2p} > \frac{n}{4}.$$

In particular, when n is a prime such that $n = 2p^i + 1$, then

$$\frac{n-1}{3} \leq \max_{k \geq 2} L(G(n, k)) \leq \frac{n-1}{2} - 1. \tag{5.20}$$

The upper bound in (5.20) is attained if and only if n is a prime of the form $2p + 1$, i.e., p is a Sophie Germain prime, and the lower bound in (5.20) is achieved if and only if n is a prime of the form $2 \cdot 3^i + 1$, where $i \geq 1$.

Remark It is noted in Friedlander, Pomerance, and Shparlinski [116, p. 1592] that Theorem 5.31 (i) holds. It was shown in Lucheta, Miller, and Reiter [247] that the upper bound in (5.20) is attained when n is a prime of the form $2p + 1$ for p an odd prime.

Chapter 6
Pseudoprimes

6.1 What Is a Pseudoprime?

In this chapter, we will introduce a special class of composite numbers—pseudoprimes that exhibit some of the same properties as prime numbers, see [208]. We give an overview of their basic characteristics. We establish their asymptotic density and show that their occurrence is much rarer than the occurrence of prime numbers. For instance, there are only three pseudoprimes (their definition will be given on the next page) that are less than 1000 whereas there exist 168 primes in the same interval. We also introduce an interesting set of pseudoprimes called Carmichael numbers and describe some algorithms for generating pseudoprimes.

To define the term "pseudoprime", we need Fermat's Little Theorem 2.17, which determines some of the basic properties of prime numbers and on which the majority of probabilistic algorithms for testing prime numbers is based. Fermat's Little Theorem thus plays a key role in number theory. It is usually presented in two equivalent versions. The first version says that if p is a prime number, then

$$a^p \equiv a \pmod{p}$$

for all integers a (in other words, p divides $a^p - a$ without remainder). According to the second version, when p is a prime number coprime with a, then

$$a^{p-1} \equiv 1 \pmod{p}. \tag{6.1}$$

Unfortunately, the converse of Fermat's Little Theorem does not hold. For an arbitrary base $a > 1$ there exists a composite number n coprime with a such that

$$a^n \equiv a \pmod{n}. \tag{6.2}$$

For instance, the composite number $n = 341 = 31 \cdot 11$ fulfills (6.2) if $a = 2$, for which $(a, n) = 1$. We can easily calculate that $2^{10} \equiv 1 \pmod{341}$. By raising this

© The Author(s), under exclusive license to Springer Nature Switzerland AG 2021
M. Křížek et al., *From Great Discoveries in Number Theory to Applications*,
https://doi.org/10.1007/978-3-030-83899-7_6

congruence to the thirty-fourth power, we get $2^{340} \equiv 1 \pmod{341}$, i.e. (6.1) holds. Multiplying now the last congruence by two, we obtain

$$2^{341} \equiv 2 \pmod{341}, \tag{6.3}$$

i.e., relation (6.2) holds as well.

For another base $a = 3$ we can similarly prove that $3^{340} \equiv 56 \pmod{341}$ and $(3, 341) = 1$, which by (6.1) means that the number 341 is composite (without factoring it into primes).

A composite number n is called a *pseudoprime to the base a*, if (6.2) holds, i.e. when n divides $a^n - a$.

Note also that if the numbers a and n are coprime, then (6.2) holds if and only if

$$a^{n-1} \equiv 1 \pmod{n}. \tag{6.4}$$

6.2 Historical Notes

The first pseudoprime number 341 to the base 2 was found in 1819 by Pierre Frédérique Sarrus (see Annales de Math. 10 (1819), 184–187 or Dickson [91, p. 92]), which showed the validity of the congruence (6.3). The property (6.3) of the number 341 was also found by the anonymous author of the article *Théorèmes et problèmes sur les nombres,* which was published in the Journal für die reine und angewandte Mathematik 6 (1830), 100–106.

Also János Bolyai (1802–1860), one of the founders of non-Euclidean geometries, informed his father (see [165]) about his discovery of the pseudoprime number 341 in a letter dated May 1855. Moreover, according to E. Kiss [165, p. 72], Bolyai was the first to show that the composite Fermat number $F_5 = 2^{32} + 1 = 641 \cdot 6700417$ is a pseudoprime number to the base 2. His proof was based on the congruence

$$2^{2^{32}} \equiv 1 \pmod{F_5}.$$

To make sure this relationship holds true, we first square the congruence $2^{32} \equiv -1 \pmod{F_5}$, cf. (1.4),
$$2^{64} \equiv 1 \pmod{F_5}.$$

Using the equality $2^{32} = 64 \cdot 2^{26}$, we can continue in squaring to get

$$2^{2^{32}} = \left(2^{64}\right)^{2^{26}} \equiv 1 \pmod{F_5}.$$

Multiplying the left- and right-hand sides of this congruence by two, we get

$$2^{F_5} = 2 \cdot 2^{2^{32}} \equiv 2 \pmod{F_5},$$

and thus F_5 is a pseudoprime to the base 2.

In 1909, Tadeusz Banachiewicz published 5 pseudoprime numbers to the base 2 smaller than 2000 and later discovered two more remaining in this interval (see [23], [91, p. 94], [352]). The complete list of these pseudoprimes has the form:

$$341 = 11 \cdot 31,$$
$$561 = 3 \cdot 11 \cdot 17,$$
$$645 = 3 \cdot 5 \cdot 43,$$
$$1105 = 5 \cdot 13 \cdot 17,$$
$$1387 = 19 \cdot 73,$$
$$1729 = 7 \cdot 13 \cdot 19,$$
$$1905 = 3 \cdot 5 \cdot 127.$$

Pseudoprimes to the base 2 were discovered first and are also most often studied. Therefore, let us agree on the following.

Convention. For simplicity, pseudoprimes to the base 2 will be called just *pseudo-prime numbers*.

Pseudoprimes to other bases are tabulated e.g. in Ribenboim [321].

Example We show that $91 = 7 \cdot 13$ is a pseudoprime to the base 3. By Fermat's Little Theorem 2.17 we have $3^6 \equiv 1 \pmod 7$, and thus by (1.4) we obtain $3^{90} \equiv 1 \pmod 7$. Furthermore, we see that $3^3 \equiv 1 \pmod{13}$, i.e., $3^{90} \equiv 1 \pmod{13}$. Altogether we get $3^{90} \equiv 1 \pmod{91}$, i.e., 91 is a pseudoprime to the base 3. In addition, it can be verified that it is the smallest pseudoprime number to this base. Another such pseudoprime is the even number 286.

Pseudoprimes were once also called almost primes (see Erdős [103]) or also *Poulet numbers* (see Duparc [98]), since they were intensively studied by Poulet in [311], where all pseudoprime numbers up to 10^8 are tabulated. See also Lehmer [226].

In a number of publications (see for example [23], [91, p. 59], [154]) it is stated that the ancient Chinese 500 years BC believed that

$$2^n \equiv 2 \pmod n \tag{6.5}$$

holds if and only if n is a prime.[1] An interesting explanation, how this almost certainly false story came about, is described in Ribenboim [321, p. 86] (see also [323, pp. 103–

---

[1] With all respect to the ancient Chinese mathematicians, they would probably hardly be able on their abacuses to find that $341 \mid (2^{341} - 2)$, i.e. that the implication $\Rightarrow$ does not hold, since the number $2^{341} - 2$ has over 100 digits.

105]). The main counter-argument is that the ancient Chinese never formulated the term "pseudoprime". This error first appeared probably in the article [154] by Jeans and was then repeated by many authors.

According to Mahnke [251], in the years 1680–1681 Leibniz also erroneously argued that congruence (6.5) is valid only if n is a prime number.

The following theorem, which is given in Sierpiński [352] and Steuerwald [382], can be used for a recursive generation of infinitely many pseudoprimes.

Theorem 6.1 *If n is an odd pseudoprime, then $2^n - 1$ is also an odd pseudoprime.*

Proof Since n is composite, there exists a positive integer m, $1 < m < n$, which divides n. Then $2^m - 1$ divides by (4.1) the number $2^n - 1$, i.e.,

$$\left(2^m - 1\right) \mid \left(2^n - 1\right),$$

and thus $2^n - 1$ is also composite. Now it is enough to show that

$$\left(2^n - 1\right) \mid \left(2^{2^n-2} - 1\right). \tag{6.6}$$

Since n is an odd pseudoprime, by congruence (6.4) we have

$$\left(2^n - 2\right)/2 = 2^{n-1} - 1 = kn$$

for some integer k. Then the number $2^n - 1$ divides $2^{kn} - 1 = 2^{(2^n-2)/2} - 1$ and relation (6.6) is satisfied, since

$$2^{2^n-2} - 1 = \left(2^{(2^n-2)/2} - 1\right)\left(2^{(2^n-2)/2} + 1\right).$$

The above completes the proof.   □

Due to this theorem, we can explicitly specify a pseudoprime, which is greater than any given integer. A similar statement for prime numbers is not known yet, even though we know that there are infinitely many prime numbers. As of 2020 the largest known prime was the Mersenne prime $2^{82589933} - 1$ having almost 25 million digits (see Sect. 4.1). Let us point out that in 1903 Theorem 6.1 was proved by Malo [255] for the case that n is prime and $2^n - 1$ is composite.

The first even pseudoprime (to the base 2)

$$161\,038 = 2 \cdot 73 \cdot 1103$$

was found by Lehmer in 1950 (see Erdős [103]). One year later Beeger proved that there exists infinitely many even pseudoprimes (see [27]).

6.3 Density and Distribution of Pseudoprimes

Although there are infinitely many pseudoprimes (see Theorem 6.1), there are far fewer of them than primes. For example, Szymiczek [391] proved that if we denote the nth pseudoprime (to the base 2) by P_n, then the series

$$\sum_{n=1}^{\infty} \frac{1}{P_n}$$

is convergent, while it is well known that the series

$$\sum_{n=1}^{\infty} \frac{1}{p_n}$$

is divergent (see Theorem 3.15), where p_n denotes the pth prime. However, later Mąkowski proved in [254] that the series

$$\sum_{n=1}^{\infty} \frac{1}{\log P_n}$$

is divergent.

Choosing randomly a number $n \leq 10^{10}$ for which $2^{n-1} \equiv 1 \pmod{n}$, the probability that n is a prime will be more than $30\,000$ times higher than that n is a pseudoprime. More details are in the article [208].

An interesting property that is common to both primes and pseudoprimes is expressed in the following theorem due to Rotkiewicz [335, 338].

Theorem 6.2 (Rotkiewicz) *Let $a, d \in \mathbb{N}$ be coprime numbers. There exist infinitely many pseudoprimes in the arithmetic progression $(a + kd)_{k=0}^{\infty}$.*

A similar result for prime numbers is given by Dirichlet's Theorem 3.10, which has the same wording as Theorem 6.2 when we omit the prefix "pseudo".

6.4 Carmichael Numbers

The enormous sparsity of pseudoprimes to a fixed base a compared to primes provides a reasonable reason to use Fermat's Little Theorem 2.17 as a primality test (see [309, p. 81]). However, there exist composite numbers n, called the *Carmichael numbers* or *absolute pseudoprimes*, that are pseudoprimes for any base a, i.e. the congruence

$$a^n \equiv a \pmod{n}$$

holds for all positive integers a.

Computer primality sieves based on Fermat's Little Theorem therefore let through all (composite) Carmichael numbers as well as primes. Therefore, appropriate primality tests have to be modified in some way. As already mentioned, the Wilson equivalence (see Theorem 3.6): "*p is prime* $\Leftrightarrow$ $(p-1)! \equiv -1 \pmod{p}$" is not suitable for primality testing, since an effective method for calculating the factorial modulo p is not known.

In 1899, Korselt proved the following necessary and sufficient condition to test whether a given composite positive integer n is an absolute pseudoprime (see [174]).

Theorem 6.3 (Korselt) *A composite positive integer n is an absolute pseudoprime if and only if n is not divisible by the square of any integer greater than one and p − 1 divides n − 1 for all prime factors p of the number n.*

Interestingly, when Korselt proved his theorem, he did not give any example of an absolute pseudoprime. From Korselt's theorem 6.3, for example, it immediately follows that

$$561 = 3 \cdot 11 \cdot 17$$

is an absolute pseudoprime, since $3 - 1 = 2, 11 - 1 = 10, 17 - 1 = 16$ and all these three numbers divide 560. On the other hand, $341 = 11 \cdot 31$ is not an absolute pseudoprime, since $31 - 1 = 30$ does not divide 340.

Robert Daniel Carmichael, who introduced the λ-function (see Sect. 2.9), independently discovered the above Korselt criterion and reformulated it in the following way even though he did not use the term the "Carmichael number" (see [62]):

Theorem 6.4 (Carmichael) *A composite number n is a Carmichael number if and only if n is not divisible by the square of any integer greater than one and $\lambda(n) \mid (n-1)$.*

In the paper [62], Carmichael also showed that every Carmichael number has at least three prime divisors and he gave there four examples which included the two smallest ones: 561 and 1105. In another paper [63], fifteen examples of Carmichael numbers are given. By [308] there exist only 2163 Carmichael numbers less than $25 \cdot 10^9$.

In [7], it is shown that there exist infinitely many Carmichael numbers. Moreover, it is proved that if we denote by $C(x)$ the number of Carmichael numbers not exceeding x, than for a sufficiently large x we have

$$C(x) > x^{2/7}.$$

In 2008, Glyn Harman [139] improved the lower bound for $C(x)$ to

$$C(x) > x^{1/3}$$

for sufficiently large x.

6.5 Mersenne and Fermat Pseudoprimes

The following theorem states that each composite *Mersenne number* $M_p = 2^p - 1$, where p is prime, is a pseudoprime (to the base 2).

Theorem 6.5 *All Mersenne numbers are either primes or pseudoprimes.*

Proof Suppose that $M_p = 2^p - 1$, where p is a prime, is a composite Mersenne number. Then p is an odd prime. It is enough to show that

$$2^{M_p - 1} \equiv 1 \pmod{M_p},$$

or equivalently

$$M_p \mid \left(2^{M_p - 1} - 1\right). \tag{6.7}$$

However, we will prove a stronger statement, namely

$$M_p \mid \left(2^{(M_p - 1)/2} - 1\right), \tag{6.8}$$

which implies (6.7).

According to Fermat's Little Theorem 2.17, we have

$$\frac{M_p - 1}{2} = 2^{p-1} - 1 \equiv 0 \pmod{p}.$$

Hence, $(M_p - 1)/2 = kp$ for some positive integer k. Since

$$\left(2^p - 1\right) \mid \left(2^{kp} - 1\right),$$

we see that (6.8) holds. □

There exist also several connections between Fermat numbers $F_m = 2^{2^m} + 1$, $m = 0, 1, 2, \ldots$, and pseudoprimes. For instance, in 1904 M. Cipolla derived that $n \mid \left(2^n - 2\right)$ for $n = 2^{2^m} + 1$, $m = 0, 1, 2, \ldots$ Five years later T. Banachiewicz proved that all Fermat numbers are either primes or pseudoprimes to the base 2 (a proof is given e.g. in [199, p. 36]).

Theorem 6.6 *All Fermat numbers are either primes or pseudoprimes. Moreover, if $2^n + 1$ is a pseudoprime, then n is a power of 2.*

Maybe just this assertion led Fermat to his incorrect conjecture that all Fermat numbers are primes.

Several authors (see [71, 152, 154, 336, 390]) used Fermat numbers to generate infinitely many pseudoprimes. The following theorem from 1904, proved in Cipolla [71], shows how Fermat numbers can be used to create infinitely many pseudoprimes that have any number of prime factors.

Theorem 6.7 (Cipolla) *If $a > b > \cdots > s > 1$ and $n = F_a F_b \cdots F_s$, then n is a pseudoprime if and only if $2^s > a$. For an arbitrarily large $m > 1$ thus there exist infinitely many pseudoprimes having at least m different prime factors.*

In 1949, Paul Erdős proved that for every $m > 1$ there exist infinitely many pseudoprimes that have exactly m different prime factors. Interestingly, the first proof of an infinite number pseudoprimes was given by the astronomer James Hopwood Jeans as early as 1897. In [154], he showed that if $a > b$ and $2^b > a$, then the product $F_a F_b$ is a pseudoprime. This result is a special case of Theorem 6.7.

6.6 Further Types of Pseudoprimes

Let N be a composite odd integer and write $N - 1 = 2^s t$, where t is odd and $s \in \{1, 2, \ldots\}$. Let $a \geq 1$ be coprime to N. Then N is a *strong pseudoprime to the base a* if

$$a^t \equiv 1 \pmod{N} \tag{6.9}$$

or

$$a^{2^r t} \equiv -1 \pmod{N} \quad \text{for some } r \in \{0, \ldots, s - 1\}.$$

Strong pseudoprimes to the base 2 will simply be called *strong pseudoprimes*.

Remark Strong pseudoprimes to the base a were first defined by Selfridge (see [419, Sect. 17]). It is evident that strong pseudoprimes to the base a are pseudoprimes to the base a. Note that

$$a^{N-1} - 1 = a^{2^s t} - 1 = \left(a^t - 1\right)\left(a^t + 1\right)\left(a^{2t} + 1\right) \cdots \left(a^{2^{s-1}t} + 1\right). \tag{6.10}$$

If N is prime, then by Fermat's Little Theorem 2.17,

$$a^{N-1} - 1 \equiv 0 \pmod{N},$$

which implies that one of the factors on the right-hand side of (6.10) is congruent to $0 \pmod{N}$. Hence, N satisfies (6.9). It is proved in Rabin [314] and Monier [277] that a composite odd integer $N > a$ can be a strong pseudoprime to the base a for at most $\phi(n)/4$ bases a such that $1 \leq a < n$ and $(a, n) = 1$. Thus, if the base a is chosen randomly, the probability that N is a strong pseudoprime to the base a is less than or equal to $\frac{1}{4}$. This gives rise to the *Miller-Rabin probabilistic primality test*:

Let N be a fixed odd integer. Choose a base a such that $(a, N) = 1$ at random and observe whether (6.9) is satisfied. If (6.9) is not satisfied, then N is composite. If (6.9) holds, then repeat the test. The probability that (6.9) will be satisfied for k independent tests if N is composite is less than or equal to 4^{-k}. If N passes k tests, where k is fairly large, we declare N to be a probable prime.

The following theorem is proved in [199, p. 138].

Theorem 6.8 *Let $F_m = 2^{2^m} + 1$, $m \in \{5, 6, \ldots\}$, be a composite Fermat number and let $M_p = 2^p - 1$ be a composite Mersenne number, where p is a prime. Then F_m and M_p both are strong pseudoprimes.*

The following theorem is proved in [308, p. 1008] and [199, p. 139].

Theorem 6.9 *If N is an odd pseudoprime, then $2^N - 1$ is a strong pseudoprime.*

Proof Since N is composite, there exists an integer c, $1 < c < N$, such that $c \mid N$. Then $2^c - 1 \mid 2^N - 1$, and $2^N - 1$ is also an odd composite integer. Since N is a pseudoprime, we have that $2^{N-1} - 1 = Nt$ for some odd integer t. Then $2(2^{N-1} - 1) = (2^N - 1) - 1 = Nt \cdot 2^1$. Clearly, $2^N \equiv 1 \pmod{2^N - 1}$. Hence,

$$2^{Nt} = 2^{2^{N-1}-1} \equiv 1^t \equiv 1 \pmod{2^N - 1},$$

and $2^N - 1$ is a strong pseudoprime. □

We have the following theorem using pseudoprimes that are not strong pseudoprimes to give a nontrivial factorization of the positive integer N.

Theorem 6.10 *Let N be a pseudoprime to the base a that is not a strong pseudoprime to the base a. Let $N - 1 = 2^s t$, where t is odd. Then there exists an integer $r \in \{0, \ldots, s - 1\}$ such that $a^{2^{r+1}t} \equiv 1 \pmod{N}$ and $a^{2^r t} \equiv c \pmod{N}$, where $c \not\equiv \pm 1 \pmod{N}$. Then $(c - 1, N) > 1$ and $(c + 1, N) > 1$.*

Proof By the definition of a pseudoprime and a strong pseudoprime, there exist such integers r, s, t, and c. Then

$$c^2 - 1 = (c - 1)(c + 1) \equiv 0 \pmod{N},$$

but

$$c \not\equiv -1 \text{ and } c \not\equiv 1 \pmod{N}.$$

The result now follows. □

Example Consider the pseudoprime 341, which is a pseudoprime but is not a strong pseudoprime. Then

$$2^{340} \equiv 1 \pmod{341}, \quad 2^{170} \equiv 1 \pmod{341}, \text{ and } 2^{85} \equiv 32 \not\equiv \pm 1 \pmod{341}.$$

We now observe that

$$(32 - 1, 341) = 31, \quad (32 + 1, 341) = 11,$$

giving that $341 = 11 \cdot 31$.

The following theorem was stated by Karel Micka.

Theorem 6.11 (Micka) *Any positive integer which is not a prime is a pseudoprime for a suitably chosen base.*

Proof By the Binomial Theorem, for an arbitrary composite number n the difference $a^n - a$ is divisible by n for $a = n + 1$,

$$(n + 1)^n - n - 1 = n^n + \binom{n}{1} n^{n-1} + \cdots + \binom{n}{n-2} n^2.$$

From this and congruence (6.2) it follows that n is a pseudoprime to the base $n + 1$.   □

The research on pseudoprimes is now quite extensive. There exist a number of different classes of pseudoprimes (see [17, 82, 158, 165, 167, 199, 299, 305, 306, 339, 364, 365, 390]), for instance, the above mentioned absolute pseudoprimes, and also Dickson, Fibonacci, Frobenius, Lehmer, and Fermat and Lucas d-pseudoprimes, superpseudoprimes, etc.

The composite odd integer N is an *Euler pseudoprime* to the base a if $(a, N) = 1$ and

$$\left(\frac{a}{N}\right) \equiv a^{(N-1)/2} \pmod{N},$$

where $a \geq 1$ and (a/N) denotes the Jacobi symbol. The Euler pseudoprimes appear in the Solovay-Strassen probabilistic primality test (see [199, p. 137]). Also strong pseudoprimes are especially useful in probabilistic primality tests (see e.g. Crandall and Pomerance [82], Wagstaff [408]).

Chapter 7
Fibonacci and Lucas Numbers

7.1 Fibonacci Numbers

In the literature, the Fibonacci numbers are usually denoted by F_n, but this symbol is already reserved for the Fermat numbers in this book. So we will denote them by K_n. The sequence of *Fibonacci numbers* $(K_n)_{n=0}^{\infty}$ starts with $K_0 = 0$ and $K_1 = 1$ and satisfies the recurrence

$$K_{n+2} = K_{n+1} + K_n \qquad \text{for all } n = 0, 1, 2, \ldots \tag{7.1}$$

Thus the first several initial terms are

$$0, 1, 1, 2, 3, 5, 8, 13, 21, 34, 55, 89, 144, 233, 377, 610, 987, 1597, 2584, \ldots.$$

These numbers first appeared as the solution to a classical problem of the reproduction of rabbits in the second extended edition of the book entitled *Liber Abaci* (1228) by the Italian mathematician Fibonacci, also called Leonardo de Pisa (cca 1180–cca 1250). They were named after Fibonacci by Édouard Lucas in the second half of the 19th century. Lucas also discovered a formula that shows how the Fibonacci numbers are hidden in the Pascal triangle (see e.g. [58, 201, 203]).

In Chap. 12 of the above-mentioned book (cf. [303, p. 404]), the main task is to establish how fast rabbits can reproduce under certain very specific and idealized conditions. Suppose that one newborn couple is placed in an empty enclosure, rabbits can mate after a month, the gestation period is also 1 month, every female gives birth to just one pair every month, and finally suppose that rabbits do not die. In equation (7.1), K_n denotes the number of rabbit pairs in the enclosure in the nth month. The first person who wrote the recurrence (7.1) for Fibonacci numbers, was Albert Girard (1595–1632), and therefore not Fibonacci (see Tanton [393]).

It would be an overwhelming task to mention all mathematical activity that Fibonacci numbers generated. It suffices to say that there is a journal *The Fibonacci*

© The Author(s), under exclusive license to Springer Nature Switzerland AG 2021
M. Křížek et al., *From Great Discoveries in Number Theory to Applications*,
https://doi.org/10.1007/978-3-030-83899-7_7

Quarterly, appearing four times a year and publishing articles primarily dedicated to the investigation of Fibonacci numbers, that there exists a *Fibonacci Association* which holds an International Conference every two years, and several books dedicated solely to the Fibonacci numbers (see e.g. [145, 153, 175, 236, 404, 405]). These numbers are so popular that 2 m high neon Fibonacci numbers from 1 to 55 are attached to a tall power plant chimney in Turku (Finland).

In this section we will give a brief overview about interesting properties of Fibonacci numbers, about completely new results and open problems. We shall see that Fibonacci numbers appear in some completely unexpected contexts.

Example If we subsequently divide the atomic numbers of noble (inert) gases ($_2$He, $_{10}$Ne, $_{18}$Ar, $_{36}$Kr, $_{54}$Xe, $_{86}$Rn) by 5π, then after rounding to integer numbers we obtain the sequence 0, 1, 1, 2, 3, 5. We can continue in this Fibonacci growth. For a nonstable noble gas $_{118}$Uuo, the so-called ununoktinium, we get 8.

Example We also encounter frequent occurrences of Fibonacci numbers in biology. For instance, spruce or pine cone seeds are divided into two types (clockwise and counterclockwise) of helices, the numbers of which are usually in the ratio 13 : 8. Similar spirals can also be seen in artichokes or pineapples. In Fig. 7.1 we see that helices (spirals) on the surface of the pineapple are formed by shapes that resemble rhombuses.

Figure 7.2 then schematically illustrates the developed surface of this pineapple. Rectangles can be numbered so that the numbers on the individual helices form arithmetic progressions with differences 8 and 13. The numbers of rhombuses that touch just one vertex have in turn the difference of 5 or 21 (in absolute value). It is remarkable that roughly every second pineapple has a mirrored helix on its surface, i.e. it has 13 counterclockwise and 8 clockwise helices. It is believed that this complex helix structure is not encoded in DNA, but that it is formed automatically during growth, so that a new rhombus is created where there is the most space.

Finally note that flowers usually have 3, 5, 8 or 13 petals. More examples of the occurrence of Fibonacci numbers in nature can be found e.g. in Koshy [175].

Example In 1876 the French mathematician Édouard Lucas derived the formula

$$\sum_{i=0}^{\lfloor n/2 \rfloor} \binom{n-i}{i} = K_{n+1},$$

which indicates how the Fibonacci numbers are hidden in Pascal's triangle (also Fermat's numbers are hidden in Pascal's triangle, see [199, p. 35]). Let us write this triangle as shown in Fig. 7.3. We see that the sums of the binomial coefficients along diagonals with a slope of $+45°$ are just equal to the numbers K_{n+1}. For example $1 + 1 = 2$, $1 + 2 = 3$, $1 + 3 + 1 = 5$, $1 + 4 + 3 = 8$, $1 + 5 + 6 + 1 = 13$, etc.

Fig. 7.1 Systems of right-handed and left-handed spirals on the surface of a pineapple

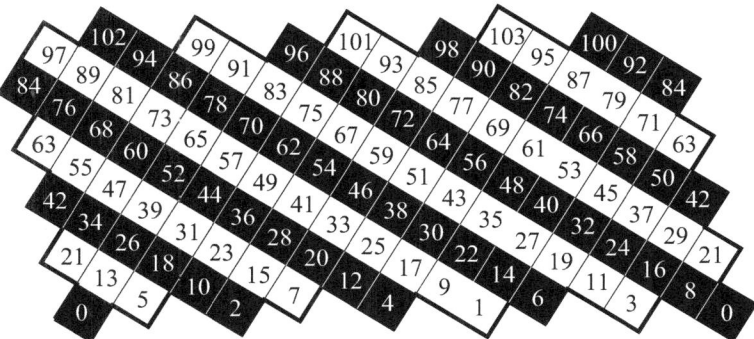

Fig. 7.2 Eight counterclockwise helices on a developed pineapple's shell. For higher resolution they are colored alternately black and white. Approximately perpendicular to this direction there are thirteen clockwise helices. (Right and left edges of the drawing must be identified.)

Fig. 7.3 Pascal's triangle in
a beveled shape

$$
\begin{array}{ccccccc}
1 & & & & & & \\
1 & 1 & & & & & \\
1 & 2 & 1 & & & & \\
1 & 3 & 3 & 1 & & & \\
1 & 4 & 6 & 4 & 1 & & \\
1 & 5 & 10 & 10 & 5 & 1 & \\
1 & 6 & 15 & 20 & 15 & 6 & 1 \\
\vdots & \vdots & \vdots & \vdots & \vdots & \vdots & \vdots
\end{array}
$$

Example One can prove that the number of subsets of the set $A = \{1, 2, \ldots, n\}$ which do not contain two consecutive numbers, is K_{n+2}. For instance, if $n = 4$ then the following subsets $\emptyset$, $\{1\}$, $\{2\}$, $\{3\}$, $\{4\}$, $\{1, 3\}$, $\{1, 4\}$, and $\{2, 4\}$ satisfy that condition. Also the number of possibilities, when making n coin tosses that two consecutive heads do not occur, is K_{n+2}. For instance, if $n = 4$ we have the following possibilities,

$$(T,T,T,T), \ (H,T,T,T), \ (T,H,T,T), \ (T,T,H,T),$$

$$(T,T,T,H), \ (H,T,H,T), \ (H,T,T,H), \ (T,H,T,H),$$

where T denotes tails and H heads. The fact that in both cases we get the same number of possibilities follows from the following assignment. Let B be a subset of A, which does not contain two consecutive numbers. Then the corresponding n-tuple consisting of T and H has H in the ith position if and only if $i \in B$.

Example Consider two panes of window glass on top of each other. The sun's rays can refract in various ways inside this double layer. In the book [236, p. 99] by Livio, it is shown that the ray can either pass through directly or is reflected inside in two ways, or is reflected twice in three ways, or is reflected three times in five ways, or is reflected four times in eight ways, etc. We get again the Fibonacci sequence.

Example The classical Fibonacci problem with rabbits can be equivalently formulated as follows. Consider a sequence of strings formed by the digits 0 and 1, which are defined recursively as follows:

The number 0 is replaced by 1 in the next string (i.e., a newborn rabbit pair matures) and the number 1 is replaced by 10 (i.e., an adult couple gives birth to one new couple). Starting from 1, we get the following sequence of strings:

$$1,$$

$$10,$$

$$101,$$

$$10110,$$

$$10110101,$$

$$1011010110110,$$

$$1011010110110101101101,$$

$$\vdots$$

It can be proved that the length of the nth string is equal to K_{n+1} for $n = 1, 2, 3, \ldots$ It can also be shown that the number of 0's in the nth string is equal to K_{n-1} and the number of 1's in the nth string is equal to K_n. In this way one may construct an infinite sequence of 0's and 1's

$$10110101101101011010110110101101 10 \ldots, \tag{7.2}$$

which is called the *golden string*. The golden string is a transcendental number if it is preceded by a decimal point.

The sequence (7.2) has a fractal character. Any arbitrary substring appearing in the golden string is contained in it infinitely many times, even though this sequence is not periodic. Hence, Fibonacci introduced for the first time a problem that leads to fractal structures.

7.2 Fibonacci Numbers and The Mandelbrot Set

The Fibonacci numbers are "hidden" even in the famous Mandelbrot's fractal set. Consider the recurrence equation

$$y_{n+1} = y_n^2 + c, \quad n = 1, 2, \ldots \tag{7.3}$$

in the complex plane, i.e., $c, y_n \in \mathbb{C}$. Hence, values y_n depend on the parameter c which will be denoted by $y_n = y_n(c)$. Then *Mandelbrot's set* $\mathcal{M}$ is the set of all parameters $c \in \mathbb{C}$ for which the sequence $(y_n(c))_{n=1}^{\infty}$ with initial condition $y_1(c) = 0$ is bounded, see the black region in Fig. 7.4 and also Mandelbrot [256], i.e.,

$$\mathcal{M} = \{c \in \mathbb{C};\ y_1(c) = 0,\ \exists C \in \mathbb{R}\ \forall n \in \mathbb{N}:\ |y_n(c)| \leq C\}.$$

Remark Already in 1906, the French mathematician Pierre Fatou described even more general objects, where the above quadratic function is replaced by a rational function, see Fatou [110] and also [111]. However, at that time there were no elec-

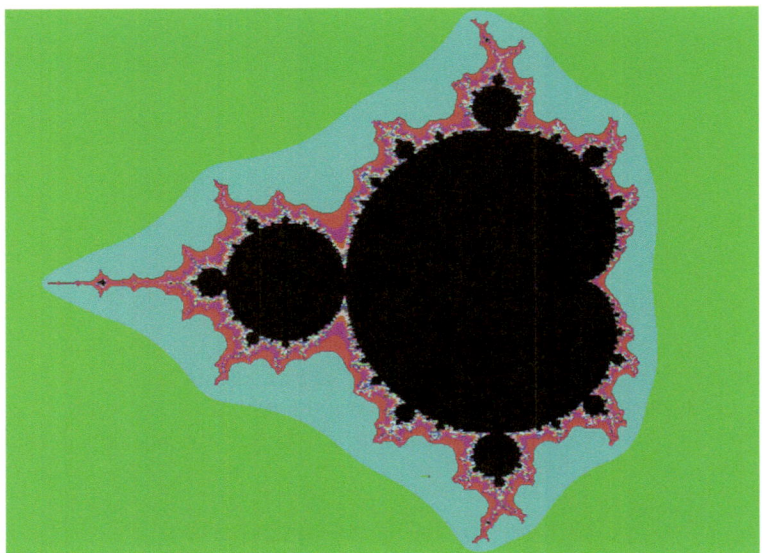

Fig. 7.4 The black color corresponds to the Mandelbrot set $\mathcal{M}$ of all complex numbers c for which the sequence $y_1(c) = 0$, $y_2(c) = c$, $y_3(c) = c^2 + c$, $y_4 = (c^2 + c)^2 + c$,... is bounded. The other colors indicated the rate of divergence of the sequence $(y_n(c))$ for $c \in [-2, 0.93] \times [-1.1\,\mathrm{i}, 1.1\,\mathrm{i}]$

tronic computers with graphical facilities and thus the beauty of fractals could not be discovered. Also Gaston Julia for $c \in \mathbb{C}$ investigated sets of the form

$$\mathcal{J}_c = \partial\{y_1(c) \in \mathbb{C};\ \exists C \in \mathbb{R}\ \forall n \in \mathbb{N} : |y_n(c)| \le C\},$$

where the symbol ∂ denotes the boundary. In [159], he proved that $c \in \mathcal{M}$ if and only if $\mathcal{J}_c$ is connected. The set $\mathcal{J}_c$ is called the *Julia set* after him.

The largest part of the Mandelbrot set $\mathcal{M}$ defined by the relation is formed by the region (called the *body* of the Mandelbrot set) bounded by the cardioid curve. In our case this is a plane curve traced by a point on the perimeter of a circle with radius $\frac{1}{4}$ that is rolling around a fixed circle of the same radius with the center at the origin. Adjacent to the body of Mandelbrot's set there are infinitely many circular areas with other circular protrusions (see [295, p. 857]). Their number is also infinite.

Multicolored images of the surroundings of the Mandelbrot fractal set can be obtained e.g. as follows: To any point $c \in \mathbb{C} \setminus \mathcal{M}$ we assign a certain color according to the "speed of divergence" of the sequence $(y_n(c))_{n=1}^{\infty}$. In Fig. 7.4, the green color indicates the region with the largest speed of divergence, blue denotes a region with slightly lower speed, red stands for an interval with even lower divergence rate, etc.

However, it is more natural to color the interior of the Mandelbrot sets $\mathcal{M}$ according to the number of accumulation points of the sequence (y_n) (see Fig. 7.5). For $k \in \mathbb{N}$ set

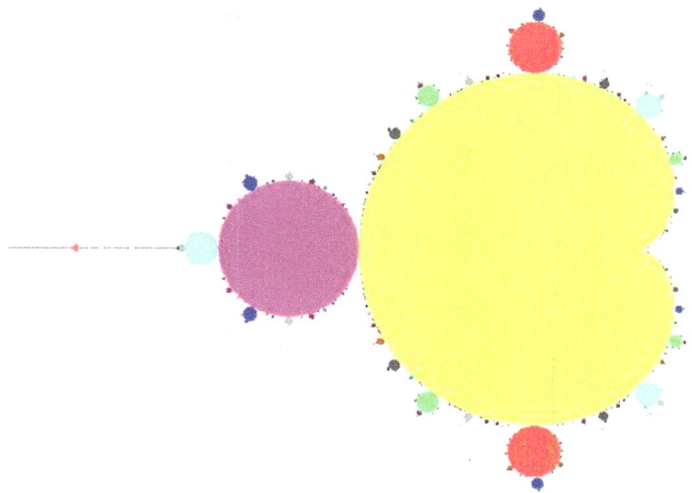

Fig. 7.5 Coloring of the Mandelbrot set $\mathcal{M}$ according to the number of accumulation points of the sequence $(y_n(c))$ for $c \in [-2, 0.93] \times [-1.1\,\mathrm{i}, 1.1\,\mathrm{i}]$

$$\mathcal{M}(k) = \{c \in \mathcal{M};\ (y_n(c))_{n=1}^{\infty} \text{ has exactly } k \text{ accumulation points}\}.$$

In Figs. 7.5 and 7.6, the yellow color indicates the set $\mathcal{M}(1)$ (when the sequence $(y_n(c))_{n=1}^{\infty}$ has a limit) and light purple indicates the set $\mathcal{M}(2)$ (i.e., when the associated the sequence has exactly 2 accumulation points). The other colors are assigned as follows: 3—light red, 4—azure blue, 5—light green, 6—light blue, 7—dark gray, 8—light gray, 9—brown, 10—violet, 11—red, 12—blue, 13—green, 14—dark blue, 15 or more accumulation points—black. The region, where the sequence $(y_n(c))$ diverges for $n \to \infty$, is colored white. These colors allow us to trace Fibonacci numbers in Mandelbrot's set.

In Fig. 7.5, we see different colored circles that surround the perimeter of the cardioid (in parentheses we indicate the corresponding number of accumulation points). Notice that the largest circle between light purple (2) and the upper, or lower, light red circle (3) is light green (5). Similarly in Fig. 7.6, the largest circle between light red (3) and light green (5) is light gray (8). The largest circle between light green (5) and light gray (8) is green (13), etc. Notice that the numbers in parentheses are the Fibonacci numbers. The proof of this remarkable property and other interesting relations between the Mandelbrot set and the Fibonacci the numbers are given in [90], [295, p. 688], [423], etc.

Example The point $c = 0$ obviously belongs to $\mathcal{M}(1)$, see the yellow region in Fig. 7.5.

Fig. 7.6 Coloring of the Mandelbrot set $\mathcal{M}$ according to the number of accumulation points of the sequence $(y_n(c))$ in the rectangle $[-0.6, 0] \times [0.45\,\mathrm{i}, 0.9\,\mathrm{i}]$

Example If $c = -1$, then $c \in \mathcal{M}(2)$, since the associated sequence $y_1 = 0$, $y_2 = -1$, $y_3 = 0$, $y_4 = -1,\ldots$ has exactly two accumulation points, see the largest light purple circle in Fig. 7.5. Its radius is $\frac{1}{4}$ and the center $(-1, 0)$.

Example Three accumulation points are obtained from the requirement $y_1 = y_4$ which leads to the relation $(c^2 + c)^2 + c = 0$ for some $c \in \mathbb{C}$. Eliminating the trivial solution, we get the cubic equation

$$c^3 + 2c^2 + c + 1 = 0$$

whose roots are $-0.122564 \pm \mathrm{i}\,0.744861$ and -1.754877, see red circles in Fig. 7.5. In Fig. 7.6, we see detail of a neighborhood of one complex root. It is colored by light red and belongs to $\mathcal{M}(3)$. Similarly the light green circle is a part of the set $\mathcal{M}(5)$, etc.

The sequence $(y_n(c))_{n=1}^{\infty}$ is called an *orbit at the point c*. It can have infinitely many accumulation points for $c \in \mathcal{M}$. Setting

$$\mathcal{M}(\infty) = M \setminus \bigcup_{k=1}^{\infty} M(k),$$

Mandelbrot's set can be expressed as the union of infinitely many disjoint sets

$$\mathcal{M} = \mathcal{M}(1) \cup \mathcal{M}(2) \cup \cdots \cup \mathcal{M}(\infty).$$

Let us mention one more remarkable connection between the fractal Mandelbort's set and theory of deterministic chaos. The Eq. (7.3) is, in fact, the standard logistic equation (see [199, p. 177]) in the canonical form, i.e. without the linear term. Now consider only real values of the parameter c. If $c \in [-\frac{3}{4}, \frac{1}{4}]$, then the sequence (y_n) has a limit. If $c \in [-\frac{5}{4}, -\frac{3}{4})$, then (y_n) has two accumulation points, etc. The points $c_2 = -\frac{3}{4}$, $c_3 = -\frac{5}{4}$, $c_4 = -1.368$, $c_5 = -1.394$, ...are called *bifurcation points* of the dynamical system (7.3). They correspond to the so-called *doubling of the period*, see Antonyuk and Stanyukovich [11]. Thus on the interval $[c_4, c_3)$ there are four accumulation points, on $[c_5, c_4)$ eight points, etc. The corresponding bifurcation branches are illustrated in Fig. 7.7. It can be showed that

$$\delta = \lim_{j \to \infty} \frac{c_j - c_{j-1}}{c_{j+1} - c_j} = 4.66920160910299067185320382047\ldots,$$

where δ is called the *Feigenbaum constant*, cf. [112]. This means that the distances $c_{j+1} - c_j$ form an almost geometrical sequence. For c smaller than the value $c_\infty = \lim_{j \to \infty} c_j \approx -1.401155$ we can obtain chaos, but also finitely many accumulation points. This means that for specific values of c the so-called windows appear, where the chaotic behavior of the solution branches becomes regular, see Fig. 7.7.

7.3 Golden Section and Lucas Numbers

Denote by

$$\alpha = \frac{1 + \sqrt{5}}{2} \quad \text{and} \quad \beta = \frac{1 - \sqrt{5}}{2} \tag{7.4}$$

the two roots of the characteristic equation[1]

$$x^2 - x - 1 = 0$$

corresponding to the difference Eq. (7.1). The number

$$\alpha = 1.6180339887\ldots$$

---

[1] The solution of the linear difference equations of the kth order is described, for example, in Henrici [143, p. 213].

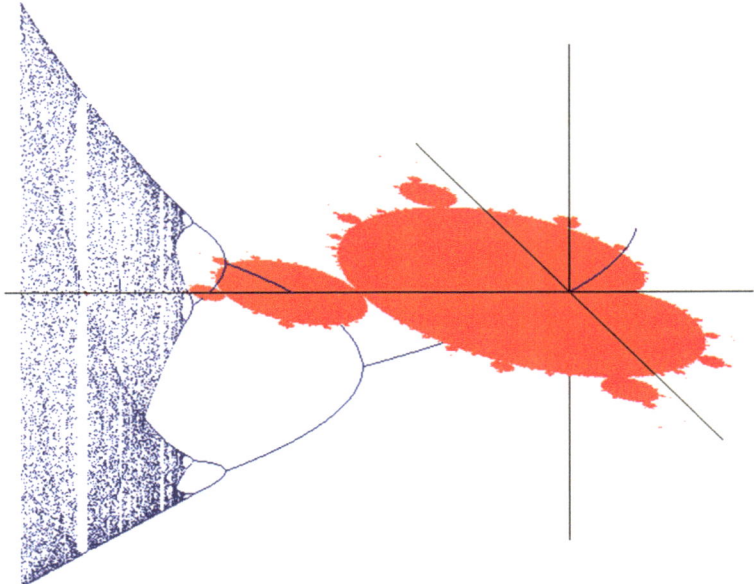

Fig. 7.7 Connection between the Mandelbrot set $\mathcal{M}$ and the chaotic behavior of the dynamical system (7.3)

is called the *golden section* or *golden ratio* (or sectio aurea in Latin). Note that the root β is the same (except for its sign) which appears in the Euclidean construction of the regular pentagon (see [199, p. 198]). The constants

$$\alpha = \alpha^2 - 1 \quad \text{and} \quad \beta = -\frac{1}{\alpha}$$

play an important role in the analysis of Fibonacci numbers because of the general formula

$$K_n = \frac{\alpha^n - \beta^n}{\alpha - \beta} = \frac{\alpha^n - \beta^n}{\sqrt{5}} \qquad \text{for all } n = 0, 1, 2, \ldots \qquad (7.5)$$

This relation is called the *Binet formula*. We have $\alpha - \beta = \sqrt{5}$ and $\alpha + \beta = 1$. Formula (7.5) was published as early as 1765 by Leonhard Euler. However, in 1843 it was rediscovered by the French mathematician Jacques Philippe Marie Binet (1786–1856). Notice that (7.5) generates only nonnegative integers even though it contains several irrational numbers. Moreover, it follows from (7.5) that K_n is the closest integer to the number $\alpha^n/\sqrt{5}$, since $|\beta| = 1/\alpha < 1$.

Moreover, since $\alpha^2 = \alpha + 1$, we have $\alpha^3 = \alpha^2 + \alpha = 2\alpha + 1$, and similarly, $\alpha^4 = 3\alpha + 2$, $\alpha^5 = 5\alpha + 3$, etc. By induction, we find that

$$\alpha^n = K_n\alpha + K_{n-1} \quad \text{for } n = 1, 2, \ldots$$

The golden ratio α represents the basic aesthetic ratio since antiquity. A rectangle whose ratio of lengths of sides yields the golden ratio is considered to be the most beautiful. Such a rectangle is created, for example, as the convex hull of two opposite edges of the regular icosahedron (see Fig. 12.3). It should be noted that the authors Livio [236] and Markowsky [259] dispute the assertion that there is general consensus that the rectangle whose ratio of length to width equals α is judged to be the most beautiful. The ratio of the lengths of a diagonal to a side of the regular pentagon is α; the diagonals themselves are divided in the ratio of the golden section, too. The radius of the circumscribed circle to the regular decagon to the length of its side is also in the ratio α. Johannes Kepler (1571–1630) described a special polyhedron which is the intersection of five differently rotated cubes with the same center. Each of its thirty congruent faces is a rhombus whose ratio of its lengths of diagonals is α. The golden section appears also in four-dimensional regular polytopes, see Table 12.1.

The golden section has a number of interesting algebraic expressions, for instance,

$$\alpha = \sqrt{1 + \sqrt{1 + \sqrt{1 + \sqrt{1 + \cdots}}}}.$$

Every real number y such that $0 < y \le 1$ can be written in the form of a continued fraction

$$y = \cfrac{1}{a_1 + \cfrac{1}{a_2 + \cfrac{1}{a_3 + \cdots}}}, \quad a_i \in \mathbb{Z},$$

which is usually denoted by $[a_1, a_2, a_3, \ldots]$. For the reciprocal value of the golden section we have

$$\alpha^{-1} = [1, 1, 1, \ldots] = \frac{1}{2}(\sqrt{5} - 1).$$

Remark By means of

$$\sigma^{-1} = [2, 2, 2, \ldots] = \sqrt{2} - 1 \quad \text{and} \quad \zeta^{-1} = [3, 3, 3, \ldots] = \frac{1}{2}(\sqrt{13} - 3)$$

we can analogously define the reciprocal values of the *silver section* $\sigma = \sqrt{2} + 1$ and the *bronze section* $\zeta = \frac{1}{2}(\sqrt{13} + 1)$, respectively.

There is an important integer sequence accompanying the sequence K_n.

$$L_n = \alpha^n + \beta^n = \alpha^n + (-1)^n \alpha^{-n} \quad \text{for all } n = 0, 1, 2, \ldots, \tag{7.6}$$

called the *Lucas sequence*. Its terms called the *Lucas numbers* also satisfy recurrence (7.1), if we replace K_n by L_n, i.e.,

$$L_{n+2} = L_{n+1} + L_n \quad \text{for all } n = 0, 1, 2, \ldots \tag{7.7}$$

with initial values $L_0 = 2$ and $L_1 = 1$. Several of its first terms are

$$2, \ 1, \ 3, \ 4, \ 7, \ 11, \ 18, \ 29, \ 47, \ 76, \ldots$$

The Lucas sequence is sometimes called the *companion sequence* of the Fibonacci sequence.

7.4 Equalities Containing Fibonacci Numbers

In 1680 Giovanni Domenico Cassini, a French mathematician and astronomer of Italian origin, found that (see Koshy [175, p. 74])

$$K_{n+1} K_{n-1} - K_n^2 = (-1)^n. \tag{7.8}$$

This identity for $n = 6$ is the basis of a remarkable geometric paradox given in Fig. 7.8 (see also the remark at the end of this section).

From relations (7.1), (7.5), (7.6), and (7.7) we can derive a number of interesting identities:

$$
\begin{aligned}
K_{n+1} + K_{n-1} &= L_n, \\
L_{n+1} + L_{n-1} &= 5K_n, \\
K_n^2 - K_{n+k} K_{n-k} &= (-1)^{n+k} K_k^2, \\
L_n^2 - L_{n+1} L_{n-1} &= (-1)^n 5, \\
L_n^2 - 5K_n^2 &= (-1)^n 4, \\
K_{n+1}^2 - K_{n-1}^2 &= K_n L_n = K_{2n}, \\
K_{n+1}^2 + K_n^2 &= K_{2n+1}, \\
K_m K_{n+1} + K_{m-1} K_n &= K_{m+n}.
\end{aligned}
$$

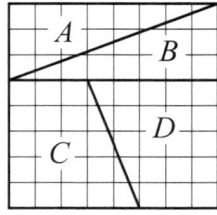
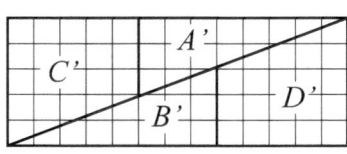

Fig. 7.8 The 8×8 square and the 13×5 rectangle which have different areas are partitioned into the domains A, B, C, D and A', B', C', D', respectively, that have seemingly the same areas

Fig. 7.9 The geometric interpretation of the relation $\sum\limits_{i=1}^{n} K_i^2 = K_n K_{n+1}$

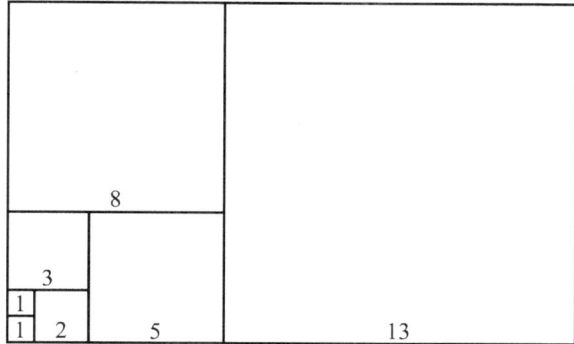

The equalities on the last three lines can be used recursively for calculating Fibonacci numbers with large indices. By induction one may prove another elegant relation

$$\begin{pmatrix} K_{n+1} & K_n \\ K_n & K_{n-1} \end{pmatrix} = \begin{pmatrix} 1 & 1 \\ 1 & 0 \end{pmatrix}^n, \quad n = 1, 2, \ldots$$

Notice that (7.8) actually expresses the equality of determinants of these matrices. More than a hundred similar relationships between the Fibonacci and Lucas numbers can be found in [145, 175, 404]. Here are some more (see also Fig. 7.9):

$$\sum_{i=1}^{n} K_i = K_{n+2} - 1, \qquad \prod_{i=0}^{n-1} L_{2^i} = K_{2^n}, \qquad \sum_{i=0}^{n} \binom{n}{i} K_i = K_{2n}.$$

The determinants of the $n \times n$ tridiagonal matrices

$$T(n) = \begin{pmatrix} 3 & 1 & & & \\ 1 & 3 & 1 & & \\ & 1 & 3 & \ddots & \\ & & \ddots & \ddots & 1 \\ & & & 1 & 3 \end{pmatrix}, \qquad U(n) = \begin{pmatrix} 1 & i & & & \\ i & 1 & i & & \\ & i & 1 & \ddots & \\ & & \ddots & \ddots & i \\ & & & i & 1 \end{pmatrix},$$

where i is the imaginary unit, are Fibonacci numbers. By induction (see [57, 385]) we can derive that

$$\det T(n) = K_{2n+2} = \prod_{k=1}^{n} \left(3 - 2\cos \frac{k\pi}{n+1} \right),$$

$$\det U(n) = K_{n+1} = \prod_{k=1}^{n} \left(1 - 2i\cos \frac{k\pi}{n+1} \right).$$

These factorizations in real and complex arithmetic are further generalized in Seibert and Trojovský [351].

The well-known Russian mathematician Matiyasevich [262, p. 40] disproved by means of Fibonacci numbers Hilbert's tenth problem concerning the solvability of Diophantine equations using a finite number of arithmetic operations (see [157, 261]). In 1985, he derived another surprising relationship (see [263]) between Fibonacci numbers and *Ludolph's number* $\pi = 3.14159\ldots$ which is transcendental (i.e., it is not a solution of an algebraic equation with integer coefficients),

$$\pi = \lim_{n \to \infty} \sqrt{\frac{6\log(K_1 K_2 \cdots K_n)}{\log[K_1, K_2, \ldots, K_n]}},$$

where $[K_1, K_2, \ldots, K_n]$ denotes the least common multiple of the Fibonacci numbers $K_1, K_2, \ldots, K_n$.

Jones in [157] proved that the set of all Fibonacci numbers is identical with the set of all nonnegative values of the following polynomial of the fifth degree in two nonnegative integer variables:

$$p(x, y) = -x^4 y - 2x^3 y^2 + x^2 y^3 + 2xy^4 - y^5 + 2y.$$

For instance, $p(1, 1) = 1$, $p(2, 3) = 3$, $p(8, 13) = 13$.

The Belgian mathematician Eugène Charles Catalan (1814–1894) introduced the so-called *Fibonacci polynomials* using the following recurrence similar to (7.1):

$$f_{n+2}(x) = x f_{n+1}(x) + f_n(x) \qquad \text{for all } n = 0, 1, 2, \ldots,$$

where $f_0(x) = 0$ and $f_1(x) = 1$. At the same time, we see that

$$f_n(1) = K_n.$$

Similarly one may define the Lucas polynomials. Their properties are presented in the monograph [175] by Koshy.

From relations (7.5) and (7.6) we can easily prove that

$$\alpha = \lim_{n \to \infty} \frac{K_{n+1}}{K_n} = \lim_{n \to \infty} \frac{L_{n+1}}{L_n}. \tag{7.9}$$

The golden section α has a number of important applications. For example, it can be used to search for the minimum value of a real continuous and generally non-differentiable function J on a closed interval $[a, b] \subset \mathbb{R}$ by the golden section search algorithm, see [205, p. 177]:

For simplicity assume that J is decreasing in the proper subinterval $[a, x^*]$ and increasing on $[x^*, b]$, i.e., J attains its minimum in $x^* \in (a, b)$. The initial conditions of the golden section algorithm are $x_0 := a$, $x_3 := b$, and

Fig. 7.10 Searching for the minimum of the functional J by the golden section algorithm

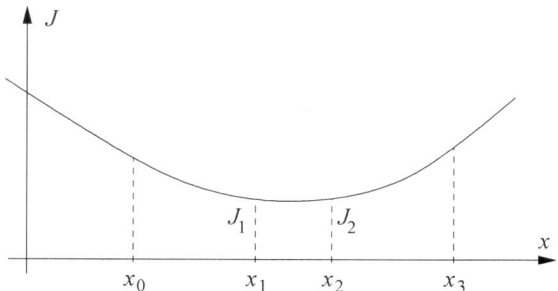

$$x_1 := \gamma x_3 - \beta x_0, \quad J_1 := J(x_1),$$
$$x_2 := \gamma x_0 - \beta x_3, \quad J_2 := J(x_2),$$

where

$$\beta = \frac{1 - \sqrt{5}}{2} = -\frac{1}{\alpha} \quad \text{and} \quad \gamma = \beta^2 = \frac{3 - \sqrt{5}}{2}.$$

If $J_1 < J_2$, we set (cf. Fig. 7.10)

$$x_3 := x_2, \quad x_2 := x_1, \quad x_1 := x_0 + x_3 - x_2, \quad J_2 := J_1, \quad J_1 := J(x_1);$$

otherwise

$$x_0 := x_1, \quad x_1 := x_2, \quad x_2 := x_0 + x_3 - x_1, \quad J_1 := J_2, \quad J_2 := J(x_2).$$

We repeat this iteration process until the difference $x_3 - x_0$ is small enough. The point x^*, in which the minimum of the functional J is attained, will be inside the closed interval $[x_0, x_3]$.

The ratios of the lengths of the respective segments during this iteration process will always be equal to α (see (7.4)). For instance, we see that

$$\frac{x_3 - x_0}{x_3 - x_1} = \frac{x_3 - x_0}{x_3 - \beta^2 x_3 + \beta x_0} = \frac{x_3 - x_0}{-\beta x_3 + \beta x_0} = \alpha.$$

Analogously we can show that

$$\frac{x_3 - x_0}{x_2 - x_0} = \frac{x_2 - x_0}{x_1 - x_0} = \frac{x_3 - x_1}{x_1 - x_0} = \alpha.$$

Due to this property it can be shown that the number of evaluations of the function J (which can sometimes be computationally very expensive) is generally the smallest possible.

Fibonacci numbers are also used in game theory, genetics, for sorting algorithms (the Fibonacci sorting algorithm), for the problem of banking deposits, for special solutions of electrical circuits, etc. These applications can be found, for example, in the books [145, 175, 350]. Note that the chromatic scale on the piano has 13 keys, 8 of which are white and $2 + 3 = 5$ black.

Remark Recall the geometric paradox of Fig. 7.8. We would like to inform you that the smallest angle in the triangle A is $\arctan\frac{3}{8} = 20.56°$, while at the triangle A' it is $\arctan\frac{5}{13} = 21.04°$ (cf. also (7.9)). This small difference is almost invisible in Fig. 7.8. The author of this paradox is the mathematician C. L. Dodgson, known as Lewis Carroll, the author of the book *Alice's Adventures in Wonderland*.

7.5 The Most Beautiful Theorems on Fibonacci and Lucas Numbers

First, let us mention Zeckendorf's completeness theorem.

Theorem 7.1 (Zeckendorf) *Every positive integer can be uniquely expressed as the sum of different Fibonacci numbers K_i for $i \geq 2$, where no two indices i are consecutive.*

Its proof is presented in Brown [46]. Based on Zeckendorf's Theorem 7.1, we can write each positive integer as a sum of Fibonacci numbers. Relation (7.1) guarantees that a representation can be given that does not contain two consecutive Fibonacci numbers.

Writing 10^{n-2} instead of K_n for any integer $n \geq 2$, we can assign to all non-negative integers the following strings consisting of zeros and ones

$$0 \mapsto \underline{0}, \quad 1 \mapsto \underline{1}, \quad 2 \mapsto 1\underline{0}, \quad 3 \mapsto 10\underline{0}, \quad 4 = 3 + 1 \mapsto 10\underline{1}, \quad 5 \mapsto 100\underline{0},$$
$$6 = 5 + 1 \mapsto 100\underline{1}, \quad 7 = 5 + 2 \mapsto 101\underline{0}, \quad 8 \mapsto 1000\underline{0}, \quad 9 = 8 + 1 \mapsto 1000\underline{1},$$
$$10 = 8 + 2 \mapsto 1001\underline{0}, \quad 11 = 8 + 3 \mapsto 1010\underline{0}, \quad 12 = 8 + 3 + 1 \mapsto 1010\underline{1}, \ldots$$

Notice that from the last (underlined) digits we can compile a string that is inverse to the golden string (7.2), i.e., zeros and ones are swapped. Is this not a small miracle?

Theorem 7.2 *Every positive integer can be expressed as the sum of different Lucas numbers L_i for $i \geq 0$.*

The proof can be found e.g. in Hoggatt [145, p. 73]. It is not difficult to show (see Koshy [175, p. 136], Vorobiev [405, p. 57]) that every two consecutive Fibonacci numbers are coprime, i.e.

$$(K_{n+1}, K_n) = 1.$$

This result can be generalized as follows:

Theorem 7.3 *For arbitrary positive integers m and n we have*

$$(K_m, K_n) = K_{(m,n)},$$
$$K_m \mid K_n \iff m \mid n, \quad m \neq 2, \tag{7.10}$$

and

$$L_m \mid L_n \iff \exists k \in \{1, 2 \ldots\} : n = (2k - 1)m, \quad m > 1. \tag{7.11}$$

Proofs of individual statements can be found, for example, in [145, pp. 39–40], [175, pp. 198–200], [405, pp. 51–58].

Theorem 7.4 *The number K_{mn} is divisible by $K_m K_n$ if and only if $(m, n) \in \{1, 2, 5\}$.*

Theorem 7.5 *Let $m > 1$ and $n > 1$. Then the number L_{mn} is divisible by $L_m L_n$ if and only if m and n are odd coprime integers.*

Proofs of the previous two theorems can be found in Jarden [153, pp. 1 and 3]. Another remarkable statement is given in Koshy [175, p. 422]:

Theorem 7.6 *If $p \geq 5$ is a prime, then $K_{p^2} \equiv p^2$ (mod 100), i.e., the last digits of the numbers p^2 and K_{p^2} are the same.*

The following statement comes from the prominent French mathematician Gabriel Lamé (1795–1870), see [175, pp. 138–140].

Theorem 7.7 *Let $2 \leq k \leq m$. The number of steps of the Euclidean algorithm needed to calculate (k, m) is at most five times the number of digits of k. Moreover, we have $k \geq K_{n+1}$ and $m \geq K_{n+2}$, if the Euclidean algorithm requires n steps.*

The Euclidean algorithm (see Sect. 1.6) to calculate the greatest common divisor requires the most steps if k and m are two consecutive Fibonacci numbers. Thus, if $n \geq 2$, then to determine (K_{n+1}, K_{n+2}) exactly n divisions are needed.

Let us finish this section by one more theorem. The proof that we will show is substantially simpler than the original proof by Halton, see [134].

Theorem 7.8 (Halton) *Let $n \geq m$ and let r be the remainder when dividing K_n by the number K_m. Then either r or $K_m - r$ is a Fibonacci number.*

Proof By (7.5) and (7.6),

$$K_n = \frac{\alpha^n - \beta^n}{\alpha - \beta} = \frac{\alpha^m - \beta^m}{\alpha - \beta}(\alpha^{n-m} + \beta^{n-m})$$
$$+ \operatorname{sgn}(n - 2m)(\alpha\beta)^{\min(m,n-m)} \frac{\alpha^{|n-2m|} - \beta^{|n-2m|}}{\alpha - \beta}$$
$$= K_m L_{n-m} + \operatorname{sgn}(n - 2m)(-1)^{\min(m,n-m)} K_{|n-2m|}.$$

If $|n - 2m| < m$, then $r = K_{|n-2m|}$ or $K_m - r = K_{|n-2m|}$. If $|n - 2m| \geq m$, then we divide $K_{|n-2m|}$ by the number K_m and continue as before until the remainder is less than K_m. □

7.6 Primes in Fibonacci and Lucas Sequences

There are over 50 known values of n for which K_n is prime (see [434]). The first 12 values of n for which K_n is prime are

$$n = 3, 4, 5, 7, 11, 13, 17, 23, 29, 43, 47, 83.$$

In addition, there exist also more than 50 known values of m for which L_m is prime (see [435]). The first 13 values of m for which L_m is prime are

$$m = 0, 2, 4, 5, 7, 8, 11, 13, 16, 17, 19, 31, 37.$$

It is conjectured that K_n and L_m are prime for infinitely many values of n and m (see Ribenboim [323, pp. 362–364]). Since $K_1 = K_2 = 1$ and $K_n > 1$ for $n > 2$, it follows from equivalence (7.10) that if K_n is prime, then $n = 4$ or n is also a prime. Similarly, it further follows from equivalence (7.11) that if L_m is prime, then $m = 0$ or m is prime or m is a power of 2.

On the other hand, $K_{19} = 37 \cdot 113$ is composite, even though its index is a prime. Drobot [96] proved the following theorem giving simple conditions for K_p to be composite.

Theorem 7.9 (Drobot) *Suppose that $p > 7$ is a prime, $p \equiv 2$ or 4 (mod 5), and $2p - 1$ is also a prime. Then $2p - 1 \mid K_p$ and $K_p > 2p - 1$ and thus composite.*

A similar statement for Mersenne primes follows from Theorem 4.24, namely, if p and $2p + 1$ are primes and $p \equiv 3$ (mod 4), then $2p + 1 \mid M_p$ and M_p is composite. Somer generalized Theorem 7.9 in [366] from the Fibonacci sequence to all second-order linear recurrences $u(a, b)$ defined by $u_{n+2} = au_{n+1} + bu_n$ with initial terms $u_0 = 0$ and $u_1 = 1$.

Example Making use of Appendix A.3 of Koshy [175], we see that

$$37 \mid K_{19} = 4181 = 37 \cdot 113,$$
$$73 \mid K_{37} = 24157817 = 73 \cdot 149 \cdot 2221,$$
$$157 \mid K_{79} = 14472334024676221 = 157 \cdot 92180471494753,$$

and

$$193 \mid K_{97} = 83621143489848422977 = 193 \cdot 389 \cdot 3084989 \cdot 361040209.$$

Now we present the following two complementary results to Drobot's Theorem 7.9 providing values of n for which n is equal to a power of 2 or a prime and L_n is composite.

Theorem 7.10 *Suppose that $p \equiv 3$ (mod 4) and $2^p - 1$ is a Mersenne prime. Then $2^p - 1 \mid L_{2^{p-1}}$. Moreover, $2^p - 1 < L_{2^{p-1}}$ when $p > 3$.*

Theorem 7.11 *Let $p \equiv 29$ (mod 30) and $2p + 1$ be primes. Then $2p + 1 \mid L_p$ and $2p + 1 < L_p$.*

These two theorems are proved in [376].

Example By the use of Appendix A.4 [175], we see that

$$2^7 - 1 = 127 \mid L_{2^6} = L_{64} = 23725150497407 = 127 \cdot 186812208641$$

and

$$59 \mid L_{29} = 1149851 = 59 \cdot 19489.$$

7.7 Prime Factors of the Fibonacci Numbers

The Fibonacci numbers

$$2, \ 3, \ 5, \ 13, \ 89, \ 233, \ 1597, \ldots$$

are primes. However, as stated earlier, up to now it is not known whether their number is infinite. Also the Lucas numbers

$$2, \ 3, \ 7, \ 11, \ 29, \ 47, \ 199, \ 521, \ldots$$

are primes and we also do not know if their number is infinite. A table of all prime factors of the Fibonacci and Lucas numbers for $n \leq 385$ is contained e.g. in Jarden [153].

An interesting observation was made by M. J. Zerger, see Koshy [175, p. 11]. He noticed that the product of the first seven prime numbers is equal to

$$2 \cdot 3 \cdot 5 \cdot 7 \cdot 11 \cdot 13 \cdot 17 = K_7 K_8 K_9 K_{10}. \tag{7.12}$$

In the next theorem (see [203]), we prove that this interesting property cannot be "extended".

Theorem 7.12 *The largest product of consecutive primes that is equal to the product of at least three consecutive Fibonacci numbers, is $K_7 K_8 K_9 K_{10}$.*

Proof Let n be a positive integer. Denote by $z(n)$ the smallest positive integer j such that $n \mid K_j$. It is known that such $z(n)$ always exists, see Ribenboim [323, p. 60]. First we show that

$$n \mid K_m \iff z(n) \mid m. \tag{7.13}$$

$\Rightarrow$: Let $n \mid K_m$ and set $k = (z(n), m)$. Then by Theorem 7.3 we have $(K_{z(n)}, K_m) = K_k$ and $k \leq z(n)$. If $k < z(n)$, then $n \mid K_k$, since $n \mid K_{z(n)}$ and $n \mid K_m$. However, this contradicts the fact that $z(n)$ is the smallest positive integer j such that $n \mid K_j$. Therefore, $k = z(n)$ and $z(n) \mid m$.

$\Leftarrow$: Let $z(n) \mid m$. Then by (7.10) we have $K_{z(n)} \mid K_m$. Since n divides by definition the number $K_{z(n)}$, we obtain $n \mid K_m$.

Now we prove our statement. First note that of three or more consecutive integer indices, at least one of these is a multiple of 3. Hence, if $3 \mid n$, then by the equivalence (7.10) we have $K_3 \mid K_n$. Thus K_n is even, since $K_3 = 2$, and the product of the left-hand side has to begin with 2.

If the example is larger than that of Zerger's (7.12), then 19 must appear on the left-hand side. The smallest Fibonacci number divisible by 19 is $K_{18} = 2584$. Hence, by (7.13) the product of the Fibonacci numbers on the right-hand side has to contain an index n divisible by $z(19) = 18$, which is a multiple of 6. However, by (7.10) we have that $K_6 \mid K_n$ and also $K_6 = 8 = 2^3$. Consequently, the product on the right-hand side can no longer be a product of consecutive primes. $\square$

Theorem 7.13 *For any $n \in \mathbb{N}$ the sequence of remainders $(K_i \pmod{n})_{i=0}^{\infty}$ is periodic.*

Proof The sequence $(K_i \pmod{n})_{i=0}^{\infty}$ contains at most n different remainders. Thus, by Dirichlet's pigeonhole principle (Theorem 1.5) there exists $j \in \{0, 1, \ldots, n^2\}$ such that the ordered pair $\langle K_j, K_{j+1} \rangle$ repeats in the sequence of the Fibonacci numbers modulo n, since there are n^2 such pairs. The sequence $(K_i \pmod{n})_{i=0}^{\infty}$ is therefore periodic for $i \geq j$.

By contradiction we show that $j = 0$. Let $j \geq 1$ be the smallest number such that the sequence $(K_i \pmod{n})_{i=0}^{\infty}$ is periodic for $i \geq j$ and let ℓ be the period of this sequence. Then $K_i \equiv K_{i+\ell} \pmod{n}$ for all $i \geq j$. In particular, $K_j \equiv K_{j+\ell} \pmod{n}$ and $K_{j+1} \equiv K_{j+1+\ell} \pmod{n}$. But then

$$K_{j-1} = K_{j+1} - K_j \equiv K_{j+1+\ell} - K_{j+\ell} = K_{j-1+\ell} \pmod{n}$$

and $(K_i \pmod{n})_{i=0}^{\infty}$ is thus periodic also for $i \geq j - 1$, which is a contradiction. Hence, $j = 0$ and the sequence $(K_i \pmod{n})_{i=0}^{\infty}$ is periodic from the very beginning. $\square$

Let n be a positive integer. Since $K_0 = 0$ and (K_i) is periodic modulo n starting with $i = 0$, we see that there exists a positive integer i such that $K_i \equiv 0 \pmod{n}$. Thus every positive integer n is a divisor of some member of the Fibonacci sequence. We will see below that the Lucas numbers (L_n) have a quite different property with respect to prime divisors.

Moreover, D. D. Wall proved that the length of the period of the sequence of remainders $(K_i \pmod{n})_{i=0}^{\infty}$ is always even for $n > 2$. For $n = 2, 3, 4$ the lengths of the minimal periods are 3, 8, and 6, respectively. One can easily verify that for $n = 24$ the length of the minimal period is also 24.

A prime factor p of K_n is called *primitive* if p does not divide K_m for any positive integer $m < n$. For example, $K_{12} = 144 = 2^4 3^2$ has no primitive prime factor, because $2 \mid K_3$ and $3 \mid K_4$. Robert Daniel Carmichael (see [64]) proved the following very general theorem. It is so nice that it is destined to be rediscovered (see e.g. Boase [34]).

Theorem 7.14 (Carmichael) *The number K_n has at least one primitive prime divisor for all $n > 12$.*

Let us point out that primitive divisors must be relatively large. If $p \neq 5$ is a primitive divisor of K_n, then $p \equiv 1 \pmod{n}$ for $p \equiv 1, 4 \pmod{5}$ and $p \equiv -1 \pmod{n}$ for $p \equiv 2, 3 \pmod{5}$, see Theorem 7.19 below. Therefore, $p \geq n - 1$.

Carmichael's Theorem 7.14 can be generalized as follows. Consider the general second order recurrence given by

$$U_{n+2} = rU_{n+1} + sU_n,$$

where r and s are integers, $rs \neq 0$, $(r, s) = 1$, and $U_0 = 0$, $U_1 = 1$. Suppose further that if γ and δ are the roots of the equation

$$x^2 - rx - s = 0,$$

then γ/δ is not a root of unity. Bilu et al. [32] proved that U_n has a primitive prime divisor for any $n > 30$.

It was also proved by Zsigmondy [425] that if $a > b$ are positive coprime integers, then $a^n - b^n$ has a primitive prime divisor for $n > 1$ unless

(i) $n = 2$ and $a + b = 2^k$ for some $k \in \mathbb{N}$, or
(ii) $n = 6$, $a = 2$, and $b = 1$.

Wayne McDaniel proved that K_n for $n \neq 1, 2, 3, 4, 6, 8, 12, 16, 24, 32$, or 48 always has at least one prime factor of the form

$$p \equiv 1 \pmod{4},$$

see [266]. We further note that it follows from Theorem 7.24 below that if $n > 1$ is odd, then any odd prime divisor of K_n is congruent to 1 modulo 4. The probability that K_n has a prime factor of the form $p \equiv 3 \pmod{4}$ is exactly $\frac{1}{2}$ (see Luca [241]). On the other hand, the density of the set of prime factors of Lucas numbers is only $\frac{2}{3}$ (for details see [221]). This means that

$$\lim_{m \to \infty} \frac{|\{n \in \mathcal{P}; \ n \leq m\}|}{\pi(m)} = \frac{2}{3},$$

where $\mathcal{P}$ is the set of all primes that divide some L_n and $\pi(m)$ denotes the number of primes p such that $p \leq m$.

7.8   Properties of Digits of Fibonacci and Lucas Numbers

Stewart in [383] deals with the question of the computational complexity to express K_n in any number system with base $b > 1$ for increasing n. Let $s_b(m)$ denote the number of digits of a positive integer m in the number system with base b. Then for any $b > 1$ there exists a constant $c = c(b)$ such that if $n > 15$ then

$$s_b(K_n) > c\frac{\log n}{\log \log n}.$$

Recall (see Sect. 4.9) that a positive integer m is called a *palindrome in the number system with base* b, if the string of its digits is the same when read from the left to right as from the right to left. For instance $K_{10} = 55$, $L_5 = 11$, and $L_{25} = 167761$ are palindromes in the decimal system. Denoting by $\mathcal{P}_b$ the set of positive integers n for which K_n or L_n is a palindrome in base $b > 1$, then the asymptotic density of the set $\mathcal{P}_b$ is zero (see Luca [242]), i.e.,

$$\lim_{m \to \infty} \frac{|\{n \in \mathcal{P}_b;\ n \le m\}|}{m} = 0.$$

However, this result is not so strong that it can be deduced that the sum of the reciprocals of the numbers from $\mathcal{P}_b$ is convergent.

7.9   Further Properties of the Fibonacci Numbers

In 1974, Good in [121] found that

$$\sum_{n=0}^{\infty} \frac{1}{K_{2^n}} = \frac{7 - \sqrt{5}}{2}.$$

This irrational number is algebraic, because it is the root of a polynomial with integer coefficients.

In 1980, Erdős and Graham in the monograph [104, p. 64] ask the question whether the sums

$$\sum_{n=0}^{\infty} \frac{1}{K_{2^n+1}} \quad \text{and} \quad \sum_{n=0}^{\infty} \frac{1}{L_{2^n}}$$

are irrational numbers. Seven years later this was positively answered in Badea [16].

It would be difficult to recall all the series containing the Fibonacci numbers that are known to be transcendental. Therefore, we restrict ourselves only to a few more examples:

$$\sum_{n=1}^{\infty} \frac{K_n}{n2^n}, \quad \sum_{n=1}^{\infty} \frac{1}{nK_{2^n}}, \quad \sum_{n=1}^{\infty} \frac{2+(-1)^n}{K_{2^n}}, \quad \sum_{n=1}^{\infty} \frac{1}{n!K_{2^n}}.$$

However, proofs of these results require a very difficult technique from the theory of transcendental numbers. There is another deep question in Erdős and Graham [104], namely, whether

$$\sum_{n=1}^{\infty} \frac{1}{K_n}$$

is an irrational number. This was first demonstrated in André-Jeannin [8] using Padé approximations, but its proof was far from being simple. A more elementary proof later appeared in Duverney [99].

For β from (7.4) and all complex numbers z such that $|z| < |\beta|$, the following equality is valid (see Schroeder [350, p. 225])

$$\sum_{i=1}^{\infty} K_i z^i = \frac{z}{1 - z - z^2}.$$

The rational function on the right-hand side is called a *generating function* of the Fibonacci sequence. In particular, for $z = \frac{1}{2}$ the above power series converges and

$$\sum_{i=1}^{\infty} \frac{K_i}{2^i} = 2.$$

7.10 Diophantine Equations

There are several works that study how to express Fibonacci numbers in the form $P(x)$, where P is a polynomial of at least the second degree with rational coefficients and x is an integer. Most Diophantine equations $K_n = P(x)$ have only finitely many integer solutions (n, x), although there are exceptions. For instance, if $n = 3m$ and m is odd, then

$$K_n = K_{3m} = K_m(5K_m^2 - 3).$$

In particular, for $P(x) = x(5x^2 - 3)$ the Diophantine equation $K_n = P(x)$ has infinitely many integer solutions (n, x), i.e., $(n, x) = (3m, K_m)$ for m odd. In fact, for $n = km$ we have

$$K_n = K_{km} = P_{k,i}(K_m), \qquad i \in \{0, 1\}, \tag{7.14}$$

where $P_{k,0}$ and $P_{k,1}$ are two polynomials with rational coefficients of degree k, and the relation (7.14) holds when $m \equiv i$ (mod 2). In [286], it is shown that these polynomials are basically the only exception.

The largest positive integer n such that $K_n = x(x+1)/2$ for some integer x is $n = 10$ (see Luo [248]). In other words, $K_{10} = 55$ is the largest Fibonacci triangular number. The largest integer n such that $K_n = x^2$ for some integer x is $n = 12$ (see [74, 75, 237]). So it is natural to ask whether all powers are in the Fibonacci sequence. Due to (7.10) it is enough to consider only prime exponents. There was a conjecture that $K_{12} = 144$ is the largest such power. H. London and R. Finkelstein proved it for the exponent 3, i.e., 8 is the largest cube in the Fibonacci sequence (see [238]), and J. McLaughlin proved it for exponents 5, 7, 11, 13, and 17 (see [268]). Finally, in 2004 the conjecture was proved for all exponents using modular techniques by Andrew Wiles to resolve Fermat's Last Theorem (see [50, 51]).

Fibonacci numbers appear quite unexpectedly when solving some Diophantine equations. Let us illustrate this by the following example. A *Diophantine quadruple* is a set of four rational numbers such that the product of any two numbers of this quadruple increased by one is the square of a rational number. Diophantus of Alexandria found one such quadruple $\langle \frac{1}{16}, \frac{33}{16}, \frac{68}{16}, \frac{105}{16} \rangle$. The first to construct a similar solution with integer values

$$\langle 1, 3, 8, 120 \rangle,$$

was Pierre de Fermat. The fact that the first three members of this quadruple are Fibonacci numbers, enabled us to reveal a much more general solution. For $k \geq 1$

$$\langle K_{2k}, K_{2k+2}, K_{2k+4}, 4K_{2k+1}K_{2k+2}K_{2k+3} \rangle$$

is always a Diophantine quadruple. If x is such that $xK_i + 1$ is a square for all $i \in \{2k, 2k+2, 2k+4\}$, then (see Dujella [97])

$$x = 4K_{2k+1}K_{2k+2}K_{2k+3}.$$

D. A. Lind sought a solution to the Diophantine equation for binomial coefficients (see [234])

$$\binom{n}{k} = \binom{n-1}{k+1} \tag{7.15}$$

and he surprisingly found an infinite class of solutions

$$\langle n, k \rangle = \langle K_{2i}K_{2i+1}, K_{2i-2}K_{2i+1} \rangle$$

for $i = 1, 2, \ldots$ There is a conjecture that the Diophantine equation

$$\binom{n}{k} = \binom{m}{\ell} \qquad \text{for all } n > m, \ 2 \leq k \leq n/2, \ 2 \leq \ell \leq m/2$$

has only finitely many integer solutions $\langle n, k, m, \ell \rangle$ except for those which form the above mentioned infinite class of solutions of (7.15) in Fibonacci numbers.

Example The solutions of (7.15) with smallest n are

$$\binom{2}{0} = \binom{1}{1} = 1, \; \binom{15}{5} = \binom{14}{6} = 3003, \; \binom{104}{39} = \binom{103}{40}, \; \binom{714}{272} = \binom{713}{273}.$$

Note that there are infinitely many trivial solutions

$$\binom{n}{0} = \binom{m}{0} = 1$$

and

$$\binom{m}{k} = \binom{\binom{m}{k}}{1}.$$

Therefore, we assumed that $k, \ell \geq 2$.

Finally, we mention Fibonacci numbers of other forms. In [240, p. 1391] by Florian Luca, it is proved that $K_{12} = 3!4!$ is the largest Fibonacci number which is the product of factorials.

It can also be shown that

$$K_1 K_2 K_3 K_4 K_5 K_6 K_8 K_{10} K_{12} = 11!$$

is the largest factorial which is the product of different Fibonacci numbers. Note that there are only finitely many Fibonacci numbers that can be expressed as the sum of at most k factorials for any fixed k, but to establish such a sum for $k \geq 3$ is computationally very difficult.

7.11 A Trick with the Number 11

Let us choose any two positive integers and let us create a sequence similar to the Fibonacci sequence by means of (7.1). If we choose e.g. 5 and 6, then the corresponding sequence will be

$$5, \; 6, \; 11, \; 17, \; 28, \; 45, \; \textbf{73}, \; 118, \; 191, \; 309, \ldots$$

In this special case, we see that the sum of the first ten numbers is equal to eleven times the seventh number (cf. Livio [236, p. 104]),

$$5 + 6 + 11 + 17 + 28 + 45 + 73 + 118 + 191 + 309 = 803 = 11 \cdot 73.$$

So we can ask: *Which number do we get when the sum of the first ten numbers formed similarly from any two initial positive integers is divided by the seventh number?* The answer will always be 11. This follows from the following theorem.

Theorem 7.15 *Let* $(f_i)_{i=1}^{10}$ *be a sequence, where* f_1 *and* f_2 *are arbitrary positive integers and* $f_{i+2} = f_{i+1} + f_i$. *Then*

$$\frac{1}{f_7} \sum_{i=1}^{10} f_i = 11.$$

Proof Put $m = f_1$ and $n = f_2$. Then we see that

$$f_1 = m,$$
$$f_2 = n,$$
$$f_3 = m + n,$$
$$f_4 = m + 2n,$$
$$f_5 = 2m + 3n,$$
$$f_6 = 3m + 5n,$$
$$f_7 = 5m + 8n,$$
$$f_8 = 8m + 13n,$$
$$f_9 = 13m + 21n,$$
$$f_{10} = 21m + 34n,$$

and thus

$$\sum_{i=1}^{10} f_i = 55m + 88n = 11 f_7.$$

This completes the proof.   □

Remark The initial values f_1 and f_2 in Theorem 7.15 can also be arbitrary complex numbers for which $f_7 \neq 0$.

7.12   Generalizations of Fibonacci and Lucas Numbers

The *tribonacci sequence* is a generalization of the Fibonacci sequence, where each term is the sum of the three preceding terms. It begins as follows: 0, 1, 1, 2, 4, 7, 13, 24, 44, 81, 149, ... The tribonacci sequence has several applications in combinatorics. For example, there are 13 ways to toss a fair coin 4 times so that tails never comes up three times in a row.

Further, we show that the solution of equation (7.3) for one specific nonzero value of c is closely related to generalized Lucas numbers. Instead of recurrence (7.7) we shall consider the recurrence (see also the previous section)

$$L_{n+2} = rL_{n+1} + sL_n \quad \text{for all } n = 0, 1, 2, \ldots$$

for given integers r and s, and some initial values L_0 and L_1. We call such numbers *generalized Lucas numbers.*

Theorem 7.16 *The solution of equation (7.3) for $c = -2$ has the form*

$$y_n = a^{2^{n-1}} + a^{-2^{n-1}}, \quad a \in \mathbb{C} \setminus \{0\}. \tag{7.16}$$

Proof From (7.16) we have

$$y_{n+1} = a^{2^n} - 2 + a^{-2^n} - 2 = \left(a^{2^{n-1}} + a^{-2^{n-1}}\right)^2 - 2 = y_n^2 - 2.$$

Consequently, the sequence defined by relation (7.16) fulfills the Eq. (7.3) for $c = -2$.

We still have to show that there are no other solutions. If the initial value of the sequence y_1 is an arbitrary complex number, then there is a nonzero number $a \in \mathbb{C}$ such that $y_1 = a + a^{-1}$. (To see this, it is enough to solve the quadratic equation $a^2 - ay_1 + 1 = 0$, both of whose solutions give $y_1 \in \mathbb{C}$.) Further, we see that the next values of the sequence (7.3) are $y_2 = a^2 + a^{-2}$, $y_3 = a^4 + a^{-4}$, etc. □

The relation (7.16) allows us to explicitly specify the sequence defined recursively as follows:

$$y_{n+1} = y_n^2 - 2, \quad y_1 = 4. \tag{7.17}$$

From (7.16) for $a = 2 + \sqrt{3}$ we get

$$y_n = \left(2 + \sqrt{3}\right)^{2^{n-1}} + \left(2 - \sqrt{3}\right)^{2^{n-1}} = \left\lfloor \left(2 + \sqrt{3}\right)^{2^{n-1}} \right\rfloor + 1,$$

where $\lfloor \cdot \rfloor$ denotes the integer part. Moreover, we have $y_n = L_{2^{n-1}}$, where the generalized Lucas numbers L_k satisfy the recurrence of a particular form (see Riesel [327, p. 131])

$$L_{k+2} = 4L_{k+1} - L_k \tag{7.18}$$

with initial conditions $L_1 = 4$ and $L_2 = 14$ (i.e. $L_k = a^k + a^{-k}$).

Let us point out that some of the generalized Lucas numbers defined by (7.18) also occur in the following Lucas-Lehmer test (cf. Theorem 4.4 and (7.17)), which is used to search for the largest known prime numbers—Mersenne primes. A modified version of Theorem 4.4 therefore reads:

Theorem 7.17 *If $p > 2$ is a prime, then the Mersenne number $M_p = 2^p - 1$ is prime if and only if $M_p \mid L_{2^{p-2}}$.*

7.13 Analogue of Fermat's Little Theorem

There is a nice analogue of Fermat's Little Theorem 2.17 involving the Fibonacci numbers.

Theorem 7.18 *Let p be an odd prime. Then*

$$K_{p-(5/p)} \equiv 0 \pmod{p},$$

where $(5/p)$ denotes the Legendre symbol.

There is a simple proof of this theorem given in [363].

Example Let $p = 11$. Then by Theorem 2.26 and the Law of Quadratic Reciprocity (see Theorem 2.27) we have

$$\left(\frac{5}{p}\right) = \left(\frac{5}{11}\right) = \left(\frac{11}{5}\right) = \left(\frac{1}{5}\right) = 1.$$

We note that, indeed,

$$K_{11-(5/11)} = K_{11-1} = K_{10} \equiv 0 \pmod{11},$$

since $K_{10} = 55$.

Theorem 7.19 *Let $p \neq 5$ be a primitive prime divisor of K_n. Then $p \equiv 1 \pmod{n}$ if $p \equiv 1$ or $4 \pmod{5}$, and $p \equiv -1 \pmod{n}$ if $p \equiv 2$ or $3 \pmod{5}$.*

Proof By the definition of a primitive factor of K_n, $z(p) = n$. Thus by Theorem 7.18 and the proof of Theorem 7.12 we have $n \mid p - (5/p)$. By the Law of Quadratic Reciprocity (see Theorem 2.27), $(5/p) = 1$ if $p \equiv 1$ or $4 \pmod{5}$, and $(5/p) = -1$ if $p \equiv 2$ or $3 \pmod{5}$. The result now follows. $\square$

The odd composite integer N is a *Fibonacci pseudoprime* if

$$K_{N-(5/N)} \equiv 0 \pmod{N},$$

where $(5/N)$ denotes the Jacobi symbol. The smallest two Fibonacci pseudoprimes are 323 and 377.

7.14 Defective Fibonacci Sequence Modulo m

Let $(S_n)_{n=0}^{\infty}$ be a sequence of integers. We say that it is *defective modulo m*, where m is a positive number, if (S_n) does not contain all residues modulo m. We let $\delta(m)$ denote the number of residues not appearing in (S_n) modulo m. We call $\delta(m)$ the *defect* of (S_n) modulo m. The sequence (S_n) is called *nondefective modulo m*, if $\delta(m) = 0$, i.e., all residues appear in (S_n) modulo m. We will consider the cases in which $(S_n)_{n=0}^{\infty}$ is $(K_n)_{n=0}^{\infty}$ or $(L_n)_{n=0}^{\infty}$.

Example Consider the Fibonacci sequence (K_n) modulo 11. Then (K_n) is periodic modulo 11 with a period of 10:

$$(K_n) \equiv (0, 1, 1, 2, 3, 5, 8, 2, 10, 1, 0, 1, 1, \ldots) \pmod{11}.$$

We observe that (K_n) is defective modulo 11 and $\delta(11) = 4$, since the residues 4, 6, 7, and 9 do not appear in (K_n) modulo 11.

The paper [379] considers for a given prime p what are the numbers that can appear in the sequence of residues and what are the numbers that do not appear in the sequence of residues. About 50 years ago, these kinds of residue sequences were considered for use in cryptography. This is no longer the case, because it was shown that such ciphers are fairly easily broken using modern computers. However, they are still of considerable interest to mathematicians and perhaps also in some other applications.

Theorems 7.20 (see Burr [55]) and 7.21 (see Avila and Chen [13]) find all moduli m for which (K_n) and (L_n) are nondefective modulo m.

Theorem 7.20 *The Fibonacci sequence (K_n) is nondefective modulo m if and only if m has one of the following forms:*

$$2 \cdot 5^k, \ 3^j \cdot 5^k, \ 4 \cdot 5^k, \ 6 \cdot 5^k, \ 7 \cdot 5^k, \ 14 \cdot 5^k,$$

where $j \geq 0$ and $k \geq 0$.

Theorem 7.21 *The Lucas sequence (L_n) is nondefective modulo m if and only if m is equal to one of the following numbers:*

$$2, \ 4, \ 6, \ 7, \ 14, \ 3^k,$$

where $k \geq 0$.

Theorems 7.20 and 7.21 were generalized in [377, 378], respectively. Theorem 7.22 below shows (see [380]) that for both the Fibonacci sequence (K_n) and Lucas sequence (L_n) the defect modulo p grows without bound, where p denotes an arbitrary prime. Noting that $\delta(p) \leq p - 1$, Theorem 7.22 also states that there exist primes p for which $\delta(p)/p$ is arbitrarily close to 1.

Theorem 7.22 *Consider the Fibonacci sequence* (K_n) *and the Lucas sequence* (L_n).
Let p *denote an arbitrary prime. Then for each sequence the following hold:*

(i) $\lim_{p \to \infty} \delta(p) = \infty$.
(ii) *For any* $\varepsilon > 0$, *there exist infinitely many primes* p *such that* $\delta(p)/p \geq 1 - \varepsilon$.

7.15 A Theorem on Fibonacci Numbers with Odd Index

The next well-known theorem follows from [56, pp. 267–268] and [138].

Theorem 7.23 *Let* m *and* n *be positive integers. Then* $m^2 + n^2$ *is divisible by a prime* $q \equiv 3 \pmod 4$ *if and only if* $q \mid (m, n)$.

Proof $\Leftarrow$: Suppose that $q \equiv 3 \pmod 4$ and $q \mid (m, n)$. Then clearly, $q \mid m^2 + n^2$.
   $\Rightarrow$: Suppose that $q \equiv 3 \pmod 4$, $q \nmid (m, n)$, and $q \mid (m^2 + n^2)$. Then $q \mid m$ if and only if $q \mid n$. Thus $(q, m) = (q, n) = 1$. By Theorem 1.3, there exist integers x and y such that $nx + qy = 1$. Thus $nx \equiv 1 \pmod q$. Noting that $m^2 + n^2 \equiv 0 \pmod q$, we have that

$$m^2 x^2 \equiv (mx)^2 \equiv -n^2 x^2 \equiv -(nx)^2 \equiv -1 \pmod q,$$

which is impossible by Theorem 2.26(iv), since $(-1/q) = -1$. $\square$

   Theorem 7.24 below appeared as a problem in [362].

Theorem 7.24 *If* $n > 1$ *is odd, then* K_n *has no prime divisor congruent to* 3 *modulo* 4.

Proof Let $n = 2k + 1$, where $k \geq 1$. From Sect. 7.4 we know that

$$K_n = K_{2k+1} = K_k^2 + K_{k+1}^2.$$

Since $(k, k + 1) = 1$, we obtain by Theorems 7.3 and 7.23 that Theorem 7.24 holds. $\square$

7.16 Fibonacci Numbers Divisible by Their Index

It is of interest to note that $12 \mid K_{12} = 144$. Smyth [359] characterized completely all the numbers n for which $n \mid K_n$ and showed that there are infinitely many such numbers.
   Some other numbers n for which $n \mid K_n$ are

$$1, \ 24, \ 36, \ 48, \ 72, \ 144, \ 288, \ 3001 \cdot 25 = 75025, \ldots$$

and
$$5^k \quad \text{for } k \geq 1.$$

Since $z(5^k) = 5^k$ for $k \geq 1$, it is easily seen from (7.13) that if $n \mid K_n$, then $5^k n \mid K_{5^k n}$ for $k \geq 1$. If also follows from (7.10) that if $n \mid K_n$, then

$$K_n \mid K_{K_n}, \quad K_{K_n} \mid K_{K_{K_n}}, \ldots$$

Other results concerning when $n \mid K_n$ are given in [8, 153, 366, 367].

Chapter 8
Further Special Types of Integers

8.1 Polygonal Numbers

For a given $k \in \mathbb{N}$ we define *k-gonal* (*polygonal* or also *figurate*) numbers $P_{k,n}$ as the sum of the first n members of the arithmetic progression $((m-1)(k-2)+1)_{m=1}^{\infty}$ with initial member 1 and difference $k-2$, i.e.,

$$P_{k,n} = 1 + (k-1) + (2k-3) + (3k-5) + \cdots + ((n-1)(k-2)+1). \quad (8.1)$$

They are usually considered only for the case $k \geq 3$, which corresponds to regular polygons.

For instance, for $k = 3, 4, 5$ and $n = 1, 2, 3, 4$ the polygonal numbers $P_{k,n}$ are represented in Fig. 8.1 by regular polygons with k sides and n dots on every side. The dots are thought of as units. The total number of dots is $P_{k,n}$.

Denote by T_n *triangular numbers* $P_{3,n}$. Similarly we can introduce the symbol S_n for *square numbers* $P_{4,n}$ or P_n for *pentagonal numbers* $P_{5,n}$. Applying the formula of the sum of the first n members of the arithmetic progression (8.1), we obtain that

$$P_{k,n} = \frac{n}{2}[1 + (1 + (n-1)(k-2))] = \frac{n}{2}[(n-1)(k-2)+2]. \quad (8.2)$$

Thus, in particular,

$$T_n = \frac{n(n+1)}{2} = 1 + 2 + \cdots + n, \quad (8.3)$$

$$S_n = n^2,$$

$$P_n = \frac{n(3n-1)}{2},$$

and for $n = 1, 2, 3, 4, 5, 6, \ldots$ we get

© The Author(s), under exclusive license to Springer Nature Switzerland AG 2021
M. Křížek et al., *From Great Discoveries in Number Theory to Applications*,
https://doi.org/10.1007/978-3-030-83899-7_8

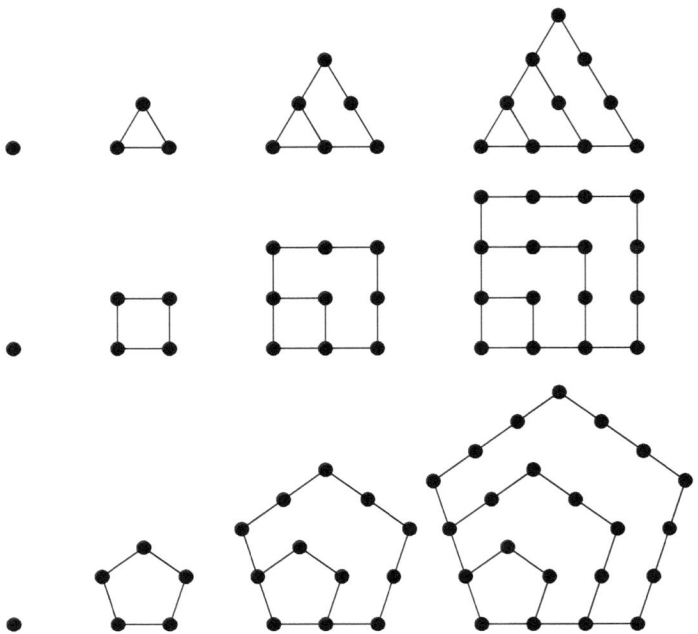

Fig. 8.1 Geometric interpretation of triangular, square, and pentagonal numbers

$$1,\ 3,\ 6,\ 10,\ 15,\ 21,\dots \quad \text{(triangular numbers)},$$
$$1,\ 4,\ 9,\ 16,\ 25,\ 36,\dots \quad \text{(square numbers)},$$
$$1,\ 5,\ 12,\ 22,\ 35,\ 51,\dots \quad \text{(pentagonal numbers)}.$$

From the middle row of Fig. 8.1 we also see that every square number is the sum of consecutive odd numbers:

$$n^2 = 1 + 3 + 5 + \cdots + (2n - 3) + (2n - 1),$$

which can also be directly proved from the formula for the sum of a finite number of terms of an arithmetic progression. Further, we easily find that

$$T_n = n + T_{n-1},$$
$$S_n = n + 2T_{n-1},$$
$$P_n = n + 3T_{n-1},$$

and generally,

$$P_{k,n} = n + (k - 2)T_{n-1}.$$

Fig. 8.2 The geometric
interpretation of
Theorem 8.1

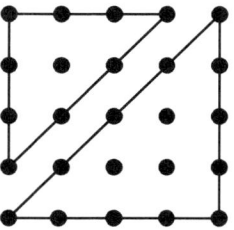

In Sects. 10.2 and 10.3, we show how the triangular numbers (8.3) are related to the
bellworks of the Prague horologe. So let us present some of their further interesting
properties. For instance Luo (in [248]) proved that the largest Fibonacci triangular
number is $K_{10} = 55$. From relation (8.3) for $n = 8$ we can easily verify that the
square number $36 = 6^2$ is at the same time a triangular number. It is known that
there are infinitely many positive integers that are simultaneously a square number
and a triangular number (see Beiler [28, p. 197]). All such numbers A_n are given by
the formula $A_n = a_n^2$, where

$$a_{n+2} = 6a_{n+1} - a_n$$

with initial terms $a_1 = 1$ and $a_2 = 6$. The first few of these numbers are $1 = 1^2 = T_1$,
$36 = 6^2 = T_8$, $1225 = 35^2 = T_{49}$, $41\,616 = 204^2 = T_{288}$.

Furthermore, we extend the definition of polygonal numbers by setting

$$P_{0,0} = P_{1,0} = P_{2,0} = \cdots = 0.$$

Hence,

$$T_0 = 0, \quad S_0 = 0 \quad \text{and} \quad P_0 = 0.$$

The following theorems hold (see Fig. 8.2):

Theorem 8.1 *The sum of two consecutive triangular numbers is the square of a
positive integer.*

Proof For an integer $n \geq 1$ we have

$$T_{n-1} + T_n = \frac{n(n-1)}{2} + \frac{n(n+1)}{2} = n^2$$

and the theorem is proved.   □

The next theorem shows that any k-gonal number is the sum of a $(k-1)$-gonal
number and a triangular number.

Theorem 8.2 *For any $k, n \in \mathbb{N}$ we have*

$$P_{k,n} = P_{k-1,n} + T_{n-1}.$$

Proof From (8.2) and (8.3) it follows that

$$P_{k,n} = \frac{n}{2}(n(k-2) + 4 - k) = \frac{n}{2}(n(k-3) + 5 - k) + \frac{n}{2}(n-1)$$
$$= P_{k-1,n} + T_{n-1},$$

which completes the proof.   □

Theorem 8.3 *For any* $k, n \in \mathbb{N}$ *we have*

$$8(k-2)P_{k,n} + (k-4)^2 = (2kn - 4n - k + 4)^2.$$

Proof Using (8.2), we arrive at

$$8(k-2)P_{k,n} + (k-4)^2 = 8(k-2)\frac{n}{2}(n(k-2) + 4 - k) + (k-4)^2$$
$$= [(2n)(k-2)]^2 + 4n(k-2)(4-k) + (4-k)^2$$
$$= (2kn - 4n - k + 4)^2,$$

whence the theorem follows.   □

In particular, for $k = 3$ we obtain

$$8T_n + 1 = (2n+1)^2,$$

and thus $8T_n + 1$ is the square of a positive integer (cf. Fig. 10.6). For $k = 5$ we get

$$24P_n + 1 = (6n-1)^2 \quad \text{and} \quad P_n = n^2 + T_{n-1}.$$

If $k = 6$, then $P_{6,n} = T_{2n-1}$.

Theorem 8.4 *Every positive integer can be expressed as the sum of k-gonal numbers with at most k terms.*

This theorem was already known to Pierre de Fermat. He claimed that he also had its proof. Such a proof was never found. Theorem 8.4 for the case of square numbers was known to Diophantus, but it was proved later by C. G. Jacobi, J. L. Lagrange (1772), and L. Euler. It is called the *Four-Squares Theorem*:

Theorem 8.5 *Every positive integer is the sum of four squares.*

In 1796, C. F. Gauss presented the proof of Theorem 8.4 for the case of triangular numbers. However, a complete proof for an arbitrary k was given by A. L. Cauchy in 1813 (see Guy [131]). For instance, by Theorem 8.4 for $k = 3$ we have

$$18 = 0 + 3 + 15 = T_2 + T_5.$$

Similarly for $k = 4$ we get

$$1634 = 1^2 + 9^2 + 16^2 + 36^2 = S_1 + S_9 + S_{16} + S_{36}.$$

For a given positive integer k denote by $g(k)$ least number of the kth powers of numbers: 0, 1, 2, 3,..., the sum of which is equal to a given natural number. Clearly, $g(1) = 1$. By Theorem 8.5 we have $g(2) = 4$ and this number cannot be reduced, since

$$7 = 1^2 + 1^2 + 1^2 + 2^2.$$

The number

$$23 = 1^3 + 1^3 + 1^3 + 1^3 + 1^3 + 1^3 + 1^3 + 2^3 + 2^3$$

can be expressed as the sum of nine cubes and this number cannot be reduced, i.e. $g(3) \geq 9$. In 1909 Wieferich [417] proved that $g(3) = 9$. Since 79 can be similarly expressed as the sum of 19 fourth powers and this number cannot be reduced, we obtain that $g(4) \geq 19$. In 1986 it was found that $g(4) = 19$, see [20]. We also know that $g(5) = 35$ (see [69]) and $g(6) = 73$ (see [302]). At present, it is known that $g(k) = 2^k - 2 + \lfloor (3/2)^k \rfloor$ for $3 \leq k \leq 471600000$, where $\lfloor r \rfloor$ stands for the integer part of a real number r, cf. [219] (see also [138]).

Let us now deal with a remarkable theorem coming from from Euler, which connects the so-called generalized pentagonal numbers and the decompositions of numbers into sums of natural numbers. Denote by $p(n)$ the number of different ways by which we can express the number n as the sum of one or more positive integers regardless of the order of particular terms (in doing so we set $p(0) = 1$). For instance $p(4) = 5$, since

$$4 = 3 + 1 = 2 + 2 = 2 + 1 + 1 = 1 + 1 + 1 + 1.$$

The *generalized pentagonal numbers* are defined by relations

$$P_n = \frac{3n^2 - n}{2} \quad \text{for } n \in \{\dots, -2, -1, 0, 1, 2, \dots\}.$$

So we see that their definition agrees with the usual definition of pentagonal numbers for positive integers n.

Theorem 8.6 (Euler) *For any positive integer n we have*

$$p(n) = p(n-1) + p(n-2) - p(n-5) - p(n-7) + p(n-12) + p(n-15) - \dots$$
$$= \sum_j (-1)^{j+1} p\left(n - \frac{3j^2 - j}{2}\right) + \sum_j (-1)^{j+1} p\left(n - \frac{3j^2 + j}{2}\right),$$

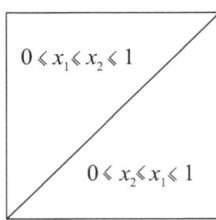 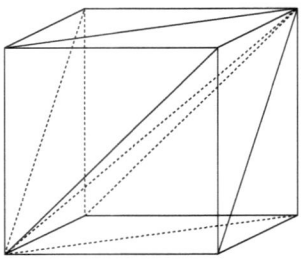

Fig. 8.3 Partition of a unit square into two triangles and of a unit cube into 6 tetrahedra

where we sum over all positive integers j for which the arguments in $p(\cdot)$ are non-negative.

A proof of the theorem is given e.g. in Hardy and Wright [138, p. 284], and Niven et al. [291, p. 456].

Similarly to polygonal numbers, we can define polyhedral numbers. For example, according to Theorem 8.1 and (8.3), we have

$$2T_{n-1} = n^2 - n,$$

which can be generalized to three-dimensional space as follows. We divide a cube into 6 tetrahedra of the same volume that have a common diagonal (see Fig. 8.3); one of them is, for example, a tetrahedron whose coordinates satisfy the inequalities $0 \leq x_1 \leq x_2 \leq x_3 \leq 1$. Therefore, we have

$$6U_{n-1} = n^3 - n,$$

where $U_n = \frac{1}{6}n(n+1)(n+2)$ are called *tetrahedral numbers*, $U_1 = 1$, $U_2 = 4$, $U_3 = 10$, etc.

Tetrahedral numbers are also called *triangular pyramidal numbers*, since they are defined as the sum of the first n triangular numbers $U_n = T_1 + T_2 + \cdots + T_n$.

In 1878, Meyl [271] proved that U_n is a square if and only if $n \in \{1, 2, 48\}$, which gives rise to the squares $1 = T_1$, $4 = 2^2 = T_1 + T_2 = 1 + 3$, and

$$19600 = 140^2 = T_1 + T_2 + \cdots + T_{48} = 1 + 3 + \cdots + 1176.$$

Using Theorem 8.1, we find that

$$140^2 = S_2 + S_4 + S_6 + \cdots + S_{48}.$$

Dividing this equality by 4, we get

$$70^2 = S_1 + S_2 + S_3 + \cdots + S_{24}. \tag{8.4}$$

One can deduce similar relations to get expressions for cubic numbers, octahedral numbers, etc. Such generalizations can be further extended to d-dimensional spaces to get polytopic numbers for d-cube, d-simplex, or d-orthoplex. For instance, 4-simplicial numbers are defined by

$$D_n = \frac{1}{4!}n(n+1)(n+2)(n+3)$$

and they represent the number of fourth derivatives of a smooth function f in n variables.

Analogously, we can investigate *square pyramidal numbers* 1, 5, 14, 30, 55, 91, 140, ... which can be obtained as particular sums of square numbers

$$S_1 + S_2 + \cdots + S_n = \frac{1}{6}n(n+1)(2n+1).$$

It is of interest that the only square pyramidal numbers that are squares occur for $n = 1$ and $n = 24$ which give the squares 1 and (see (8.4))

$$4900 = 70^2 = 1^2 + 2^2 + 3^2 + \cdots + 24^2.$$

This is known as the cannonball problem, which was proposed in 1875 by Lucas [245] as follows:

A square pyramid of cannonballs contains a square number of cannonballs only when it has 24 *cannonballs along its base.*

A correct complete solution to Lucas' question was only given in 1918 by Watson [412] whose proof made use of Jacobian elliptic functions. We give an accessible elementary solution of the cannonball problem based on the paper by William Anglin [10], which makes use of a result by De Gang Ma [249], properties of the Pell equation

$$X^2 - 3Y^2 = 1,$$

and Fermat's method of infinite descent.

Theorem 8.7 *The only integer values of n for which the square pyramidal number $\frac{1}{6}n(n+1)(2n+1)$ is a nonzero square are $n = 1$ and $n = 24$.*

We will first prove Theorem 8.7 for the case in which n is even. The proof will utilize the auxiliary results presented in Theorems 8.8 and 8.9 given below.

Theorem 8.8 *There are no positive integers x such that $2x^4 + 1$ is a square.*

Proof Assume to the contrary that there exist positive integers x and y such that $2x^4 + 1 = y^2$. Then $y = 2s + 1$ for some $s \in \mathbb{N}$, and

$$x^4 = 2s(s+1).$$

Suppose that s is odd. Then s and $2(s + 1)$ are coprime, and for some positive integers t and u, $s = t^4$ and $2(s + 1) = u^4$. This yields that $2(t^4 + 1) = u^4$ with t odd and u even. Then we have that

$$2(1 + 1) \equiv 0 \pmod{8},$$

which is impossible. Therefore, s must be even.

Its follows that $2s$ and $s + 1$ are coprime, and there exist integers t and u, both greater than 1, such that $2s = t^4$ and $s + 1 = u^4$. Let v be a positive integer such that $t = 2v$. Let a be the positive integer such that $u^2 = 2a + 1$. Then

$$\frac{t^4}{2} + 1 = s + 1 = u^4,$$

which implies that

$$2v^4 = \frac{u^4 - 1}{4} = a(a + 1).$$

Noting that $u^2 = 2a + 1$, we observe that $u^2 \equiv 1 \pmod{4}$, which implies that a is even. Since $2v^4 = a(a + 1)$, it now follows from the fact that a and $a + 1$ are coprime that there exist positive integers b and c such that $a = 2b^4$ and $a + 1 = c^4$. Then

$$2b^4 + 1 = \left(c^2\right)^2.$$

However,

$$c^2 \leq c^4 = a + 1 < 2a + 1 = u^2 \leq u^4 = s + 1 < 2s + 1 = y.$$

Theorem 8.8 now follows by Fermat's method of infinite descent.   □

Theorem 8.9   *There exists exactly one positive integer x, namely 1, such that $8x^4 + 1$ is a square.*

Proof   Assume that $8x^4 + 1 = (2s + 1)^2$, where $x \geq 1$. Then

$$2x^4 = s(s + 1).$$

Suppose that s is even. Since s and $s + 1$ are coprime, there exist positive integers t and u such that $s = 2t^4$ and $s + 1 = u^4$. Then

$$2t^4 + 1 = s + 1 = u^4,$$

and by Theorem 8.8, $t = 0$, and hence $x = 0$. Thus, we only need to consider the case in which s is odd.

It now follows that there exist integers t and u such that $s = t^4$ and $s + 1 = 2u^4$, i.e.,

$$t^4 + 1 = 2u^4. \tag{8.5}$$

Since t is odd, $t^4 + 1 \equiv 2 \pmod 4$, which implies by (8.5) that u is odd. Squaring both sides of (8.5) yields that

$$4u^8 - 4t^4 = t^8 - 2t^4 + 1 = (t^4 - 1)^2,$$

and thus,

$$(u^4 - t^2)(u^4 + t^2) = \left(\frac{t^4 - 1}{2}\right)^2,$$

an integer square. Since u^4 and t^2 are coprime, it follows that both $\frac{1}{2}(u^4 - t^2)$ and $\frac{1}{2}(u^4 + t^2)$ are integer squares. We now observe that

$$(u^2 - t)^2 + (u^2 + t)^2 = 4\frac{u^4 + t^2}{2} = A^2 \tag{8.6}$$

and

$$\frac{(u^2 - t)^2 (u^2 + t)^2}{2} = \frac{u^4 - t^2}{2} = B^2,$$

where $A > 0$. From this and (8.6), if $B \neq 0$, then the triangle with sides $u^2 - t$, $u^2 + t$, $\sqrt{2(u^4 + t^2)}$ is a Pythagorean triangle whose area is an integer square. This is impossible by Theorem 2.8. Thus $B = 0$ and $u^2 = t$. Since $t^4 + 1 = 2u^4$ by (8.5), we obtain $t^4 - 2t^2 + 1 = 0$, which yields that $t^2 = 1$. It now follows that $s = t^4 = 1$ and $8x^4 + 1 = (2s + 1)^2 = 9$. Hence, $x = 1$. □

Proof of Theorem 8.7 for the case in which n is even. First, let us suppose that $x(x + 1)(2x + 1) = 6y^2$, where x is a nonnegative even integer. We will later discard the trivial solution in which $x = 0$. Then $x + 1$ is odd. Since $x, x + 1$, and $2x + 1$ are coprime in pairs and both $x + 1$ and $2x + 1$ are odd, it follows that $x + 1$ and $2x + 1$ are either squares or triples of squares. Thus, $x + 1 \not\equiv 2 \pmod 3$ and $2x + 1 \not\equiv 2 \pmod 3$. Hence, $x \equiv 0 \pmod 3$. Since x is even, $x \equiv 0 \pmod 6$ and $x + 1 \equiv 2x + 1 \equiv 1 \pmod 6$. Therefore, for some nonnegative integers f, g, and h, we have $x = 6g^2$, $x + 1 = f^2$, and $2x + 1 = h^2$. Then

$$6g^2 = h^2 - f^2 = (h - f)(h + f). \tag{8.7}$$

Since h and f are both odd, 4 is a factor of $(h - f)(h + f)$, and thus $4 \mid g^2$, Hence, by (8.7),

$$6\left(\frac{g}{2}\right)^2 = \frac{h - f}{2}\frac{h + f}{2}. \tag{8.8}$$

We note that $\frac{1}{2}(h - f)$ and $\frac{1}{2}(h + f)$ are coprime, because h^2 and f^2 are coprime. Thus, by (8.8), we have the following two cases:

Case (i). One of $\frac{1}{2}(h - f)$ and $\frac{1}{2}(h + f)$ has the form $6A^2$ and the other has the form B^2, where A and B are nonnegative integers. Then $f = \pm(6A^2 - B^2)$ and by (8.8),

$$\frac{h - f}{2}\frac{h + f}{2} = 6A^2B^2 = \frac{6g^2}{4} \Leftrightarrow g^2 = 4A^2B^2 \Leftrightarrow g = 2AB. \qquad (8.9)$$

Since $6g^2 + 1 = x + 1 = f^2$, we obtain from (8.9) that

$$24A^2B^2 + 1 = 6g^2 + 1 = (6A^2 - B^2)^2 \qquad (8.10)$$

or

$$(6A^2 - 3B^2)^2 - 8B^4 = 1. \qquad (8.11)$$

By Theorem 8.9, $B = 0$ or 1. If $B = 0$, then by (8.9), $x = 6g^2 = 24A^2B^2 = 0$. If $B = 1$, then $A = 1$ by (8.11), and $x = 24A^2B^2 = 24$.

Case (ii). One of $\frac{1}{2}(h - f)$ and $\frac{1}{2}(h + f)$ has the form $3A^2$ and the other has the form $2B^2$, where A and B are nonnegative integers. Then $f = \pm(3A^2 - 2B^2)$ and by (8.9), $g = 2AB$. By a similar argument as in Case (i), we have that

$$24A^2B^2 + 1 = (3A^2 - 2B^2)^2$$

or

$$(3A^2 - 6B^2)^2 - 2(2B)^4 = 1.$$

By Theorem 8.8, $B = 0$ and hence, $x = 24A^2B^2 = 0$.

Consequently, when n is an even positive integer, the square pyramidal number $\frac{1}{6}n(n + 1)(2n + 1)$ is a square if and only if $n = 24$. $\qquad \square$

We now treat the remaining case of Theorem 8.7 in which n is odd. To accomplish this, we first examine the solutions of the Pell equation $X^2 - 3Y^2 = 1$. Let $\gamma = 2 + \sqrt{3}$ and $\delta = 2 - \sqrt{3}$. Note that $\gamma\delta = 1$. For n a nonnegative integer, let

$$v_n = \frac{\gamma^n + \delta^n}{2} \quad \text{and} \quad u_n = \frac{\gamma^n - \delta^n}{2\sqrt{3}}.$$

Then v_n and u_n are integers and it is well known that $\langle v_n, u_n \rangle$ is the nth nonnegative integer solution of $X^2 - 3Y^2 = 1$ for $n \geq 0$ (see Robbins [330, pp. 273–274], Table 91 of Beiler [28], and [28, pp. 252–255]). The proof of the case in which n is odd given later will make use of Theorems 8.10–8.18 given below.

Theorem 8.10 *Both the sequences $(v_n)_{n=0}^{\infty}$ and $(u_n)_{n=0}^{\infty}$ satisfy the recursion relation*

$$w_{n+2} = 4w_{n+1} - w_n \qquad (8.12)$$

for $n \geq 0$.

Proof We observe that γ and δ are the roots of the quadratic equation

$$x^2 - 4x + 1 = 0.$$

Thus, $\gamma^2 = 4\gamma - 1$ and $\delta^2 = 4\delta - 1$. It now follows that $\gamma^{n+2} = 4\gamma^{n+1} - \gamma^n$ and $\delta^{n+2} = 4\delta^{n+1} - \delta^n$ for $n \geq 0$. Hence, the sequences $(\gamma^n)_{n=0}^{\infty}$ and $(\delta^n)_{n=0}^{\infty}$ both satisfy the recursion relation (8.12). It now follows from the definitions of v_n and u_n and induction that both the sequences $(v^n)_{n=0}^{\infty}$ and $(u^n)_{n=0}^{\infty}$ also satisfy the relation (8.12). □

Theorem 8.11 *Let $n \geq 0$. Then $v_{-n} = v_n$ and $u_{-n} = -u_n$.*

Proof By inspection, $v_0 = 1$, $v_1 = 2$ and $u_0 = 0$, $u_1 = 1$. The result now follows by induction upon making use of the recursion relation (8.12). □

Theorem 8.12 *For m a positive integer, $v_{2m} = v_m^2 - 1 = 6u_m^2 + 1$ and $u_{2m} = 2u_m v_m$.*

Proof The equalities follow from the definitions of v_m and u_m and from the facts that $\gamma\delta = 1$ and $v_m^2 - 3u_m^2 = 1$. □

Theorem 8.13 *Let $m, n \geq 0$. Then we have that $v_{m+n} = v_m v_n + 3u_m u_n$ and $u_{m+n} = u_m v_n + u_n v_m$. Further, if $m - n \geq 0$ then $v_{m-n} = v_m v_n - 3u_m u_n$ and $u_{m-n} = u_m v_m - u_n v_m$.*

Proof This follows from the definitions of v_n and u_n and from Theorem 8.11. □

Theorem 8.14 *If k and m are nonnegative integers, then the following hold:*

$$\text{(i)} \qquad v_{(2k+1)m} \equiv 0 \pmod{v_m}, \qquad\qquad (8.13)$$

$$\text{(ii)} \qquad u_{2km} \equiv 0 \pmod{v_m}, \qquad\qquad (8.14)$$

$$\text{(iii)} \qquad v_{2km} \equiv (-1)^k \pmod{v_m}. \qquad\qquad (8.15)$$

Proof We first prove parts (i) and (ii) by induction.

(i) Clearly (8.13) holds for $k = 0$. Assume that (8.13) holds also for $k \geq 0$. Then by Theorems 8.12 and 8.13, we have that

$$v_{(2k+3)m} = v_{(2k+1)m} v_{2m} + 3u_{(2k+1)m} u_{2m}$$
$$= v_{(2k+1)m} v_{2m} + 6u_{(2k+1)m} u_m v_m \equiv 0 \pmod{v_m}.$$

(ii) Since $u_0 = 0$, we see by Theorem 8.12 that (8.14) is satisfied for $k = 0$ and $k = 1$. Now assume that (8.14) holds for $k \geq 1$. Then by Theorems 8.12 and 8.13, we see that

$$u_{(2k+2)m} = u_{2km} v_{2m} + u_{2m} v_{2km} = u_{2km} v_{2m} + 2u_m v_m v_{2km} \equiv 0 \pmod{v_m}.$$

(iii) Suppose first that k is odd. Then by Theorem 8.12 and part (i) of this theorem,

$$v_{2km} = 2v_{km}^2 - 1 \equiv -1 \pmod{v_m}.$$

Now suppose that k is even. Then by Theorem 8.12 and part (ii) of this theorem,

$$v_{2km} = 6u_{km}^2 + 1 \equiv 1 \pmod{v_m}.$$

Thus, (8.15) holds.

$\square$

Theorem 8.15 *Let k, m, and n be nonnegative integers such that $2km - n$ is nonnegative. Then*

$$v_{2km \pm n} \equiv (-1)^k v_n \pmod{v_m}.$$

Proof By Theorems 8.13 and 8.14 (ii) and (iii),

$$v_{2km \pm n} = v_{2km} \pm 3u_{2km} u_n \equiv (-1)^k v_n \pmod{v_m}$$

which proves the theorem.

$\square$

Let us now consider the first several values of v_n. Beginning with $n = 0$, we have 1, 2, 7, 26, 97, 362, 1351, ...Considering these values modulo 5, we obtain 1, 2, 2, 1, 2, 2, ...By Theorem 8.10, we see that (v_n) is purely periodic modulo 5 with a period length of 3. If we consider the values of (v_n) modulo 8, we get 1, 2, 7, 2, 1, 2, 7, 2, ...Again by Theorem 8.10, we find that (v_n) is purely periodic modulo 8 with period length 4 and that v_n is odd if and only if n is even. Using the laws of quadratic reciprocity for the Jacobi symbol given in Theorem 2.28 (vi)–(viii), we have the following two theorems by our above remarks.

Theorem 8.16 *Let n be even. Then v_n is an odd nonmultiple of 5 and $\left(\frac{5}{v_n}\right) = 1$ if and only if $n \equiv 0 \pmod{3}$.*

Theorem 8.17 *Let n be even. Then v_n is odd and $\left(\frac{-2}{v_n}\right) = 1$ if and only if $n \equiv 0 \pmod{4}$.*

Theorem 8.18 below was first proved by Ma in [249].

Theorem 8.18 *For $n \geq 0$, v_n has the form $4M^2 + 3$ only when $v_n = 7$.*

Proof Suppose that $v_n = 4M^2 + 3$. Then $v_n \equiv 3$ or 7 (mod 8), and from the sequence of values of (v_n) modulo 8, we observe that n has the form $8k \pm 2$. Suppose that $n \neq 2$, which implies that $v_n \neq 7$. Then we can write v_n in the form $2r2^s \pm 2$, where r is odd and $s \geq 2$. By Theorem 8.15,

$$v_n = v_{2r2^s \pm 2} \equiv (-1)^r v_2 \pmod{v_{2^s}}.$$

Since r is odd and $v_2 = 7$, it follows that $4M^2 = v_n - 3 \equiv -10 \pmod{v_{2^s}}$ and thus,

$$\left(\frac{-2}{v_{2^s}}\right)\left(\frac{5}{v_{2^s}}\right) = \left(\frac{-10}{v_{2^s}}\right) = \left(\frac{4M^2}{v_{2^s}}\right) = 1. \tag{8.16}$$

However, by Theorems 8.17 and 8.16,

$$\left(\frac{-2}{v_{2^s}}\right) = 1 \quad \text{and} \quad \left(\frac{5}{v_{2^s}}\right) = -1,$$

which contradicts (8.16). Thus $n = 2$ and $v_n = 7$. □

Proof of Theorem 8.7 for the case in which n is odd. Suppose now that x is a positive odd integer such that $x(x + 1)(2x + 1) = 6y^2$ for some integer y. Since $x, x + 1$, and $2x + 1$ are coprime in pairs, it follows that x is either a square or triple a square, and thus $x \not\equiv 2 \pmod 3$. Furthermore, since $x + 1$ is even, we see that $x + 1$ is either double a square or six times a square, and consequently, $x + 1 \not\equiv 1 \pmod 3$. Hence, $x \equiv 1 \pmod 6$, $x + 1 \equiv 2 \pmod 6$, and $2x + 1 \equiv 3 \pmod 6$. Therefore, for some positive integers a, b, and c, we have $x = a^2$, $x + 1 = 2b^2$, and $2x + 1 = 3c^2$. From this we get

$$6c^2 + 1 = 4x + 3 = 4a^2 + 3. \tag{8.17}$$

We also observe that

$$(6c^2 + 1)^2 - 3(4bc)^2 = 12c^2(3c^2 + 1 - 4b^2) + 1$$
$$= 12c^2(2x + 1 + 1 - 2(x + 2)) + 1 = 1. \tag{8.18}$$

Hence, by (8.17), (8.18), and Theorem 8.18, $6c^2 + 1 = 7$. Thus $c = 1$ and $x = 1$. Therefore, $n = 1$ is the only positive odd integer such that $\frac{1}{6}n(n + 1)(2n + 1)$ is a square. □

We can now conclude that the only positive integers n, for which the square pyramidal number $\frac{1}{6}n(n + 1)(2n + 1)$ is a square, are $n = 1$ and $n = 24$ which give the square numbers 1 and $4900 = 70^2$, respectively.

8.2 Perfect Numbers

We have already dealt with even perfect numbers in Theorems 4.7 and 4.8, where we showed how closely they are related to Mersenne primes. Recall that the natural number n is called *perfect* if it equals the sum of all its divisors less than n. For example,

$$6 = 1 + 2 + 3,$$
$$28 = 1 + 2 + 4 + 7 + 14,$$
$$496 = 1 + 2 + 4 + 8 + 16 + 31 + 62 + 124 + 248,$$
$$8128 = 1 + 2 + 4 + 8 + 16 + 32 + 64 + 127 + 254 + 508 + 1016 + 2032 + 4064.$$

These four perfect numbers were already known to the Neo-Pythagorean Nicomachus of Gerasa (around the year 100 AD). Due to Euclid's Theorem 4.7 and Euler's Theorem 4.8 we can easily determine other perfect numbers

$$33\,550\,336, \quad 8\,589\,869\,056, \quad 137\,438\,691\,328, \quad 2\,305\,843\,008\,139\,952\,128, \ldots$$

So far, we know exactly as many perfect numbers as there are Mersenne prime numbers. Notice that for the sum of the reciprocals of all divisors of perfect numbers we have

$$\frac{1}{1} + \frac{1}{2} + \frac{1}{3} + \frac{1}{6} = 2,$$
$$\frac{1}{1} + \frac{1}{2} + \frac{1}{4} + \frac{1}{7} + \frac{1}{14} + \frac{1}{28} = 2,$$

etc. Now will prove this property for any perfect number. As in Theorem 4.7, we will use the symbol $\sigma(n)$ to indicate the sum of all positive divisors n (including n), i.e.,

$$\sigma(n) = \sum_{d \mid n} d. \tag{8.19}$$

Theorem 8.19 *A positive integer n is perfect if and only if*

$$\sum_{d \mid n} \frac{1}{d} = 2.$$

Proof By the definition of a perfect number we have

$$2n = \sigma(n) = \sum_{d \mid n} d = \sum_{d \mid n} \frac{n}{d} = n \sum_{d \mid n} \frac{1}{d},$$

which proves the theorem. □

As a direct consequence, we obtain the following statement (see Burton [56, p. 204]):

Theorem 8.20 *If n is perfect, $m < n$, and $m \mid n$, then m is not perfect.*

From Euclid's Theorem 4.7 and Euler's Theorem 4.8 another somewhat surprising property of even perfect numbers can be derived (cf. Fig. 10.5 below):

Theorem 8.21 *All even perfect numbers are triangular.*

Proof From Theorems 4.7 and 4.8 we obtain the following connection between even perfect numbers n and the Mersenne primes M_p,

$$n = 2^{p-1}(2^p - 1) = \frac{2^p}{2}(2^p - 1) = 1 + 2 + \cdots + (2^p - 1). \tag{8.20}$$

From this and (8.3) we see that n a triangular number. □

Now we will prove several other simple properties of perfect numbers.

Theorem 8.22 *If n is perfect, then n is not a power of a prime.*

Proof Suppose to the contrary that $n = p^k$, where p is a prime and k a positive integer. Then

$$\sigma(n) = 1 + p + p^2 + \cdots + p^k = 1 + p(1 + p + \cdots + p^{k-1}),$$

and thus, $p \nmid \sigma(n)$. Consequently, $\sigma(n) \neq 2p^k$. □

For any coprime positive integers r and s it follows from the definition (8.19) that

$$\sigma(rs) = \sigma(r)\sigma(s). \tag{8.21}$$

We will use this relationship several times.

Theorem 8.23 *If n is perfect, then n is not a square.*

Proof Assume to the contrary that $n = m^2$. Then

$$n = \prod_{i=1}^{r} p_i^{k_i},$$

where $2 \mid k_i$ for $i = 1, 2, \ldots, r$. From this and (8.21) by induction we get

$$\sigma(n) = \sigma\left(\prod_{i=1}^{r} p_i^{k_i}\right) = \prod_{i=1}^{r} \sigma\left(p_i^{k_i}\right).$$

However, we see that

$$\sigma(p_i^{k_i}) = 1 + p_i + p_i^2 + \cdots + p_i^{k_i} \tag{8.22}$$

is odd for $i = 1, 2, \ldots, r$, since k_i is even. Therefore, (8.22) contains an odd number of terms and $\sigma(n)$ is also an odd number. Hence, $\sigma(n) \neq 2n$. □

Theorem 8.24 *If n is an even perfect number of the form (8.20), then n can be written in the binary system as follows:*

$$111\ldots1000\ldots0$$

with p ones and p − 1 zeros.

Proof This immediately follows from the expression (8.20) as a special triangular number.   □

Theorem 8.25 *The last digit of the number n given by (8.20) is either 6, or 8. If the last digit is 8, then n the last two digits are 28.*

Proof First we show that $n \equiv 6$ or 8 (mod 10). If $p = 2$, then by (8.20) we have $n = 6$. Thus we will assume that p is odd, i.e., $p - 1$ is even. Now we can easily check that $2^{p-1} \equiv 6$ (mod 10) if $p - 1 = 4k$, and $2^{p-1} \equiv 4$ (mod 10) if $p - 1 = 4\ell + 2$ for some positive integers k and ℓ.

If $2^{p-1} \equiv 6$ (mod 10), then by (8.20) we have

$$n = 2^{p-1}(2^p - 1) \equiv 6(2 \cdot 6 - 1) \equiv 6 \cdot 11 \equiv 6 \quad (\text{mod } 10).$$

If $2^{p-1} \equiv 4$ (mod 10), then similarly we get

$$n = 4(2 \cdot 4 - 1) \equiv 4 \cdot 7 \equiv 8 \quad (\text{mod } 10).$$

The rest about the last two digits is proved in Burton [56, pp. 203–204].   □

In Heath's Theorem 4.10 we showed that every even perfect number except for 6 can be expressed as the sum of the third powers of odd numbers. On the other hand, according to Mąkowski [253], we have that

$$28 = 2^2(2^3 - 1) = 3^3 + 1^3$$

is the only even perfect number which is of the form $m^3 + 1$, where m is a positive integer.

Theorem 8.26 (Mąkowski) *The only even perfect number of the form $m^3 + 1$ is 28.*

Proof All even perfect numbers are of the form (8.20). Hence, for a prime $p \geq 3$ we have $2^{p-1} \equiv 1$ (mod 3), $2^p - 1 \equiv 1$ (mod 3) which gives $2^{p-1}(2^p - 1) \equiv 1$ (mod 3). If $2^{p-1}(2^p - 1) = m^3 + 1$, then m is divisible by 3 and m is odd, i.e. $m = 3(2y + 1)$ and

$$2^{p-1}(2^p - 1) = (m + 1)(m^2 - m + 1).$$

We see that

$$(m + 1, m^2 - m + 1) = (m + 1, -2m + 1) = (m + 1, 3) = (3(2y + 1) + 1, 3) = 1.$$

Since $m > 2$, we have $m^2 - m + 1 > m + 1 > 1$. However, the only expression of an even perfect number as the product of two coprime positive integers greater than 1 is given by relation (8.20). Therefore, $m + 1 = 2^{p-1}$ and $m^2 - m + 1 = 2^p - 1$. Subtracting $2m + 2 = 2^p$ from the last equation, we get $m^2 - 3m - 1 = -1$. Hence, $m(m - 3) = 0$, and from this we obtain $m = 3$. □

As a consequence we get the following statement.

Theorem 8.27 *The only even perfect number of the form $m^m + 1$ is 28.*

Proof If $m^m + 1$ is an even perfect number, then m is divisible by 3, i.e., $m = 3k$ and $[(3k)^k]^3 + 1$ is even perfect number. By the foregoing theorem $(3k)^k = 3$, and thus $k = 1$, $m = 3$, and $m^m + 1 = 28$. □

From this another theorem follows.

Theorem 8.28 *There does not exist an even perfect number of the form*

$$m^{m^{m^{\cdot^{\cdot^{\cdot^m}}}}} + 1.$$

This result is further extended in [356] to numbers of the form $a^{m^{m^{\cdot^{\cdot^{\cdot^m}}}}} + 1$. In this paper, the authors showed that 28 is the only perfect number of the form $a^n + b^n$, where $n \geq 2$ and $(a, b) = 1$, which generalizes Mąkowski's Theorem 8.26.

There are many open problems concerning even perfect numbers (see Sect. 4.1 and [346]).

8.3 Deficient and Abundant Numbers

For a positive integer n we set

$$s(n) = \sigma(n) - n = \sum_{\substack{d|n \\ d<n}} d, \tag{8.23}$$

where $\sigma(n)$ is defined by (8.19). The value $s(n)$ is the sum of the so-called *aliquot parts,* i.e., it is the sum of all divisors of the number n that are less than n. The perfect number is thus fully characterized by the relation (see Fig. 8.4)

$$s(n) = n.$$

A positive integer n is called *deficient* or *abundant,* if

$$s(n) < n \quad \text{or} \quad s(n) > n, \text{ respectively.}$$

According to this definition the numbers

$$1, 2, 3, 4, 5, 7, 8, 9, 10, 11, 13, 14, 15, 16, 17, 19, \ldots \quad \text{are deficient and}$$
$$12, 18, 20, 24, 30, 36, 40, 42, 48, 54, 60, \ldots \quad \text{are abundant.}$$

For instance, $14 = 2 \cdot 7$ is deficient, since

$$s(14) = 1 + 2 + 7 = 10 < 14.$$

On the other hand, the number $12 = 2^2 3$ is abundant, because

$$s(12) = 1 + 2 + 3 + 4 + 6 = 16 > 12.$$

The smallest abundant odd number is 945.

The symbol $\sigma(n)$ will again indicate the sum of all positive divisors of n.

As already mentioned in Sect. 4.1, we do not know yet if there are infinitely many even perfect numbers. Moreover, no odd perfect number is known. In contrast to that, we have the following theorem.

Theorem 8.29 *There exist infinitely many even deficient numbers, odd deficient numbers, even abundant numbers, and odd abundant numbers.*

Proof We shall distinguish the following four cases:

(a) All numbers 2^k for $k \geq 1$ are even and deficient, since

$$s(2^k) = 2^k - 1 < 2^k.$$

(b) Let p be an odd prime and $k \geq 1$. Then p^k is odd and deficient, since

$$s(p^k) = 1 + p + p^2 + \cdots + p^{k-1} = \frac{p^k - 1}{p - 1} < \frac{p^k}{p - 1} < p^k.$$

(c) All numbers divisible by 12 are even and abundant, since

$$\frac{s(n)}{n} > \frac{1}{2} + \frac{1}{3} + \frac{1}{4} > 1.$$

(d) Let $n = 945 \, k$, where $k \geq 1$ is an arbitrary odd integer, which is not divisible by 3, 5, and 7. Since $945 = 3^3 \cdot 5 \cdot 7$, we see that $(945, k) = 1$, and thus by (8.21) we have

$$\sigma(n) = \sigma(945)\sigma(k) = \sigma(3^3)\sigma(5)\sigma(7)\sigma(k) = 40 \cdot 6 \cdot 8 \cdot \sigma(k) \geq 1920\,k > 2 \cdot 945\,k,$$

since $\sigma(k) \geq k$. Hence, every number of the form $n = 945\,k$ is abundant by (8.23) and obviously odd. $\qquad\square$

Let us point our that Claude Gaspard Bachet de Méziriac provided a relatively lengthy proof of Euclid's Theorem 4.7 with the following addition:

If $2^q - 1$ is composite, then the number $n = 2^{q-1}(2^q - 1)$ is abundant (see Porubský [309, p. 54]). Denoting by $A(x)$ the number of abundant numbers not exceeding x, then

$$0.241x < A(x) < 0.314x,$$

which means that most of the positive integers are deficient.

A positive integer n is called *weird*, if it is abundant and if it cannot be expressed as the sum of some of its proper divisors.

Remark We see that 70 is an abundant number, since

$$1 - 2 + 5 + 7 + 10 + 14 + 35 = 74 > 70.$$

Moreover, no subset of $\{1, 2, 5, 7, 10, 14, 35\}$ sums to the number 70. It can be shown that 70 is the smallest weird number and that there are only 7 weird numbers less than 10 000. All of them are even. For the time being it is not known whether there exists an odd weird number.

8.4 Amicable Numbers

The positive integers m and n are called *amicable*, if each of them is equal to the sum of all positive divisors of the second number except for itself, i.e., if

$$\sum_{\substack{d\mid m \\ d < m}} d = n \quad \text{and} \quad \sum_{\substack{d\mid n \\ d < n}} d = m.$$

Another equivalent expression of the amicable numbers m and n is given by the equality

$$\sigma(m) = m + n = \sigma(n). \tag{8.24}$$

Hence, every perfect number is amicable to itself. It is easy to verify that $220 = 2^2 5 \cdot 11$ and $284 = 2^2 71$ are amicable numbers; the first one is abundant and the second deficient, and we have

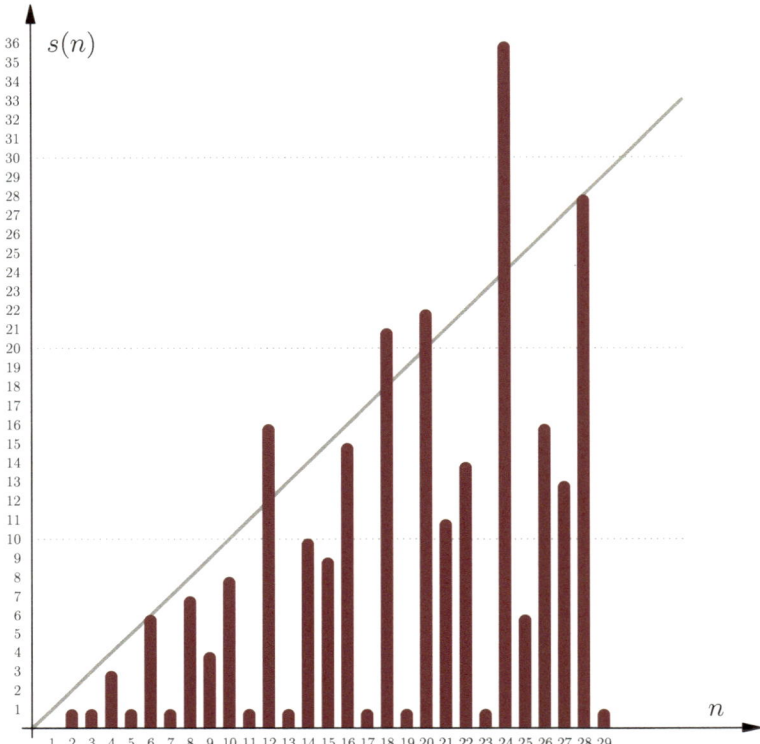

Fig. 8.4 Graph representing the dependence of $s(n)$ on n. The diagonal corresponds to perfect numbers. The triangular region above and below the diagonal corresponds to abundant and deficient numbers, respectively

$$220 = 1 + 2 + 4 + 71 + 142,$$
$$284 = 1 + 2 + 4 + 5 + 10 + 11 + 20 + 22 + 44 + 55 + 110.$$

This is the smallest known pair of such numbers that are not perfect. The Greek scholar Iamblichus (250–330 AD) ascribed knowledge of this pair already to the Pythagoreans (see also Burton [56, p. 213]). Extending the graph from Fig. 8.4 to 284, we would obtain a straight line connecting the function values $s(220)$ and $s(284)$ with slope (-1).

The first explicit rule for expressing special amicable numbers was found by the famous Arabic mathematician Thābit ibn Qurra from Baghdad already in the 9th century (see Porubský [309]):

Theorem 8.30 (Thabit) *If* $p = 3 \cdot 2^{k-1} - 1$, $q = 3 \cdot 2^k - 1$, *and* $r = 9 \cdot 2^{2k-1} - 1$ *are primes for* $k \geq 2$, *then* $2^k pq$ *and* $2^k r$ *are amicable numbers.*

Proof Setting $m = 2^k pq$ and $n = 2^k r$, we have

$$\begin{aligned}
m + n &= 2^k(3 \cdot 2^{k-1} - 1)(3 \cdot 2^k - 1) + 2^k(9 \cdot 2^{2k-1} - 1) \\
&= 2^k(9 \cdot 2^{2k-1} - 3 \cdot 2^{k-1} - 3 \cdot 2^k + 1) + 2^k(9 \cdot 2^{2k-1} - 1) \\
&= 9 \cdot 2^{3k-1} - 3 \cdot 2^{2k-1} - 3 \cdot 2 \cdot 2^{2k-1} + 2^k + 9 \cdot 2^{3k-1} - 2^k \\
&= 2 \cdot 9 \cdot 2^{3k-1} - 9 \cdot 2^{2k-1} = 9 \cdot 2^{2k-1}(2^{k+1} - 1).
\end{aligned} \tag{8.25}$$

Since 2^k, p, and q are coprime numbers for $k \geq 2$, we can use relation (8.21),

$$\sigma(m) = \sigma(2^k)\sigma(p)\sigma(q). \tag{8.26}$$

Since p and q are primes, we obtain $\sigma(p) = p + 1$ and $\sigma(q) = q + 1$. Next, we use the formula for the sum of a geometric series

$$\sigma(2^k) = 1 + 2 + 2^2 + \cdots + 2^k = \frac{2^{k+1} - 1}{2 - 1} = 2^{k+1} - 1.$$

Hence, altogether from (8.26) and (8.25) we get

$$\sigma(m) = (2^{k+1} - 1)(3 \cdot 2^{k-1})(3 \cdot 2^k) = (2^{k+1} - 1)9 \cdot 2^{2k-1} = m + n. \tag{8.27}$$

Since r is also a prime, we find that

$$\sigma(n) = \sigma(2^k r) = \sigma(2^k)\sigma(r) = (2^{k+1} - 1)9 \cdot 2^{2k-1} = m + n. \tag{8.28}$$

Therefore, from relation (8.27) and (8.28) we obtain that (8.24) holds, and thus m and n are amicable numbers. $\qquad\square$

Numbers of the form $3 \cdot 2^k - 1$ are called *Thabit numbers*. If $3 \cdot 2^k - 1$ is a prime number for some $k \geq 0$, it is called the *Thabit prime number*, for example,

$$2, \ 5, \ 11, \ 23, \ 47, \ 191, \ 383, \ 6143, \ 786431, \ 51539607551.$$

We observe that Thabit's Theorem 8.30 for $k = 2$ yields the amicable numbers 220 and 284 (see Borho [36]). However, it is not known whether Thabit used the theorem for $k > 2$. The second smallest pair of amicable numbers {1184, 1210} was found in 1866 by the student Nicolo Paganini, who was only 16 years old. However, Theorem 8.30 cannot be used in this particular case, because it gives only a sufficient condition for the existence of amicable numbers. It is said that Pierre de Fermat rediscovered Thabit's Theorem 8.30 and found the pair {17 296, 18 416}, which corresponds to the case $k = 4$. But before that, the same pair was discovered by the Arab mathematician Ibn al-Bannā' al-Marrākushī in Morocco at the turn of the 13th and 14th centuries (see Porubský [309, p. 60]). Also Fermat's contemporary René

Descartes found a pair of relatively large amicable numbers {9363584, 9437056} which corresponds to the case $k = 7$ of Theorem 8.30.

The smallest pairs of amicable numbers are as follows:

{220, 284},   {1184, 1210},   {2620, 2924},   {5020, 5564},   {6232, 6368},

{10744, 10856},   {12285, 14595},   {17296, 18416},   {63020, 76084},

{66928, 66992},   {67095, 71145},   {69615, 87633},   {79750, 88730}, ...

Leonhard Euler found 60 such pairs. He also generalized (see [108]) Thabit's Theorem 8.30 to the case in which $p = 2^{k-\ell} f - 1$, $q = 2^k f - 1$, and $r = 2^{2k-\ell} f^2 - 1$ are primes, $f = 2^\ell + 1$ and $k > \ell \geq 1$. His proof is quite similar to the proof of Theorem 8.30. The case $\ell = 1$ actually corresponds to Thabit's Theorem 8.30. For $\ell = 7$ and $k = 8$ we get the pair

$$\{2172649216, 2181168896\}.$$

For $\ell = 11$ and $k = 40$ we get another pair (see te Riele [326])

$$m = 272491804039370655778575224081940584 8576,$$

$$n = 272491804039618485630625803878723590 5536.$$

Notice that the ratio $m : n$ is about $1 - 2^{-40}$. Thanks to supercomputer facilities, millions of such pairs are currently known, the largest of these pairs has over 24 000 digits as of 2020.

Theorem 8.31 *If m and n are amicable numbers, then*

$$\left(\sum_{d|m}\frac{1}{d}\right)^{-1} + \left(\sum_{d|n}\frac{1}{d}\right)^{-1} = 1.$$

Proof By (8.24) we see that

$$m + n = \sigma(m) = \sum_{d|m} d = \sum_{d|m}\frac{m}{d} = m\sum_{d|m}\frac{1}{d},$$

$$m + n = \sigma(n) = \sum_{d|n} d = \sum_{d|n}\frac{n}{d} = n\sum_{d|n}\frac{1}{d}.$$

Hence,

$$\sum_{d|m}\frac{1}{d} = \frac{m+n}{m} \quad \text{and} \quad \sum_{d|n}\frac{1}{d} = \frac{m+n}{n},$$

and from this it follows that

$$\left(\sum_{d|m}\frac{1}{d}\right)^{-1} + \left(\sum_{d|n}\frac{1}{d}\right)^{-1} = \frac{m}{m+n} + \frac{n}{m+n} = 1.$$

The above completes the proof.   □

We will now mention some open problems:

(1) *Are there infinitely many pairs of amicable numbers?*
(2) *Is there an amicable pair of coprime numbers?*
(3) *It there an amicable pair of different parities (i.e. m odd and n even)?*
(4) *Is there an amicable pair, one of which is a square?*

8.5   Cunningham Numbers

At the beginning of the 20th century a large project was started dealing with the factorization of *Cunningham numbers,* i.e. numbers of the form $b^n \pm 1$, where b is "small" (usually $b \le 12$) and n is "large".

If $b \ge 2$ and $b^n + 1$ is a prime, then necessarily $n = 2^m$. This follows from relation (4.8), where we replace the base 2 with b. Note that $(b^n - 1)/(b - 1)$ is prime only if n is a prime, where $b \ge 2$. (Note further that $b - 1 = 1$ when $b = 2$.) Similarly, we can replace the base 2 with b in (4.1).

At present, every month many prime divisors of Cunningham numbers are found by computers. The tables of their divisors for $b \le 12$ are the contents of the book [45].

However, the biggest benefit of Cunningham's project is that it initiated the effort to develop new highly effective methods for primality testing, searching for prime decompositions, etc. Generalized Cunningham numbers of the form $b^n \pm a^n$ are also investigated.

8.6   Cullen Numbers

We do not know yet whether the sequence of Fermat numbers F_m contains an infinite number of composite numbers. On the other hand, it is proved in the sequence

$$C_n = n2^n + 1$$

there are infinitely many composite numbers for $n = 1, 2, \ldots$ (see Crandall and Pomerance [82, p. 71]).

In 1905, the Reverend James Cullen proved that C_n is composite for $n \in \{2, 3, \ldots, 100\}$ except for one case $n = 53$ (see Cullen [83]), and therefore the numbers C_n are called *Cullen numbers*. A year later A. J. Cunningham found that the prime number 5591 divides C_{53} and proved that C_n is composite for all $n \in \{2, 3, \ldots, 200\}$ except for the case $n = 141$, which he could not decide. About fifty years later Robinson in 1957 showed that C_{141} is a prime number.

Until the end of 2020, it was known that C_n are primes only for

$$n = 1, \ 141, \ 4713, \ 5795, \ 6611, \ 18496, \ 32292, \ 32469, \ 59656,$$
$$90825, \ 262419, \ 361275, \ 4811899, \ 1354828, \ 6328548, \ 6679881.$$

These numbers are called *Cullen primes*. It is not known yet whether there are infinitely many Cullen primes.

The Cullen numbers C_n are similar to the Fermat numbers in that $C_n - 1$ is divisible by high powers of two for large n and that Cullen's primes are probably very sparse (see [199, p. 157]). Generalized Cullen numbers of the form $nb^n + 1$ are studied as well.

8.7 Other Special Types of Integers

In 1917, A. J. Cunningham and H. J. Woodall dealt with properties of the numbers

$$W_n = n2^n - 1,$$

that are at present called the *Woodall numbers*. Sometimes they are also called *Cullen numbers of the second kind.*

There exist infinitely many composite Woodall numbers. For

$$n = 2, \ 3, \ 6, \ 30, \ 75, \ 81, \ 115, \ 123, \ 249, \ 362, \ 384, \ 462, \ 512, \ 751, \ 822, \ldots$$

the numbers W_n are prime (the so-called *Woodall primes*).

If for any $n \in \mathbb{N}$ the number

$$k2^n - 1$$

is composite, then k is called a *Riesel number*. There is a conjecture that $k = 509203$ is the smallest Riesel number. Their definition is similar to that of Sierpiński numbers: A natural number k is said to be a *Sierpiński number* if the sequence $(k2^n + 1)_{n=1}^{\infty}$ contains only composite numbers. The smallest known Sierpiński number $k = 78557$ was found by John L. Selfridge in 1962. As of November 2018, there are only five candidates $k = 21181, 22699, 24737, 55459,$ and 67607 which have not been eliminated as possible Sierpiński numbers. For more details about these numbers see [199].

The *Harmonic numbers* are given by

$$H_\alpha(n) = \sum_{t=1}^{n} \frac{1}{t\alpha^t},$$

for instance,

$$H_1(n) = 1 + \frac{1}{2} + \cdots + \frac{1}{n}.$$

Chapter 9
Magic and Latin Squares

9.1 Magic Squares

Millions of people around the world dealing with recreational mathematics investigate various magic and Latin squares (including Sudoku) for their mathematical beauty and also pleasure. These activities are also highly recommended by math teachers and pedagogues, since logical thinking is developed.

By a *magic square of order n* we shall understand an $n \times n$ square matrix (array) $M = (m_{ij})$ which contains n^2 distinct positive integer numbers so that the sum of all numbers in any row, in any column, and even on both diagonals is the same. We will denote this sum throughout this chapter by the symbol s. It is actually the *trace* of a matrix, i.e., $s = \sum_i m_{ii}$.

The case $n = 1$ is trivial and the case $n = 2$ is overdetermined, so no magic square of order 2 exists. Therefore, from now on, we shall consider only $n \geq 3$. The ancient Chinese already knew the following magic square (see Fig. 9.1) that by legend was engraved on a tortoise's shell around 2200 BC,

$$\begin{bmatrix} 4 & 9 & 2 \\ 3 & 5 & 7 \\ 8 & 1 & 6 \end{bmatrix}. \tag{9.1}$$

In Europe, Albrecht Dürer in his famous painting *Melencolia I* from 1514 depicted the magic square of order 4, which in the middle of the bottom line shows the year of creation of this painting,

$$\begin{bmatrix} 16 & 3 & 2 & 13 \\ 5 & 10 & 11 & 8 \\ 9 & 6 & 7 & 12 \\ 4 & 15 & 14 & 1 \end{bmatrix}. \tag{9.2}$$

Also Michael Stifel (1487–1567) investigated magic squares. In his treatise *Arithmetica Integra* (1544) he built a 16×16 magic square. Claude Gaspard Bachet de

© The Author(s), under exclusive license to Springer Nature Switzerland AG 2021
M. Křížek et al., *From Great Discoveries in Number Theory to Applications*,
https://doi.org/10.1007/978-3-030-83899-7_9

Fig. 9.1 The Lo Shu magic
square

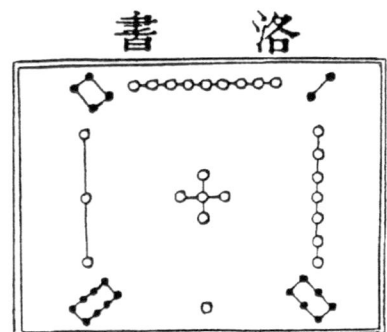

Méziriac (1581–1638) is also remembered in connection with construction of magic squares. He described the "diamond method" for constructing odd ordered magic squares in his book *Problèmes Plaisants* in 1624. According to the Oxford English Dictionary, the words "magic square" first appeared in 1704 in a technical lexicon.

A magic square of order n is said to be *normal*, if it contains consecutive integers $1, 2, \ldots, n^2$, see (9.1) and (9.2). A constructive proof of the next theorem is given, for instance, in [9, 301, 310].

Theorem 9.1 *For any $n \geq 3$ there exists a normal magic square of order n.*

Normal magic squares of order n are, of course, non-uniquely determined. For instance the transposition of matrix (9.1) or its rotation about the angle 90° is again a normal magic square. However, these operations cannot convert the following two magic squares into each other (notice that they have the same diagonals with a positive slope and the sum $s = 11 + 12 + 13 + 14 + 15 = 65$)

$$
\begin{bmatrix}
17 & 24 & 1 & 8 & 15 \\
23 & 5 & 7 & 14 & 16 \\
4 & 6 & 13 & 20 & 22 \\
10 & 12 & 19 & 21 & 3 \\
11 & 18 & 25 & 2 & 9
\end{bmatrix},
\quad
\begin{bmatrix}
23 & 6 & 19 & 2 & 15 \\
10 & 18 & 1 & 14 & 22 \\
17 & 5 & 13 & 21 & 9 \\
4 & 12 & 25 & 8 & 16 \\
11 & 24 & 7 & 20 & 3
\end{bmatrix}.
\tag{9.3}
$$

The French mathematician Bernard Frénicle de Bessy (1605–1675), a friend of Pierre de Fermat, found that there are 880 magic squares of order $n = 4$, which cannot be transformed by a finite number of matrix transposes and rotations of 90° to another magic square. For $n = 3$ a similar number is only 1 and there are 275 305 224 different magic squares of order $n = 5$ counted up to rotations and reflections, see Ratliff [317]. A similar result for the case $n \geq 6$ is unknown. Nevertheless, the following upper bound is proved in Ward [410, p. 111].

Theorem 9.2 *The number of normal magic squares of order $n \geq 3$ is at most*

$$
\frac{(n^2)!}{8(2n+1)!}.
$$

Theorem 9.3 *The sum of each row (column and both the diagonals) of a normal square of order n is equal to $s = \frac{1}{2}n(n^2 + 1)$.*

Proof Since a magic square of order n has n lines, we obtain

$$s = \frac{1}{n}\sum_{k=1}^{n^2} k = \frac{1}{n}\frac{n^2(n^2+1)}{2} = \frac{1}{2}n(n^2+1),$$

which proves the theorem. $\qquad\Box$

For instance, for the magic squares from (9.1), (9.2), and (9.3) we get $s = 15$, $s = 34$, and $s = 65$, respectively. These sums are called the *magic constants*.

When looking for general magic squares for $n \geq 3$, we have a large amount of arbitrariness, since we need to find n^2 different natural numbers satisfying a system of $2n + 2$ linear Diophantine equations. Consequently, people just for fun have been constructing various magic squares with some interesting properties. For instance,

$$\begin{bmatrix} 96 & 64 & 37 & 45 \\ 39 & 43 & 98 & 62 \\ 84 & 76 & 25 & 57 \\ 23 & 59 & 82 & 78 \end{bmatrix} \quad \text{and} \quad \begin{bmatrix} 69 & 46 & 73 & 54 \\ 93 & 34 & 89 & 26 \\ 48 & 67 & 52 & 75 \\ 32 & 95 & 28 & 87 \end{bmatrix}$$

is a pair of magic squares in which the first square has digits in the reverse order from the second square.

Another bizarre magic square was constructed by the Argentine mathematician Rodolfo Marcelo Kurchan. The digits of its entries are all different and also the row (column and diagonal) sum $s = 4\,129\,607\,358$ has the same property,

$$\begin{bmatrix} 1037956284 & 1036947285 & 1027856394 & 1026847395 \\ 1026857394 & 1027846395 & 1036957284 & 1037946285 \\ 1036847295 & 1037856294 & 1026947385 & 1027956384 \\ 1027946385 & 1026957384 & 1037846295 & 1036857294 \end{bmatrix}.$$

Dénes and Keedwell in their book [87] show a *multiplicative magic square*

$$\begin{bmatrix} 162 & 207 & 51 & 26 & 133 & 120 & 116 & 25 \\ 105 & 152 & 100 & 29 & 138 & 243 & 39 & 34 \\ 92 & 27 & 91 & 136 & 45 & 38 & 150 & 261 \\ 57 & 30 & 174 & 225 & 108 & 23 & 119 & 104 \\ 58 & 75 & 171 & 90 & 17 & 52 & 216 & 161 \\ 13 & 68 & 184 & 189 & 50 & 87 & 135 & 114 \\ 200 & 203 & 15 & 76 & 117 & 102 & 46 & 81 \\ 153 & 78 & 54 & 69 & 232 & 175 & 19 & 60 \end{bmatrix}.$$

The product and sum of all numbers in any row, any column and in both the diagonals is equal to the same number $2\,058\,068\,231\,856\,000$ and $s = 840$, respectively.

Magic squares of the same order can be multiplied by each other as matrices (see Thompson [400]), even though we generally do not get magic squares.

Theorem 9.4 *If M is a magic square of order 3 and k an odd natural number, than M^k is also a magic square.*

The proof proceeds by induction, but we will not do it. It is based on standard algebraic properties of matrices and the following fact. From the definition of a magic square M it follows directly that an eigenvalue of M is the trace s and the corresponding (right and left) eigenvector is $(1, 1, \ldots, 1)^\top$. For two magic squares M_1 and M_2 of the same order, one can show that the matrix $M_1 M_2$ has the same eigenvector (also $(M_1 M_2)^\top$ has the same eigenvector). The corresponding eigenvalue is $s_1 s_2$, where s_i is the trace of M_i for $i = 1, 2$.

Example Let M be the magic square from (9.1) with $s = 15$. It is easy to verify that M^2 is not a magic square, since all its diagonal and nondiagonal entries are equal to 59 and 83, respectively. However,

$$M^3 = \begin{bmatrix} 1149 & 1029 & 1197 \\ 1173 & 1125 & 1077 \\ 1053 & 1221 & 1101 \end{bmatrix}$$

is a magic square whose trace is equal to

$$s^3 = 15^3 = 3375.$$

Theorem 9.4 can be generalized to magic squares of order 4 and 5 under several restrictive conditions on off diagonal entries of M, see Thompson [400].

Theorem 9.5 *Let $2s$ be divisible by n. If each entry of a magic square of order n is subtracted from $2s/n$ and the result is always positive, then we again get a magic square.*

Proof Let $M = (m_{ij})$ be a given magic square. Since $2s$ is divisible by n, we find that

$$\overline{M} = \left(\frac{2s}{n} - m_{ij} \right) \tag{9.4}$$

contains only natural numbers. From (9.4) we observe that the row, column, and diagonal sums of $\overline{M}$ are equal to $2s - s = s$.   □

The magic square (9.4) satisfying the assumptions of Theorem 9.5 is called a *complementary magic square*. According to Theorem 9.3, the above assumption $n \mid 2s$ is satisfied for any normal magic square. In this case $\overline{M}$ is again a normal square.

9.2   The Existence of Prime Number Magic Squares

According to the Green-Tao Theorem 3.11 there exist arbitrarily long arithmetical progressions of finite length containing only prime numbers (Klazar [168]). We use this deep mathematical result to prove the existence of magic squares of an arbitrary order $n \geq 3$ containing only primes.

It is easy to verify that the following arithmetic progression of length $k = 10$

$$199, \ 409, \ 619, \ 829, \ 1039, \ 1249, \ 1459, \ 1669, \ 1879, \ 2089 \qquad (9.5)$$

consists solely of primes. An even longer arithmetic progression of primes is of the form

$$(223\,092\,870\,j + 2\,236\,133\,941)_{j=0}^{15}. \qquad (9.6)$$

Notice that the primes in progression (9.6) are very large. The main reason is the strong irregularities in the distribution of the prime numbers.

The Green-Tao Theorem 3.11 is a special case of the Erdős-Turán conjecture which claims that:

Let the sum of the reciprocal values from a subset $B \subset \{1, 2, 3, \dots\}$ be infinite. Then B contains an arithmetic progression of an arbitrarily long finite length.

To date, this conjecture has not been proved in full generality. Green and Tao have proved only the special case in which $B = P$ is the set of primes, since

$$\sum_{p \in P} \frac{1}{p} = \infty.$$

Theorem 9.6 *The sum of two magic squares of the same order is a magic square if all its n^2 entries are distinct.*

  Multiplying all entries of a magic square by the same positive integer, we again get a magic square.

  Moreover, adding the same integer to each entry of a magic square also gives a magic square if all entries are positive.

The proof is obvious. Theorem 9.6 is used for calculation of modules or cones of magic squares, see [24, 410].

Remark The sum of the following magic squares (cf. (9.2))

$$\begin{bmatrix} 16 & 3 & 2 & 13 \\ 5 & 10 & 11 & 8 \\ 9 & 6 & 7 & 12 \\ 4 & 15 & 14 & 1 \end{bmatrix} + \begin{bmatrix} 16 & 2 & 3 & 13 \\ 5 & 11 & 10 & 8 \\ 9 & 7 & 6 & 12 \\ 4 & 14 & 15 & 1 \end{bmatrix} = \begin{bmatrix} 32 & 5 & 5 & 26 \\ 10 & 21 & 21 & 16 \\ 18 & 13 & 13 & 24 \\ 8 & 29 & 29 & 2 \end{bmatrix}$$

is not a magic square, since its entries are not distinct as required by Theorem 9.6.

By a *prime number magic square* we understand a magic square containing only primes. The proof of the following theorem is constructive.

Theorem 9.7 *For any $n \geq 3$ there exist infinitely many prime number magic squares of order n.*

Proof First we show that for a given $n \geq 3$ there exists at least one prime number magic square. By the Green-Tao Theorem 3.11 there exist n^2 primes $p_1, p_2, \ldots, p_{n^2}$ that form an arithmetic progression. Denote by $a = p_1$ the initial term of this progression and by d the difference $d = p_2 - p_1$. By Theorem 9.1 there exists a normal magic square of order n containing consecutive integers $1, 2, \ldots, n^2$. If we multiply it by the number d, we get again a magic square (cf. Theorem 9.6). Now if we add the number $a - d$ to all its entries, we also get a magic square (see Theorem 9.6). In other words, we replace the numbers $1, 2, \ldots, n^2$ in the same order by the primes $p_1, p_2, \ldots, p_{n^2}$, and thus we get the required magic square.

By a contradiction argumentation we show now that there must be infinitely many such prime number magic squares of a given order $n \geq 3$. So let $n \geq 3$ be fixed and assume that there are only $m \geq 1$ prime number magic squares of order n, where m is fixed. By the Green-Tao Theorem 3.11 there exists an arithmetic progression of length $n^2 + m$ containing only primes. This enables us to construct $m + 1$ different prime number magic squares from progressions

$$p_1, p_2, \ldots, p_{n^2},$$
$$p_2, p_3, \ldots, p_{n^2+1},$$
$$\vdots$$
$$p_{m+1}, p_{m+2}, \ldots, p_{n^2+m},$$

which is a contradiction.   □

Now we shall present two examples to show how prime number magic squares can be constructed.

Example In the proof of Theorem 9.7 we choose $a = 199$ and $d = 210$ which corresponds to progression (9.5). By means of (9.1) we obtain the following prime number magic square

$$\begin{bmatrix} 829 & 1879 & 409 \\ 619 & 1039 & 1459 \\ 1669 & 199 & 1249 \end{bmatrix},$$

for which $s = 3117$. We can also choose $a = 409$ in the same progression (9.5) and get another prime number magic square.

Example Let us consider the normal square (9.2) and in the proof of Theorem 9.7 set $a = 2\,236\,133\,941$ and $d = 223\,092\,870$. In this way we obtain by (9.6) the following prime number magic square

$$\begin{bmatrix} 5582526991 & 2682319681 & 2459226811 & 4913248381 \\ 3128505421 & 4243969771 & 4467062641 & 3797784031 \\ 4020876901 & 3351598291 & 3574691161 & 4690155511 \\ 2905412551 & 5359434121 & 5136341251 & 2236133941 \end{bmatrix}.$$

9.3 Further Prime Number Magic Squares

Let us look at further prime number squares, which are not constructed by means of the Green-Tao Theorem 3.11. Rudolf Ondrejka (1928–2001) discovered the following prime number magic square

$$\begin{bmatrix} 17 & 89 & 71 \\ 113 & 59 & 5 \\ 47 & 29 & 101 \end{bmatrix}$$

whose entries do not form an arithmetic progression if we compare them by size.

Another example shows that

$$\begin{bmatrix} 37 & 83 & 97 & 41 \\ 53 & 61 & 71 & 73 \\ 89 & 67 & 59 & 43 \\ 79 & 47 & 31 & 101 \end{bmatrix}$$

is a magic square containing consecutive primes (see Table 13.1) which evidently do not form an arithmetic progression. It is surprising that a similar magic square of order 3 with consecutive primes and the smallest possible sum s contains much larger numbers:

$$\begin{bmatrix} 1480028201 & 1480028129 & 1480028183 \\ 1480028153 & 1480028171 & 1480028189 \\ 1480028159 & 1480028213 & 1480028141 \end{bmatrix}.$$

Relatively small consecutive primes (see Table 13.1) are contained in the following magic square of order 6,

$$\begin{bmatrix} 251 & 389 & 311 & 449 & 347 & 353 \\ 313 & 359 & 293 & 373 & 379 & 383 \\ 397 & 271 & 419 & 263 & 401 & 349 \\ 269 & 317 & 367 & 421 & 283 & 443 \\ 439 & 307 & 277 & 337 & 409 & 331 \\ 431 & 457 & 433 & 257 & 281 & 241 \end{bmatrix}.$$

Further, let us introduce an *apocalyptic prime number magic square* of order 6 which was discovered by Allan W. Johnson (cf. [300]),

$$\begin{bmatrix} 3 & 107 & 5 & \mathbf{131} & 109 & 311 \\ 7 & 331 & 193 & 11 & \mathbf{83} & 41 \\ 103 & 53 & 71 & 89 & 151 & \mathbf{199} \\ \mathbf{113} & 61 & 97 & 197 & 167 & 31 \\ 367 & \mathbf{13} & 173 & 59 & 17 & 37 \\ 73 & 101 & \mathbf{127} & 179 & 139 & 47 \end{bmatrix},$$

whose sum of numbers in each row, column, and on both main diagonals is equal to the *apocalyptic number* $s = 666$ (which is also called the *beastly number*, since it is the number of the beast). The same sum is also given by the so-called broken or interrupted diagonals in both directions (one of them is in bold), for instance,

$$113 + 13 + 127 + 131 + 83 + 199 = 666,$$
$$3 + 101 + 173 + 197 + 151 + 41 = 666.$$

The apocalyptic number 666 has a lot of special representations. For example, it is a sum of squares of consecutive primes

$$666 = 2^2 + 3^2 + 5^2 + 7^2 + 11^2 + 13^2 + 17^2.$$

It is the largest triangular number that has all its digits the same, i.e.,

$$666 = 1 + 2 + 3 + \cdots + 36 = T_{36},$$

and its digits are perfect numbers. It is also a *Smith number* whose sum of digits is equal to the sum of the digits of all prime factors of the number $666 = 2 \cdot 3 \cdot 3 \cdot 37$, that is,

$$6 + 6 + 6 = 2 + 3 + 3 + 3 + 7.$$

The number 666 is also the sum of two palindromic primes $666 = 313 + 353$,

$$666 = 1^6 - 2^6 + 3^6, \quad 666 = 6 + 6 + 6 + 6^3 + 6^3 + 6^3,$$

etc. Moreover, the value of the Euler function is $\phi(666) = 6 \times 6 \times 6$.

9.4 Construction of $3 \times 3 \times 3$ Prime Number Magic Cube

Magic squares can be generalized to rectangles, stars, cubes or even hypercubes, see e.g. Dénes and Keedwell [87] , Pickover [301, p. 238]. For instance, in Trenkler [403] a simple necessary and sufficient condition for the existence of a magic d-dimensional hypercube of order n is introduced. By the Green-Tao Theorem 3.11 we can prove the existence of a prime number magic hypercube.

Now, we show how a prime number $3 \times 3 \times 3$ magic cube can be constructed concretely.

Example Consider a $3 \times 3 \times 3$ cube whose front, middle, and back layers have the form

$$\begin{bmatrix} 8 & 24 & 10 \\ 12 & 7 & 23 \\ 22 & 11 & 9 \end{bmatrix}, \quad \begin{bmatrix} 15 & 1 & 26 \\ 25 & 14 & 3 \\ 2 & 27 & 13 \end{bmatrix}, \quad \begin{bmatrix} 19 & 17 & 6 \\ 5 & 21 & 16 \\ 18 & 4 & 20 \end{bmatrix}. \tag{9.7}$$

Each column sum is equal to $s = 42$ and also the sum of all three numbers in each of both kinds of horizontal rows is s. The middle layer forms a magic square, i.e., the sum of each of its diagonals is s. From all three second rows and all three second columns we can also construct 3×3 magic squares with the same sum of s:

$$\begin{bmatrix} 12 & 7 & 23 \\ 25 & 14 & 3 \\ 5 & 21 & 16 \end{bmatrix} \quad \text{and} \quad \begin{bmatrix} 24 & 1 & 17 \\ 7 & 14 & 21 \\ 11 & 27 & 4 \end{bmatrix},$$

respectively. Moreover, we also see that all four spatial diagonals of the cube give the same sum:

$$s = 8 + 14 + 20 = 22 + 14 + 6 = 9 + 14 + 19 = 10 + 14 + 18.$$

Since the cube (9.7) contains all the integers $1, 2, \ldots, 27$, it is called a *normal magic cube*.

A normal $3 \times 3 \times 3$ magic cube is not uniquely determined up to rotations and reflections as a 3×3 magic square is. For instance, the cube

$$\begin{bmatrix} 2 & 13 & 27 \\ 22 & 9 & 11 \\ 18 & 20 & 4 \end{bmatrix}, \quad \begin{bmatrix} 16 & 21 & 5 \\ 3 & 14 & 25 \\ 23 & 7 & 12 \end{bmatrix}, \quad \begin{bmatrix} 24 & 8 & 10 \\ 17 & 19 & 6 \\ 1 & 15 & 26 \end{bmatrix}$$

is not isomorphic to that in (9.7).

Coincidentally, the longest known prime number arithmetic progression was found in September 2019 (see [428]):

$$(18\,135\,696\,597\,948\,930\,j + 224\,584\,605\,939\,537\,911)_{j=0}^{26}. \tag{9.8}$$

We can use it to construct a prime magic cube of order 3 so that in the normal magic cube (9.7) we will gradually replace the numbers 1, 2, ..., 27 by the above prime numbers (cf. the proof of Theorem 9.7). Using (9.8), we can also construct several concrete prime number magic squares of order 5.

9.5 Latin Squares

The notion of the Latin square was introduced by Leonhard Euler in 1782 as follows: A *Latin square of order n* is an array with n rows and n columns, where in each row and each column, each entry of some set of n elements appears exactly once, see [107]. Entries of Latin squares can be different objects, e.g., playing cards, chess figurines, people, letters or natural numbers,

$$
\begin{bmatrix}
1 & 4 & 3 & 6 & 5 & 2 \\
6 & 1 & 5 & 4 & 2 & 3 \\
5 & 3 & 1 & 2 & 6 & 4 \\
4 & 6 & 2 & 1 & 3 & 5 \\
3 & 2 & 4 & 5 & 1 & 6 \\
2 & 5 & 6 & 3 & 4 & 1
\end{bmatrix}
\quad \text{and} \quad
\begin{bmatrix}
A & B & F & C & E & D \\
B & C & A & D & F & E \\
C & D & B & E & A & F \\
D & E & C & F & B & A \\
E & F & D & A & C & B \\
F & A & E & B & D & C
\end{bmatrix}.
\tag{9.9}
$$

A famous problem that stood at the birth of the Latin square notion concerns 36 officers. It was formulated by Euler in 1779 in the following way:

Six different regiments have six officers, each one belonging to different ranks. Can these 36 officers be arranged in a square formation so that each row and column contains one officer of each rank and one of each regiment?

Euler thought that this problem has no solution, but he could not prove it. As of 1900, Gaston Tarry (1843–1913) proved by a systematic calculation that this problem really has no solution (see [161]). It is worth mentioning that many works devoted to Latin squares concerned card-related games. For example, a task of arranging 16 bridge cards was presented by Jacques Ozanam at the end of the 17th century (see Fig. 9.2).

Let L_1 and L_2 be two Latin squares, where $L_1 = (a_{ij})$, $L_2 = (b_{ij})$, $i, j = 1, 2, \ldots, n$. Then L_1 and L_2 are called *orthogonal,* if all pairs (a_{ij}, b_{ij}), $i, j = 1, 2, \ldots, n$, are mutually distinct.

The introduction of the notion of orthogonality has its origin in Euler's problem about 36 officers. Its solution was to find two mutually orthogonal Latin squares of order 6. If the Latin squares $L_1 = (a_{ij})$ and $L_2 = (b_{ij})$ are orthogonal, then in the set of pairs $E = (a_{ij}, b_{ij})$, each of these pairs occurs just once. Such a square is then called the *Euler square,* (see e.g. Fig. 9.2). It is easy to find that there are no orthogonal Latin squares of order 2. The non-existence of orthogonal squares of order 6 is actually the reason for the unsolvability of the problem about 36 officers. Compare, for example, the entries for $i = 5$, $j = 1$ and $i = 2$, $j = 6$ in Latin squares (9.9).

$$\begin{bmatrix} A\clubsuit & K\spadesuit & Q\heartsuit & J\diamondsuit \\ Q\diamondsuit & J\heartsuit & A\spadesuit & K\clubsuit \\ J\spadesuit & Q\clubsuit & K\diamondsuit & A\heartsuit \\ K\heartsuit & A\diamondsuit & J\clubsuit & Q\spadesuit \end{bmatrix} .$$

Fig. 9.2 Example of two orthogonal Latin squares, i.e., the Euler square. In each row and in each column (even on both diagonals) exactly one of the ranks: A, K, Q, J is located and at the same time exactly one of the suits: hearts $\heartsuit$, clubs $\clubsuit$, spades $\spadesuit$, and diamonds $\diamondsuit$

Euler conjectured that if $n \equiv 2$ (mod 4), then there are no orthogonal squares of order n. But he was wrong. In 1960, an original proof was published (see [40]) that orthogonal Latin squares exist for all natural numbers n except for the cases $n = 2$ and $n = 6$. However, to a given Latin square there may not exist an orthogonal Latin square, in general. This was proved by Mann in [257]. A generalization of this statement can be found in the article [294]. In the theory of orthogonality of Latin squares there are two basic problems: the problem of the existence of an orthogonal Latin square to a given Latin square and the related problem of its construction.

When Arthur Cayley (1821–1895) dealt with multiplication tables of finite groups, he found that these tables are Latin squares. On the other hand, a Latin square need not be the multiplication table of a finite group. Later it turned out that Latin squares can be used as a means of determining the existence of finite projective planes of different orders. Therefore, now we will briefly mention these planes which resemble classical infinite projective planes.

A *finite projective plane of order* $n \in \{1, 2, 3, \dots\}$ is a set consisting of $n^2 + n + 1$ points and $n^2 + n + 1$ lines which fulfill the following axioms:

(A1) Each point lies on $n + 1$ lines.
(A2) Each line contains $n + 1$ points.
(A3) Any two different points lie on exactly one line.
(A4) Each two different lines always intersect in exactly one point.

There are other equivalent definitions of a finite projective plane.

The above system of axioms of the finite projective plane turns out to be most appropriate, as it aptly characterizes the duality between lines and points. If we swap the words "point" and "line" in the system of axioms (A1)–(A4) and if we exchange at the same time the words "lie" and "intersect", then the system of axioms does not change.

The consequence of this is that if we exchange these terms in a true statement concerning a projective plane of order n, then we get again a true statement. It can be easily found that the smallest finite projective plane for $n = 1$ is a triangle (see Fig. 9.3). A less trivial example of a finite projective plane of order $n = 2$ is called the *Fano plane* which has, by the above definition, 7 points and 7 lines. An interesting thing about this plane is that it is not possible to place all its "lines" on regular straight lines. Therefore, one of its "lines" is mapped on a circle (see Fig. 9.3).

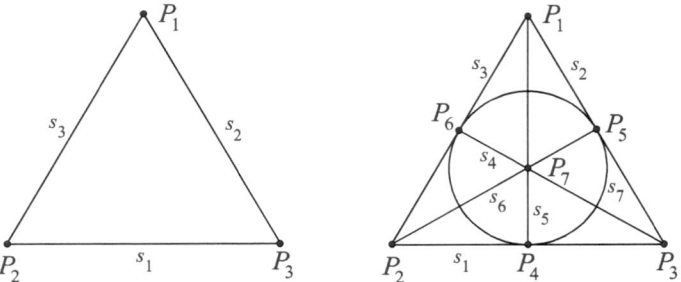

Fig. 9.3 Finite projective planes for $n = 1$ (triangle) and for $n = 2$ (the Fano plane)

It turns out that the existence of a finite projective plane of order n is closely related to the existence of $n - 1$ mutually orthogonal squares of order n. Let us still note that if there exist $n - 1$ mutually orthogonal Latin squares of order n, then this system of squares is called *complete*. Now let us state the following theorem whose proof can be found in Rybnikov [343].

Theorem 9.8 *Let $n \geq 2$. A projective finite plane of order n exists if and only if there exists a complete system of $n - 1$ mutually orthogonal Latin squares of order n.*

In 1938, Raj Chandra Bose (cf. [40]) showed why projective planes of order 6 do not exist. Because there do not exist orthogonal Latin squares of order 6, the non-existence of a finite projective plane of order 6 also follows from Theorem 9.8. The existence of a finite projective plane of order n is given by the following theorem (see Rybnikov [343]).

Theorem 9.9 *If n is a power of a prime, then there exists a finite projective plane of order n.*

It is well-known that for $n = 2$, 3, 4, 5, 7, and 8 a finite projective plane is uniquely determined. However, for $n = 9$ there exist 4 mutually non-isomorphic finite projective planes. In 1949, the following theorem was proved (see [47]).

Theorem 9.10 (Bruck-Ryser) *Let $n \equiv 1, 2 \pmod 4$. If there exists a finite projective plane of order n, then there exist integers x and y such that $n = x^2 + y^2$.*

However, this condition is not sufficient, since $10 \equiv 2 \pmod 4$, $10 = 1^2 + 3^2$, and the non-existence of a finite projective plane of order 10 was proved by extraordinarily extensive calculations on computers. The Bruck-Ryser Theorem 9.10 also does not state anything about the case $n = 12$ (investigated in [388]). On the other hand, Theorem 9.10 implies the non-existence of a finite projective plane, e.g., for $n = 14$, since $14 \equiv 2 \pmod 4$ and the number 14 cannot be expressed as the sum of two squares.

Further, notice that the Fermat Christmas Theorem 3.12 is a special case of Bruck-Ryser Theorem 9.10, since primes of the form $p \equiv 1 \pmod 4$ (for which a finite projective plane of order p exists by Theorem 9.9) can be uniquely written as the sum of two squares. For a further generalization of Theorem 9.10 see [365].

Various practical applications of Latin squares can be found when encoding secret documents, when planning experiments in operational research, or when cultivating several varieties of agricultural crops on small square fields with different qualities of the soil. There are also very significant medical applications; another example involves finding the outline of diamonds in stone using X-rays (cf. e.g. Batenburg and Sijbers [25]). An 8×8 array compiled from all possible nucleotide triplets (adenine, cytosine, guanine, and uracil) can also be interpreted as a Latin square (see [161]). Latin squares are used also when organizing golf, bridge or tennis tournaments. A curious example of the Latin square 9×9 is also exemplified by every final position of the popular puzzle Sudoku, see Herzberg and Murty [144].

9.6 Sudoku

Recall that the Sudoku square is divided into 9 blocks of type 3×3. In the final Sudoku position, each block, each row, and each column must contain exactly one of the digits $1, 2, \ldots, 9$, see [113, 140]. For example the 9×9 Latin square consisting of 3×3 blocks

$$\begin{bmatrix} 1\ 6\ 8 & 9\ 2\ 4 & 5\ 7\ 3 \\ 9\ 2\ 4 & 5\ 7\ 3 & 1\ 6\ 8 \\ 5\ 7\ 3 & 1\ 6\ 8 & 9\ 2\ 4 \\ 8\ 1\ 6 & 4\ 9\ 2 & 3\ 5\ 7 \\ 4\ 9\ 2 & 3\ 5\ 7 & 8\ 1\ 6 \\ 3\ 5\ 7 & 8\ 1\ 6 & 4\ 9\ 2 \\ 6\ 8\ 1 & 2\ 4\ 9 & 7\ 3\ 5 \\ 2\ 4\ 9 & 7\ 3\ 5 & 6\ 8\ 1 \\ 7\ 3\ 5 & 6\ 8\ 1 & 2\ 4\ 9 \end{bmatrix}$$

represents a final position of Sudoku. Its central block is the normal magic square (9.1). However, the other 8 surrounding blocks are not magic squares, even though their row and column sums are 15, but their diagonal sums are not always 15.

The word "Sudoku" is the Japanese abbreviation of the sentence *Suuji wa dokushin ni kagiru* which means the number must remain alone. The Sudoku game first appeared in May 1979 in the edition *Dell Pencil Puzzles & Word Games* on page 6 under the name *Number Place*. It was invented by the architect Howard Garns (see [86, 117]) who unfortunately died a few years later. So he did not experience its huge boom in the 21st century. Sudoku owes its success mainly due to Wayne Gould from New Zealand, who became acquainted with this game during his stay in Japan. Later, he wrote over a time span of 6 years (until 1997) a program that automatically generates a Sudoku problem giving a single solution. The digits $1, 2, \ldots, 9$ are gradually and completely randomly added to the 9×9 grid so that the rules of the final

Sudoku position are met. As soon as the program finds out that there is exactly one solution, no more digits are added and the Sudoku problem is done.

Bertram Felgenhauer from the Technical University of Dresden and Frazer Jarvis from the University of Sheffield (GB) derived (see [114]) that the number of all possible positions of the Sudoku game is

$$9! \, 2^{13} \, 3^4 = 6\,670\,903\,752\,021\,072\,936\,960 \approx 6.671 \cdot 10^{21}.$$

Eliminating all "equivalent" configurations modulo all symmetries (i.e., permutations of numbers $1, 2, \ldots, 9$, reflections, rotations, permutations of the first three rows,..., permutations of the last three columns, permutations of 3×9 rectangular blocks, etc.), the total number of different Sudoku positions is only $5\,472\,730\,538$, which is less than the total number of people on Earth. For comparison note that the total number of Latin squares of order 9 is approximately $5.525 \cdot 10^{27}$, see [22, 144].

Now we focus on the uniqueness of the Sudoku solution. Consider the problem

$$\begin{bmatrix} 9 & 2 & 6 & 5 & 7 & 1 & 4 & 8 & 3 \\ 3 & 5 & 1 & 4 & 8 & 6 & 2 & 7 & 9 \\ 8 & 7 & 4 & 9 & 2 & 3 & 5 & 1 & 6 \\ 5 & 8 & 2 & 3 & 6 & 7 & 1 & 9 & 4 \\ 1 & 4 & 9 & 2 & 5 & 8 & 3 & 6 & 7 \\ 7 & 6 & 3 & 1 & x & y & 8 & 2 & 5 \\ 2 & 3 & 8 & 7 & y & x & 6 & 5 & 1 \\ 6 & 1 & 7 & 8 & 3 & 5 & 9 & 4 & 2 \\ 4 & 9 & 5 & 6 & 1 & 2 & 7 & 3 & 8 \end{bmatrix}.$$

We see that it has two solutions $x = 4$, $y = 9$ and $x = 9$, $y = 4$. From this it follows that even 77 given digits do not guarantee that the solution is unique. However, at least 78 given digits are enough for the uniqueness.

On the other hand, there are tens of thousands of examples, where only 17 given digits in the initial position lead to a unique solution (for comparison, 17 is the number of digits on the two main diagonals). It is known that there is no unique solution of a Sudoku problem with 16 given digits, see [267]. Let us introduce an example with 16 digits which has exactly two solutions (cf. Delahaye [86]):

$$\begin{bmatrix} 1 & . & . & . & . & . & . & . & 5 \\ . & . & . & . & 3 & . & . & . & . \\ . & . & 2 & . & 4 & . & . & . & . \\ . & . & . & . & . & . & . & . & . \\ . & 3 & 4 & . & . & . & 7 & . & . \\ . & . & . & 2 & . & 6 & . & . & 1 \\ 2 & . & . & . & 5 & . & . & . & . \\ . & 7 & . & . & . & . & . & 3 & . \\ . & . & . & . & . & 1 & . & . & . \end{bmatrix}.$$

This problem cannot have a unique solution, since the grid does not contain the digits 8 and 9. Assuming the uniqueness, we easily come to a contradiction, since it is enough to exchange 8 and 9. A necessary condition for the uniqueness can be formulated as follows.

Theorem 9.11 *If a Sudoku problem has exactly one solution, then at least 8 different digits have to be given.*

This theorem is a direct consequence of much stronger statements on chromatic polynomials in graph theory [144]. There exist many algorithms to solve the Sudoku problem not only in the two-dimensional 9×9 case, but also in other grids and higher-dimensional cases, see [86, 164].

Chapter 10
The Mathematics Behind Prague's Horologe

10.1 Prague Clock Sequence

The astronomical clock in the Old Town Square of Prague is an astrolabe (see Fig. 10.1) controlled by a sophisticated clockwork mechanism. The origin of its mathematical model is attributed to Joannes Andreae (1375–1456), called *Šindel* (see Horský [148], Horský and Procházka [149]). He invented this model more than 600 years ago. In honor of this great achievement we introduced in [217] the term "Šindel sequence", which we will investigate in the next sections (for a similar approach see also Weintraub [414]). The clock was realized by the skilled clockmaker Nicholas from Kadaň in 1409. Over the centuries its construction has been renovated several times. We will demonstrate the ability of the clockmakers of that time by studying the construction of a device for the stabilization of the bell strikes.

The bellworks of the Prague horologe contains a large gear with 24 slots (the first two slots are joined into one) at increasing distances along its circumference, see Figs. 10.2 and 10.3. This arrangement allows for the periodic repetition of 1–24 strokes of the bell each day. Part of the bellwork also consists of a small auxiliary gear whose circumference is divided by 6 slots into segments of arc lengths 1, 2, 3, 4, 3, 2 (see Figs. 10.2 and 10.3). These numbers constitute a period which repeats after each revolution and their sum is

$$s = 15.$$

At the beginning of every hour a catch rises, both gears start to revolve and the bell chimes. The gears stop when the catch simultaneously falls back into the slots on both gears. The bell strikes

$$1 + 2 + \cdots + 24 = 300$$

times every day. Since the triangular number $T_{24} = 300$ is divisible by $s = 15$, the small gear is always at the same position at the beginning of each day.

© The Author(s), under exclusive license to Springer Nature Switzerland AG 2021
M. Křížek et al., *From Great Discoveries in Number Theory to Applications*,
https://doi.org/10.1007/978-3-030-83899-7_10

Fig. 10.1   Astronomical dial of the Prague horologe in 2019

Fig. 10.2   A detail of the bellworks of the astronomical clock. The catch is in the slot between the segments corresponding to 8 and 9 h on the large gear

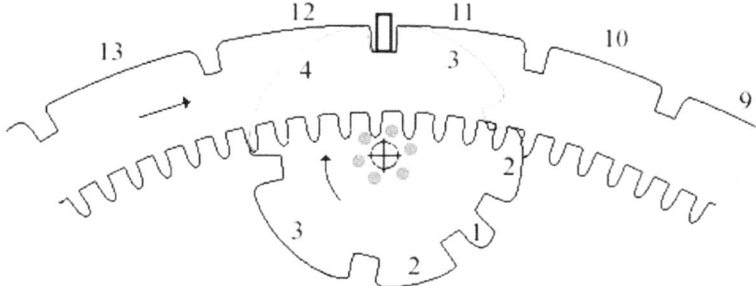

Fig. 10.3 The number of bell strokes is denoted by the numbers …, 9, 10, 11, 12, 13, …along the circumference of the large wheel. The small auxiliary gear placed behind it is divided by slots into segments of arc lengths 1, 2, 3, 4, 3, 2. The catch is indicated by a small rectangle on the top

The large gear has 120 interior teeth which drop into a pin gear with 6 little horizontal bars that surround the center of the small gear (see Figs. 10.2 and 10.3). The large gear revolves one time per day and therefore, the small gear revolves 20 times per day with approximately 4 times greater circumferential speed, since its circumference is 5 times smaller. Thus, the small gear makes the regulation of strokes sufficiently precise despite the wearing out of the slots on the large gear. Moreover, one stroke of the bell at one a.m. after midnight is due only to the movement of the small gear. The reason is that there is no tooth between the first and second slot of the large gear. It would be so thin that it would break soon.

In other similar devices there are not even any teeth corresponding to 1,2,3, and 4 strokes of the bell, see e.g. Leopold [229, p. 26] and Weintraub [415, p. 311]. This demonstrates that the small gear is not a mere mathematical curiosity, but an essential component of the bellworks (see also Fig. 10.4).

When the small gear revolves it generates by means of its slots a periodic sequence whose particular sums correspond to the number of strokes of the bell at each hour (see [437, A028356]):

$$
\underbrace{1\ 2\ 3\ 4}\ \underbrace{3\ 2}_{5}\ \underbrace{1\ 2\ 3}_{6}\ \underbrace{4\ 3}_{7}
$$

$$
\underbrace{2\ 1\ 2\ 3}_{8}\ \underbrace{4\ 3\ 2}_{9}\ \underbrace{1\ 2\ 3\ 4}_{10}\ \underbrace{3\ 2\ 1\ 2\ 3}_{11}\ \underbrace{4\ 3\ 2\ 1\ 2}_{12}
$$

$$
\underbrace{3\ 4\ 3\ 2\ 1}_{13}\ \underbrace{2\ 3\ 4\ 3\ 2}_{14}\ \underbrace{1\ 2\ 3\ 4\ 3\ 2}_{15}\ \dots \tag{10.1}
$$

In the next section we show that we could continue in this way until infinity. However, not all periodic sequences have such a nice summation property. For instance, we immediately find that the period 1, 2, 3, 4, 5, 4, 3, 2 could not be used for such a purpose, since for 6 bell strikes we have $6 < 4 + 3$. Also the period 1, 2, 3, 2 could not be used, since for 4 bell strikes we see that $2 + 1 < 4 < 2 + 1 + 2$.

Fig. 10.4 This figure
adapted from Leopold [229,
p. 26] shows another
bellwork than that used in
Prague's horologe. It
demonstrates that the small
auxiliary gear is not a mere
mathematical curiosity, but
an essential component of
bellworks at that time, since
there are no teeth of the large
gear corresponding to 1, 2, 3,
and 4 strokes of the bell, see
the lowest large gap

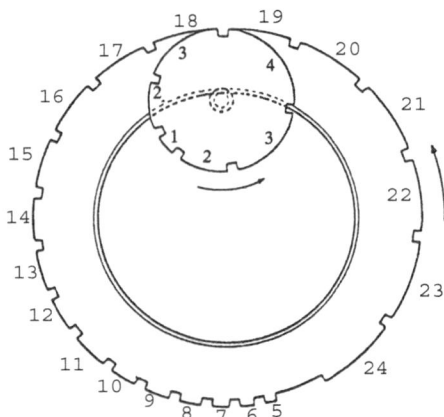

The Prague Astronomical Clock is probably the oldest still functioning clockwork
containing such an ingenious device illustrated in Fig. 10.3, see Horský [148, p. 78]).
Due to the beautiful summation property discussed above, we shall call the sequence

$$1, \ 2, \ 3, \ 4, \ 3, \ 2, \ 1, \ 2, \ 3, \ 4, \ 3, \ 2, \ldots$$

in (10.1) the *Prague clock sequence*, cf. [358, A028355, A028356].

In the following sections we prove a number of mathematical theorems concerning
a generalization of this sequence. We will actually be interested in how to design an
irregular auxiliary gear for different general values of the sum s.

10.2 Connection with Triangular Numbers and Periodic Sequences

In this section we show how the *triangular numbers*

$$T_k = 1 + 2 + \cdots + k = \frac{k(k+1)}{2}, \quad k = 0, 1, 2, \ldots, \tag{10.2}$$

are related to the Prague clock sequence. We shall look for all periodic sequences of
natural numbers that have a similar summation property as the sequence in (10.1),
i.e., that could be used in the construction of the small gear depicted in Fig. 10.3.

A sequence $(a_i)_{i=1}^{\infty}$ of positive integers is called *periodic,* if there exists $p \in \mathbb{N}$
such that

$$\forall i \in \mathbb{N} : a_{i+p} = a_i. \tag{10.3}$$

The finite sequence $a_1, \ldots, a_p$ is called a *period* and p the *period length*. The smallest p satisfying (10.3) is called the *minimal period length* and the associated sequence $a_1, \ldots, a_p$ is called the *minimal period*.

Let $(a_i) \subset \mathbb{N}$ be a periodic sequence. We say that the triangular number T_k for $k \in \mathbb{N}$ is *achievable* by (a_i), if there exists a positive integer n such that

$$T_k = a_1 + a_2 + \cdots + a_n. \tag{10.4}$$

The periodic sequence (a_i) is said to be a *Šindel sequence,* if T_k is achievable by (a_i) for all $k \in \mathbb{N}$, i.e.,

$$\forall k \in \mathbb{N} \ \exists n \in \mathbb{N} : \ T_k = \sum_{i=1}^{n} a_i. \tag{10.5}$$

The triangular number T_k on the left-hand side is equal to the sum $1 + \cdots + k$ of hours on the large gear, while the sum on the right-hand side expresses the corresponding rotation of the small gear, see Fig. 10.3. For the kth hour, we get

$$k = T_k - T_{k-1} = \sum_{i=m+1}^{n} a_i, \tag{10.6}$$

where $T_{k-1} = \sum_{i=1}^{m} a_i$. Since $a_i > 0$, the number n in (10.5) depending on k is uniquely determined. From (10.2) and (10.4) we also see that $a_1 = 1$ when (a_i) is a Šindel sequence.

The next theorem shows that condition (10.5) with infinitely many $k \in \mathbb{N}$ can be replaced by a much weaker condition containing only a finite number of k's. This enables us to perform only a finite number of arithmetic operations to check whether a given period $a_1, \ldots, a_p$ yields a Šindel sequence. From now on let

$$s = \sum_{i=1}^{p} a_i \tag{10.7}$$

denote the sum of the period.

Theorem 10.1 *A periodic sequence (a_i) for s odd is a Šindel sequence, if T_k is achievable by (a_i) for $k = 1, 2, \ldots, \frac{1}{2}(s-1)$.*

Proof The case $s = 1$ is trivial. So let $s \geq 3$ be odd and suppose that

$$\forall k \in \{1, 2, \ldots, \tfrac{1}{2}(s-1)\} \ \exists n \in \mathbb{N} : \ T_k = \sum_{i=1}^{n} a_i. \tag{10.8}$$

According to (10.7), we have

$$1 + 2 + \cdots + (s - 1) = \frac{s - 1}{2} \sum_{i=1}^{p} a_i, \tag{10.9}$$

where p is the period length and $\frac{1}{2}(s - 1)$ is an integer. For the corresponding sequence

$$\underbrace{a_1, a_2, \ldots, a_p}_{s}, \underbrace{a_1, a_2, \ldots, a_p}_{s}, \ldots, \underbrace{a_1, a_2, \ldots, a_p}_{s}, \tag{10.10}$$

formula (10.9) expresses that the period $a_1, a_2, \ldots, a_p$ in (10.10) is repeated $\frac{1}{2}(s - 1)$ times.

Assuming (10.8), we have to verify equality (10.4) for all $k \geq \frac{1}{2}(s + 1)$. For $k = s - 1$, which is even, we obtain by (10.2), (10.9), and (10.3) that

$$T_k = T_{s-1} = \frac{k}{2} \sum_{i=1}^{p} a_i = \sum_{i=1}^{pk/2} a_i,$$

i.e., $n = \frac{1}{2} pk$ in (10.4) and the number T_{s-1} is achievable.

Suppose now that $k = s - 1 - k'$, where $1 \leq k' \leq \frac{1}{2}(s - 3)$ and $s > 3$. By assumption (10.8), there exists $n' \in \mathbb{N}$ such that

$$\frac{k'(k' + 1)}{2} = \sum_{i=1}^{n'} a_i. \tag{10.11}$$

From (10.2) we observe that

$$T_k = T_{s-1-k'} = \frac{(s - 1 - k')(s - k')}{2} = \frac{s(s - 1 - 2k')}{2} + \frac{k'(k' + 1)}{2}. \tag{10.12}$$

Since s is odd and $1 \leq k' \leq \frac{1}{2}(s - 3)$, it follows that $m = s - 1 - 2k'$ is an even positive integer. Thus, by (10.12), (10.7), (10.11), and (10.3) we have

$$T_k = \frac{s - 1 - 2k'}{2} \sum_{i=1}^{p} a_i + \sum_{i=1}^{n'} a_i = \sum_{i=1}^{pm/2+n'} a_i.$$

Next, let $k = qs + k'$ with $q \in \mathbb{N}$ and $0 \leq k' < s$. Then by (10.2) and (10.7) we find that

$$T_k = \frac{(qs + k')(qs + k' + 1)}{2} = sj + \frac{k'(k' + 1)}{2} = \sum_{i=1}^{pj} a_i + T_{k'},$$

Fig. 10.5 Schematic
illustration of the triangular
number T_7. Notice that
$T_3 = 6$ and $T_7 = 28$ are
perfect numbers,
cf. Theorem 8.21. The bullets
in the kth row indicate the
number of strokes at the kth
hour, see (10.6). The
numbers denote lengths of
segments on the small gear

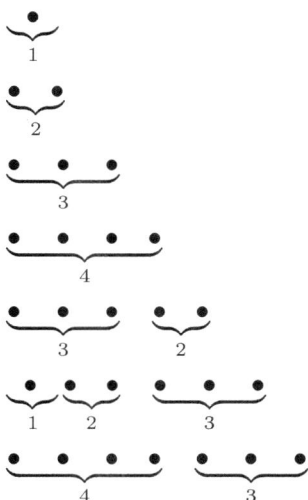

where $j = q(qs + 1)/2 + qk'$ is integer and $T_{k'} = 0$ for $k' = 0$. By our earlier observation in this proof, $T_{k'} = \sum_{i=1}^{n'} a_i$ for some $n' \in \mathbb{N}$ when $0 < k' < s$. □

Remark The number $\frac{1}{2}(s - 1)$ in (10.8) cannot be reduced if p is the minimal period length associated with s. To see this, it is enough to consider the periodic sequence (a_i) with the minimal period $1, 2, 2, 1, 4, 1, 4$ and $s = 15$. Then the triangular numbers $T_1, \ldots, T_6$ are achievable by (a_i), but T_7 is not.

Example The power of Theorem 10.1 can be demonstrated on the Prague clock sequences (10.1) for $s = 15$. It is enough to check (10.5) only for $k \leq \frac{1}{2}(s - 1) = 7$ (see Fig. 10.5 and the first row of (10.1)). The achievability of further rows of (10.1) of all $k > 7$ follows from Theorem 10.1.

Similarly, we can verify by inspection the assumptions of Theorem 10.1 for other periods:

1, 2 with $p = 2$ and $s = 3$,
1, 2, 2 with $p = 3$ and $s = 5$,
1, 2, 3, 1 with $p = 4$ and $s = 7$,
1, 2, 3, 3 with $p = 4$ and $s = 9$,
1, 2, 2, 1, 4, 1, 4, 1, 4, 1, 4 with $p = 11$ and $s = 25$.

There are also Šindel sequences with s even. We can construct one, e.g., for the period 1, 2, 1, 1, 1:

$$1\ 2\ \underbrace{1\ 1\ 1}_{3}\ \underbrace{1\ 2\ 1}_{4}\ \underbrace{1\ 1\ 1\ 2}_{5}\ \underbrace{1\ 1\ 1\ 1\ 2}_{6}\ldots \qquad (10.13)$$

However, the factor $\frac{1}{2}(s - 1)$ appearing on the right-hand side of (10.9) is not an integer. Therefore, the particular terms expressing the number $s = 6$ in (10.13) are not in the same order as the given period. Later in Theorem 10.6, we show how to find a Šindel sequence for a given s.

Theorem 10.2 *A periodic sequence (a_i) for an even s in (10.7) is a Šindel sequence, if T_k is achievable by (a_i) for $k = 1, 2, \ldots, s - 1$.*

Proof Let $s \geq 2$ in (10.7) be even and let

$$\forall k \in \{1, 2, \ldots, s - 1\} \quad \exists n \in \mathbb{N} : T_k = \sum_{i=1}^{n} a_i. \qquad (10.14)$$

From (10.7) and (10.3) we get

$$T_{2s-1} = (2s - 1) \sum_{i=1}^{p} a_i = \sum_{i=1}^{(2s-1)p} a_i.$$

Now suppose that $k = 2s - 1 - k'$, where $1 \leq k' \leq s - 1$. According to hypothesis (10.14), there exists $n' \in \mathbb{N}$ such that

$$\frac{k'(k' + 1)}{2} = \sum_{i=1}^{n'} a_i.$$

Then by (10.2) we obtain

$$T_k = T_{2s-1-k'} = \frac{(2s - 1 - k')(2s - k')}{2} = s(2s - 1 - 2k') + \frac{k'(k' + 1)}{2}.$$

Consequently,

$$T_k = (2s - 1 - 2k') \sum_{i=1}^{p} a_i + \sum_{i=1}^{n'} a_i = \sum_{i=1}^{pm+n'} a_i,$$

where $m = 2s - 1 - 2k'$.

The rest of the proof for $k \geq 2s - 1$ is similar to that of Theorem 10.1. $\qquad \square$

Remark The number $s - 1$ appearing in (10.14) is the smallest possible for the minimal period, if p is the minimal period length corresponding to s. To see this, consider the periodic sequence (a_i) with the minimal period 1, 2, 1 and $s = 4$. Then the triangular numbers T_1 and T_2 are achievable by (a_i), but T_3 is not.

10.3 Necessary and Sufficient Condition for the Existence of a Šindel Sequence

Let $n \geq 2$ and a be given integers. Recall that if the quadratic congruence

$$x^2 \equiv a \pmod{n}$$

has a solution x, then a is called a *quadratic residue modulo n*. Otherwise, a is called a *quadratic nonresidue modulo n*.

Before we state a necessary and sufficient condition for the existence of a Šindel sequence, which will be based on the concept of quadratic residue, we prove an important property that will be used later in the proof of Theorems 10.4 and 10.9.

Theorem 10.3 *If f and h are nonnegative integers, then $8f + 1$ is a quadratic residue modulo 2^h.*

Proof If $h \in \{0, 1, 2, 3\}$, then $8f + 1 \equiv 1 \pmod{2^h}$, and thus the right-hand side is a quadratic residue, since $3^2 \equiv 1 \pmod{2^h}$.

Now assume that $h \geq 4$. Let f and c be nonnegative integers such that $0 \leq c \leq 2^{h-3} - 1$ and $f \equiv c \pmod{2^{h-3}}$, i.e., $2^{h-3} \mid (f - c)$. Since the number $8 \cdot 2^{h-3} = 2^h$ divides $(8f + 1) - (8c + 1) = 8(f - c)$, we get $8f + 1 \equiv 8c + 1 \pmod{2^h}$. Altogether we have 2^{h-3} residues $8\ell + 1$ modulo 2^h which are mutually incongruent for $0 \leq \ell \leq 2^{h-3} - 1$, since $0 \leq 8\ell + 1 < 2^h$. Therefore, without loss of generality we may assume that $8f + 1$ belongs to the set

$$A = \{0 \cdot 8 + 1, \ 1 \cdot 8 + 1, \ 2 \cdot 8 + 1, \ \ldots, \ (2^{h-3} - 1)8 + 1\},$$

which has 2^{h-3} elements. In the book [291, p. 105] it is proved that the set $\{1, 3, 5, \ldots, 2^h - 1\}$ of odd remainders is identical with $\{(-1)^i 5^j\}$ modulo 2^h, where $0 \leq i \leq 1$ and $0 \leq j \leq 2^{h-2} - 1$. Clearly, $5^{2j} = (5^j)^2$ is a quadratic residue modulo 2^h for nonnegative integer j. According to our previous considerations, 2^{h-3} odd quadratic residues 5^{2j} modulo 2^h are mutually incongruent for $0 \leq j \leq 2^{h-3} - 1$, since $0 \leq 2j \leq 2^{h-2} - 1$. Further, we see that

$$5^2 = 25 = 3 \cdot 8 + 1.$$

Thus, by the Binomial Theorem we have

$$5^{2j} = (3 \cdot 8 + 1)^j$$
$$= 3^j 8^j + \binom{j}{1} 3^{j-1} 8^{j-1} + \cdots + \binom{j}{j-1} 3 \cdot 8 + 1$$
$$\equiv 1 \pmod 8$$

for $0 \leq j \leq 2^{h-3} - 1$. Since both the sets A and $\{5^{2j} \mid 0 \leq j \leq 2^{h-3} - 1\}$ have the same number of elements, the theorem is proved. □

Theorem 10.4 *A periodic sequence (a_i) is a Šindel sequence if and only if for any $n \in \{1, \ldots, p\}$ and any $j \in \{1, 2, \ldots, a_n - 1\}$ with $a_n \geq 2$ the number*

$$w = 8(a_1 + a_2 + \cdots + a_n - j) + 1$$

is a quadratic nonresidue modulo s.

Proof $\Longleftarrow$: Let a periodic sequence (a_i) not be a Šindel sequence. According to (10.5), there exist positive integers ℓ and m such that $a_m \geq 2$ and

$$\sum_{i=1}^{m-1} a_i < T_\ell < \sum_{i=1}^{m} a_i,$$

i.e., there exists a natural number $j \leq a_m - 1$ such that

$$T_\ell = \sum_{i=1}^{m} a_i - j. \tag{10.15}$$

Let $n \in \{1, \ldots, p\}$ be such that $n \equiv m \pmod{p}$. Then by (10.2), (10.15), (10.7), and (10.3) it follows that

$$(2\ell + 1)^2 = 4\ell^2 + 4\ell + 1 = 8T_\ell + 1$$
$$= 8\Big(\sum_{i=1}^{m} a_i - j\Big) + 1 \equiv 8\Big(\sum_{i=1}^{n} a_i - j\Big) + 1 \pmod{s},$$

i.e., $8\big(\sum_{i=1}^{n} a_i - j\big) + 1$ is a square modulo s. Thus the condition given in Theorem 10.4 is sufficient for (a_i) to be a Šindel sequence.

$\Longrightarrow$: Let (a_i) be a Šindel sequence with $s = 2^c d$, where $c \geq 0$ and d is odd. Suppose to the contrary that there exist positive integers n, j, and x such that $n \leq p$, $a_n \geq 2$, $j \leq a_n - 1$, $x \leq s$, and

$$w = 8\Big(\sum_{i=1}^{n} a_i - j\Big) + 1 \equiv x^2 \pmod{s}. \tag{10.16}$$

According to Theorem 10.3 and (10.16), there exists y such that

$$x^2 \equiv w \pmod{d},$$
$$y^2 \equiv w \pmod{2^{c+3}}. \tag{10.17}$$

By the Chinese Remainder Theorem 1.4 there exists an integer $u \geq 3$ (we want to avoid the case $w = u = 1$ in (10.16)) such that $u \equiv x \pmod{d}$ and $u \equiv y \pmod{2^{c+3}}$. Thus, by (10.17),

$$u^2 \equiv x^2 \equiv w \pmod{d},$$
$$u^2 \equiv y^2 \equiv w \pmod{2^{c+3}}.$$

Since $(d, 2^{c+3}) = 1$, we see that

$$u^2 \equiv w \pmod{2^{c+3}d}. \tag{10.18}$$

Clearly, u is odd, since w is odd. So let $u = 2\ell + 1$, where $\ell \geq 1$. Then, by (10.18) we get $u^2 = 4\ell^2 + 4\ell + 1 = w + 2^{c+3}dg$ for some integer g. Hence, since $u \geq 3$, we find by (10.2), (10.18), and (10.16) that

$$T_\ell = \frac{u^2 - 1}{8} = \frac{w - 1}{8} + 2^c dg \equiv \sum_{i=1}^{n} a_i - j \pmod{s}.$$

Consequently, there exists a natural number m such that $m \equiv n \pmod{p}$ and

$$T_\ell = \sum_{i=1}^{m} a_i - j,$$

which contradicts the assumption that (a_i) is a Šindel sequence. □

As a byproduct of the proof of Theorem 10.4, we get the well-known result (see Burton [56, p. 15] and Fig. 10.6):

Theorem 10.5 (Plutarch) *A natural number r is triangular if and only iff $8r + 1$ is a square.*

Proof If $r = T_\ell = \frac{1}{2}\ell(\ell + 1)$, then $8r + 1 = 4\ell^2 + 4\ell + 1 = (2\ell + 1)^2$.
On the other hand, if $8r + 1$ is a square, then $8r + 1 = (2\ell + 1)^2$ for some integer ℓ, since $8r + 1$ is odd. Then

$$r = \frac{(2\ell + 1)^2 - 1}{8} = \frac{4\ell(\ell + 1)}{8} = \frac{\ell(\ell + 1)}{2} = T_\ell,$$

which proves the theorem. □

Remark In Theorem 10.4, we require that

$$w = 8 \left(\sum_{i=1}^{n} a_i - j \right) + 1$$

be a quadratic nonresidue modulo s for various values of n and j when (a_i) is a Šindel sequence. A sufficient condition for this to occur is that w be a quadratic

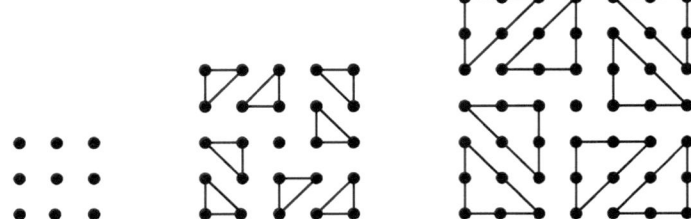

Fig. 10.6 The early Pythagoreans knew that if r is a triangular number, then $8r + 1$ is a square. This result is mentioned as early as about 100 AD in *Platonic Questions* by the Greek historian Plutarch, see [395, p. 4]

nonresidue modulo q for some odd prime q dividing s. To see that this condition is not necessary, consider the periodic sequence (a_i) given in Examples after Theorem 10.1 with $p = 11$, $s = 25$, and the period $1, 2, 2, 1, 4, 1, 4, 1, 4, 1, 4$. Then

$$8\left(\sum_{i=1}^{5} a_i - 2\right) + 1 = 65,$$

which is a quadratic nonresidue modulo 25, but is a quadratic residue modulo 5. Note that 5 is the only odd prime dividing $s = 25$.

Remark Consider the sequence (a_i) with period $1, 2, 1, 1, 1, \ldots, 1$. Notice that

$$w = 8\left(\sum_{i=1}^{2} a_i - 1\right) + 1 = 17.$$

By Theorem 10.4 and the Law of Quadratic Reciprocity (see Theorem 2.27) one sees that if s is an odd prime and

$$s \equiv 1, 2, 4, 8, 9, 13, 15, 16 \pmod{17}, \tag{10.19}$$

(see Fig. 2.6), then w is a quadratic residue modulo s. Thus, (a_i) is not a Šindel sequence (quadratic residues including zero are marked by white circles in Fig. 2.6). However, since $s = 15$ is not a prime, the sequence with period $1, 2, 1, 1, \ldots, 1$ is a Šindel sequence. Other patterns of the period of periodic sequences (a_i) can be investigated similarly.

10.4 Construction of the Primitive Šindel Sequence

First, we introduce the notion of a composite and a primitive Šindel sequence. Then we will introduce a theorem, which guarantees the existence of a single primitive Šindel sequence for a given s (see (10.7)).

A Šindel sequence (a_i') with the minimal period $p + 1$ is called *composite,* if there exists a Šindel sequence (a_i) and $\ell \in \mathbb{N}$ such that

$$
\begin{aligned}
a_i &= a_i', \quad i = 1, \ldots, \ell - 1,\\
a_\ell &= a_\ell' + a_{\ell+1}',\\
a_i &= a_{i+1}', \quad i = \ell + 1, \ldots, p.
\end{aligned}
$$

Example The period 1, 2, 3, 2, 2, 3, 2 derived from the period 1, 2, 3, 4, 3, 2 of the sequence (10.1) produces a composite Šindel sequence. In other words, the astronomical clock would also work with the small auxiliary gear (with one extra slot) corresponding to this composite Šindel sequence.

A Šindel sequence (a_i) is called *primitive* if it is not composite.

Example By inspection, we can verify that all the sequences from Examples after Theorem 10.1 are primitive.

The proof of the next theorem contains an explicit algorithm for finding a primitive Šindel sequence for a given s (see [218]).

Theorem 10.6 *For any $s \in \mathbb{N}$ there exists a unique primitive Šindel sequence (a_i) such that (10.7) holds for one of its not necessarily minimal period lengths p.*

Proof Let $1 \le b_1 < b_2 < \cdots < b_t \le s$ be all the integers such that $8b_n + 1$ is a square modulo s for $n = 1, \ldots, t$. We observe that $b_1 = 1$ and $b_t = s$. Now we choose the period as follows: $a_1 = b_1$ and $a_n = b_n - b_{n-1}$ for $n = 2, 3, \ldots, t$. Then

$$
\forall n \in \{1, 2, \ldots, t\} : \quad b_n = \sum_{i=1}^{n} a_i.
$$

We claim that (a_i) is a Šindel sequence. Note that if $n \in \{1, \ldots, t\}$, $a_n \ge 2$, and $j \in \{1, 2, \ldots, a_n - 1\}$, then

$$
b_{n-1} < \sum_{i=1}^{n} a_i - j < b_n.
$$

Therefore, $8(\sum_{i=1}^{n} a_i - j) + 1$ is a quadratic nonresidue modulo s, since

$$8b_1 + 1, \ldots, 8b_t + 1$$

are all the quadratic residues modulo s of the form $8j + 1$ for $j \in \{1, \ldots, s\}$. It now follows from Theorem 10.4 that (a_i) is a Šindel sequence.

Furthermore, one observes that (a_i) is a primitive Šindel sequence having a period length $p = t$ and satisfying (10.7). It is also clear by construction that (a_i) is the unique primitive Šindel sequence fulfilling (10.7) for some period length p.   □

The sequence $1, 1, 1, \ldots$ is called a *trivial Šindel sequence*.

Theorem 10.7 *The primitive Šindel sequence (a_i) is trivial if and only if $s = 2^h$ for $h \geq 0$.*

Proof $\Longleftarrow$: By the above construction of the period in Theorem 10.6, the primitive Šindel sequence corresponding to s is nontrivial if and only if there exists a positive integer $f \leq s$ such that $8f + 1$ is a quadratic nonresidue modulo s. By Theorem 10.3, the number $8f + 1$ is always a quadratic residue modulo $s = 2^h$ for $h \geq 0$. Hence, the primitive Šindel sequence corresponding to $s = 2^h$ is the trivial Šindel sequence.

$\Longrightarrow$: Conversely, assume that s has an odd prime divisor q. Let d be a quadratic nonresidue modulo q. Suppose that $8z \equiv 1 \pmod{q}$, i.e., 8 is invertible modulo q. Thus one sees that if z is the inverse of 8 modulo q and $f \equiv z(d - 1) \pmod{q}$, then $8f + 1 \equiv d \pmod{q}$. It now follows that the primitive Šindel sequence corresponding to s is nontrivial.   □

From Theorems 10.6 and 10.7 we immediately get the following statement.

Theorem 10.8 *Let (a_i) be a periodic sequence with the minimal period length p and $s = 2^m$, where m is a nonnegative integer. Then (a_i) is a Šindel sequence if and only if (a_i) is trivial.*

Further, we prove a surprising theorem, which guarantees that for an arbitrarily large natural k we can always find a Šindel sequence that will contain k.

Theorem 10.9 *For any $k \in \mathbb{N}$ there exist $\ell \in \mathbb{N}$ and a Šindel sequence (a_i) such that $a_\ell = k$.*

Proof The number $8r + 1$ is a square if and only if r is triangular (see Theorem 10.5). Let $k = T_k - T_{k-1}$ be given (see (10.6)). This it suffices by the proof of Theorem 10.6 to find a natural number $s \geq T_k$ such that $8(T_{k-1} + j) + 1$ is not a quadratic residue modulo s for $j = 1, 2, \ldots, k - 1$.

For a fixed $j \in \{1, \ldots, k - 1\}$ let

$$8(T_{k-1} + j) + 1 = \prod_{i=1}^{v} p_i^{\alpha_i}$$

be a prime power factorization (see Theorem 2.2). Since $8(T_{k-1} + j) + 1$ is not a square, some α_i is odd. Without loss of generality, we can assume that α_1 is odd. Let

c_1 be a quadratic nonresidue modulo p_1. By the Chinese Remainder Theorem 1.4 and Dirichlet's Theorem 3.10 on the infinitude of primes in arithmetic progressions, one can find a prime $q_j \geq T_k$ such that $q_j \equiv 1 \pmod 4$, $q_j \equiv c_1 \pmod{p_1}$, and $q_j \equiv 1 \pmod{p_i}$ for $i \in \{2, \ldots, v\}$.

Since $q_j \equiv 1 \pmod 4$, it follows from the properties of the Legendre symbol and the Law of Quadratic Reciprocity (see Theorems 2.26 and 2.27) that

$$\left(\frac{p_1}{q_j}\right) = \left(\frac{q_j}{p_1}\right) = \left(\frac{c_1}{p_1}\right) = -1$$

and

$$\left(\frac{p_i}{q_j}\right) = \left(\frac{q_j}{p_i}\right) = \left(\frac{1}{p_i}\right) = 1 \quad \text{for } i = 2, 3, \ldots, v,$$

where $\left(\frac{p}{q}\right)$ is the Legendre symbol for odd primes p and q. Since the Jacobi symbol is multiplicative (see (2.52)), we see that

$$\left(\frac{8(T_{k-1} + j) + 1}{q_j}\right) = \prod_{i=1}^{v}\left(\frac{p_i}{q_j}\right)^{\alpha_i} = (-1)^{\alpha_1}\prod_{i=2}^{v} 1^{\alpha_i} = -1,$$

and thus $8(T_{k-1} + j) + 1$ is a quadratic nonresidue modulo q_j. Now we simply let s be the product of the distinct q_j's for $j \in \{1, \ldots, k - 1\}$.   □

Example The period 1, 2, 3, 4, 5, 3, 3, 7, 2, 3, 3, 9 with minimal period length $p = 12$ and $s = 45$ yields a primitive Šindel sequence (a_i) with a large value of $a_{12} = 9$ relative to s (see Theorem 10.6).

10.5   Which Šindel Sequence Is the Most Beautiful?

The proof of Theorem 10.6 contains a numerical algorithm for creating primitive Šindel sequences. The following table shows the periods of these sequences for $s = 1, \ldots, 25$. By computer it can be verified that no primitive Šindel sequence for $s \leq 1000$ and $s \neq 15$ has such a nice "palindromic" character as the Prague clock sequence (10.1), which was used to construct the bellworks of the Prague horologe. This sequence is generated by rotating the small auxiliary gear as shown in Figs. 10.2 and 10.3.

Table 10.1 shows the values of all primitive Šindel sequences for $s = 1, \ldots, 25$. From this table we observe the property guaranteed by Theorem 10.7, namely that trivial primitive Šindel sequences appear only when $s = 2^h$ for some $h \geq 0$. The structure of sequences corresponding to $s = 13$ and $s = 19$ when $a_2 = 1$ is discussed in the Remark to the relation (10.19). If the bell rang

$$1 + 2 + \cdots + 12 = 78 = 6 \times 13$$

Table 10.1 Primitive Šindel sequences for $s = 1, \ldots, 25$

s	Periods of primitive Šindel sequences
1	1
2	1 1
3	1 2
4	1 1 1 1
5	1 2 2
6	1 2 1 2
7	1 2 3 1
8	1 1 1 1 1 1 1 1
9	1 2 3 3
10	1 2 2 1 2 2
11	1 2 1 2 4 1
12	1 2 1 2 1 2 1 2
13	1 1 1 3 2 2 3
14	1 2 3 1 1 2 3 1
15	**1 2 3 4 3 2**
16	1 1 1 1 1 1 1 1 1 1 1 1 1 1 1 1
17	1 1 1 1 2 4 1 4 2
18	1 2 3 3 1 2 3 3
19	1 1 1 3 1 2 1 5 2 2
20	1 2 2 1 2 2 1 2 2 1 2 2
21	1 2 3 1 3 3 2 6
22	1 2 1 2 4 1 1 2 1 2 4 1
23	1 2 2 1 3 1 3 2 5 1 1 1
24	1 2 1 2 1 2 1 2 1 2 1 2 1 2 1 2
25	1 2 2 1 4 1 4 1 4 1 4

times for each half day (see Weintraub [415, p. 313]), then the small auxiliary wheel would probably be chosen so that $s = 13$. Other options are $s \in \{2, 3, 6, 26, 39\}$. However, their corresponding primitive Šindel sequences are not as "nice" as the Prague clock sequence (10.1).

We also see from Table 10.1 that there does not exist a primitive Šindel sequence with the minimal period for $s < 25$ even. Now we will prove this property for an arbitrarily large s.

Theorem 10.10 *There exists a primitive Šindel sequence whose period is the minimal period if and only if s given by (10.7) is odd.*

Proof $\implies$: Let $s = 2^c d$, where $c \geq 1$ and d is odd. Since $8f + 1$ is a quadratic residue modulo 2^c for all nonnegative integers f by Theorem 10.3, we have by the Chinese Remainder Theorem 1.4 that $8f + 1$ is a square modulo s if and only if

$8f + 1$ is a square modulo d. It now follows from the construction given in the proof of Theorem 10.6 that the primitive Šindel sequence corresponding to $s = 2^c d$ has the same period, not necessarily minimal, as the period of the primitive Šindel sequence corresponding to $s = d$. Hence, we see that for s even, the primitive Šindel sequence corresponding to s does not have the associated period as its minimal period.

$\Longleftarrow$: Now let s be odd. If $s = 1$, the result is trivial. So assume that $s \geq 3$ and let (a_i) be the unique primitive Šindel sequence (see Theorem 10.6) corresponding to s and having period length p. Let p' be the minimal period length of the sequence $\{a_i\}$ and let

$$s' = \sum_{i=1}^{p'} a_i.$$

Suppose to the contrary that $p' < p$, i.e., $s' < s$. For $k \in \mathbb{N}$, we let $w_k = \sum_{i=1}^{k} a_i$. To obtain a contradiction, it suffices by the proof of Theorem 10.6 to find a positive integer $n \leq p$ such that $8w_n + 1$ is a quadratic nonresidue modulo s. To accomplish this, we need only find a divisor f of s such that $8w_n + 1$ is a quadratic nonresidue modulo f.

Since $(8, s) = 1$, there exists a unique integer b such that $0 \leq b \leq s - 1$ and $8b + 1 \equiv 0 \pmod{s}$. Simply let $b \equiv -z \pmod{s}$, where z is the inverse of 8 modulo s. Since 0 is a square modulo s, we see by the construction in the proof of Theorem 10.6 that $8w_k + 1 \equiv 0 \pmod{s}$ for some $k \in \{1, 2, \ldots, p\}$. Let m be a natural number such that $m \leq p'$ and $m \equiv k \pmod{p'}$. Then $8w_m + 1 \equiv 0 \pmod{s'}$. Since $s' < s$, there exists an odd prime q such that $q \mid \frac{s}{s'} = \frac{p}{p'}$. First, suppose that $q \nmid s'$. Consider the q integers

$$8w_m + 1, 8w_{m+p'} + 1, 8w_{m+2p'} + 1, \ldots, 8w_{m+(q-1)p'} + 1. \tag{10.20}$$

Noticing that

$$(8w_{m+jp'} + 1) - (8w_{m+ip'} + 1) = 8(j - i)s' \tag{10.21}$$

for $0 \leq i < j \leq q - 1$ and that $(8s', q) = 1$, we find that the q numbers in (10.20) are incongruent modulo q. Let e be a quadratic nonresidue modulo q. Then $8w_{m+jp'} + 1 \equiv e \pmod{q}$ for some $j \in \{0, 1, 2, \ldots, q - 1\}$, which is a contradiction, since $q \mid s$ and $m + jp' \leq m + (q - 1)p' \leq p$.

Finally, we treat the remaining case in which $q^\alpha \| s'$ for some odd prime q and integer $\alpha \geq 1$, where $q^\alpha \| s'$ means that $q^\alpha \mid s'$ but $q^{\alpha+1} \nmid s'$. Then $8w_m + 1 \equiv 0 \pmod{q^\alpha}$. By (10.21) and the fact that $q^\alpha \| s'$, we see that the q integers in (10.20) are congruent to

$$0, q^\alpha, 2q^\alpha, \ldots, (q - 1)q^\alpha \pmod{q^{\alpha+1}} \tag{10.22}$$

in some order. To complete the proof, it suffices to demonstrate that at least one of the q numbers in (10.22) is a quadratic nonresidue modulo $q^{\alpha+1}$. If α is odd, then clearly q^α is a quadratic nonresidue modulo $q^{\alpha+1}$.

Now suppose that α is a positive even integer, $q^\alpha \| r$, and r is a quadratic residue modulo $q^{\alpha+1}$. Then

$$r \equiv \left(aq^{\alpha/2}\right)^2 = a^2 q^\alpha \quad (\text{mod } q^{\alpha+1})$$

for some integer a such that $q \nmid a$. If

$$a^2 q^\alpha \equiv h q^\alpha \quad (\text{mod } q^{\alpha+1}),$$

then

$$a^2 \equiv h \quad (\text{mod } q).$$

Let u be a quadratic nonresidue modulo q and suppose that $0 \le u \le q - 1$. Then $u q^\alpha$ is a quadratic nonresidue modulo $q^{\alpha+1}$. □

The next statement follows immediately from Theorem 10.10 and its proof.

Theorem 10.11 *Let (a_i) be a primitive Šindel sequence for which the sum of the period $s = 2^c d$, where $c \ge 0$ and d is odd. Then the period of (a_i) is the concatenation, repeated 2^c times, of the primitive Šindel sequence with $s = d$.*

Example Consider the primitive Šindel sequences (a_i) and (b_i) for which the sums of the periods s are respectively equal to 3 and 12. Then their periods are $1, 2$ and $1, 2, 1, 2, 1, 2, 1, 2$, respectively. See Table 10.1 for further illustrations of Theorem 10.11.

10.6   Peculiar Šindel Sequences

Let (a_i) be a primitive Šindel sequence with period length ℓ. Suppose that $a_i = i$ for $i = 1, 2, \ldots, \ell - 1$. Then (a_i) is called a *peculiar primitive Šindel sequence*.

Clearly, if (a_i) is a peculiar primitive Šindel sequence then $a_\ell = s - T_{\ell-1} > 0$. In particular, if $s = 1$ or $s = 3$ then $a_\ell = \ell$, i.e., $a_i = i$ for $i = 1, \ldots, \ell$ (cf. Table 10.1). The following necessary and sufficient condition is given in [213].

Theorem 10.12 *A primitive Šindel sequence with sum of the period s is peculiar if and only if $s \in \{1, 2, 3, 5, 7, 9\}$.*

Proof One immediately sees from Table 10.1 that a primitive Šindel sequence is peculiar if $s = 1, 2, 3, 5, 7$, or 9.

Conversely, from here on we assume that $s \notin \{1, 2, 3, 5, 7, 9\}$ and let (a_i) be a primitive Šindel sequence. First assume that s is even. By Theorem 10.6, the period of (a_i) is not equal to its minimal period. Since $a_1 = 1$, we find that (a_i) is not increasing for $i = 1, 2, \ldots, \ell - 1$. Hence, (a_i) is not peculiar.

Now assume that s is odd. Let m be the largest integer such that $T_m \leq s$. Since $s \geq 11$, we see that $m \geq 4$. We wish to show that (a_i) is not peculiar if $s \geq 11$. Our result will follow if we can show that $\ell - 1 \geq m + 1$ if $T_m < s$ and $\ell - 1 \geq m$ if $T_m = s$. Thus it suffices to demonstrate that

$$T_{\ell-1} = \frac{\ell(\ell - 1)}{2} \geq s. \tag{10.23}$$

Let $b_n = \sum_{i=1}^{n} a_i$ for $n = 1, 2, \ldots, \ell$. By the proof of Theorem 10.6, a_n is in the period of the primitive Šindel sequence for $n \in \{1, 2, \ldots, \ell\}$ if and only if $8b_n + 1$ is a quadratic residue modulo s. Since $(8, s) = 1$, we observe that the period length ℓ is equal to $Q(s)$, where $Q(s)$ denotes the number of quadratic residues modulo s. Clearly, if $s > 1$ then $Q(s) \geq 2$, since 0 and 1 are quadratic residues modulo s.

We further prove by induction that (a_i) is not peculiar. Our induction is on the number of distinct prime divisors of s.

I. From here on in the proof, p will denote an odd prime. Suppose first that $s = p^c$, where $c \geq 1$. Since $(2, s) = 1$, it follows by Hensel's Lemma (see [291, p. 87]) that if t is a nonzero quadratic residue modulo p and $a \equiv t \pmod{p^c}$, then a is a quadratic residue modulo p^c. Noting that there are $\frac{1}{2}(p - 1)$ nonzero quadratic residues modulo p and including the quadratic residue 0 modulo p^c, we find that

$$Q(p^c) \geq \frac{p^{c-1}(p - 1)}{2} + 1 \tag{10.24}$$

for $c \geq 1$ with equality if $c = 1$ or 2.

Let $s = p^k$, $s \geq 11$, and let us write $T(\ell)$ instead of T_ℓ for convenience. Then by (10.23), it suffices to show that

$$T(\ell - 1) = T(Q(p^k) - 1) \geq \frac{p^{k-1}(p - 1)(p^{k-1}(p - 1) + 2)}{8} \geq p^k. \tag{10.25}$$

(1) Suppose that $k = 1$ and $p \geq 11$. Then (10.25) will be satisfied if

$$\frac{(p - 1)(p + 1)}{8} - p \geq 0,$$

which is equivalent to $p^2 - 8p - 1 \geq 0$ and implies that

$$p \geq \frac{8 + \sqrt{68}}{2} > 8.$$

Thus, (a_i) is not peculiar if $s = p \geq 11$ and $k = 1$.

(2) Now suppose that $k \geq 2$. Then (10.25) will be satisfied if

$$p^{2k-2}(p - 1)^2 \geq 6p^k + 2p^{k-1}.$$

First consider the case in which $p \geq 5$. Since $k \geq 2$, we get $2k - 2 \geq k$. Thus,

$$p^{2k-2}(p-1)^2 \geq p^k(5-1)^2 > 8p^k > 6p^k + 2p^{k-1},$$

as desired.

(3) Finally, suppose that $p = 3$ and $k \geq 3$. Then $2k - 2 \geq k + 1$. Therefore,

$$p^{2k-2}(p-1)^2 \geq 3p^k(3-1)^2 > 6p^k + 2p^{k-1},$$

and (a_i) is not peculiar in this case, too. Thus when $r = 1$ and $s \geq 11$, (a_i) is not peculiar.

II. Let the prime power factorization of s be given by

$$s = \prod_{i=1}^{r} p_i^{k_i},$$

where $p_1 < p_2 < \cdots < p_r$ and $k_i \geq 1$. Since a is a quadratic residue modulo s if and only if a is a quadratic residue modulo $p_i^{k_i}$ for $i = 1, 2, \ldots, r$, it follows by the Chinese Remainder Theorem 1.4 and (10.24) that

$$\ell = Q(s) = \prod_{i=1}^{r} Q(p_i^{k_i}) \geq \prod_{i=1}^{r} \frac{p_i^{k_i-1}(p_i - 1) + 2}{2}. \tag{10.26}$$

We further prove by induction that (a_i) is not peculiar when $r \geq 2$. We will show that if $(p, s) = 1$, $c \geq 1$, and

$$T(Q(s) - 1) \geq s, \tag{10.27}$$

then

$$T(Q(p^c)(Q(s) - 1)) \geq p^c s. \tag{10.28}$$

The number $p^c s$ has at least two prime divisors and by (10.2) and (10.28) we obtain that

$$T(Q(p^c s) - 1) = T(Q(p^c)Q(s) - 1) > T(Q(p^c)(Q(s) - 1)) \geq p^c s. \tag{10.29}$$

According to (10.27)–(10.29), we obtain from a nonpeculiar primitive Šindel sequence having sum of the period s and satisfying (10.23) another nonpeculiar primitive Šindel sequence with sum of the period $p^c s$ and also satisfying (10.23).

Assuming the assertion above, it will follow from our earlier results for $r = 1$ that when $r = 2$, then (a_i) is not peculiar except possibly when $s \in \{15, 21, 35, 45, 63\}$. However, in Table 10.2, we observe that when s takes any of these values, then

Table 10.2 Special values of s for which the primitive Šindel sequence is not peculiar

s	$Q(s)$	$T(Q(s) - 1)$
$15 = 3 \cdot 5$	$6 = 2 \cdot 3$	$T_5 = 15$
$21 = 3 \cdot 7$	$8 = 2 \cdot 4$	$T_7 = 28$
$35 = 5 \cdot 7$	$12 = 3 \cdot 4$	$T_{11} = 66$
$45 = 3^2 \cdot 5$	$12 = 4 \cdot 3$	$T_{11} = 66$
$63 = 3^2 \cdot 7$	$16 = 4 \cdot 4$	$T_{15} = 120$

$T(Q(s) - 1) \geq s$ and the primitive Šindel sequence with sum of the period s is indeed not peculiar.

Thus, if $r = 2$, it will follow that the primitive Šindel sequence with sum of the period s is not peculiar. Continuing by induction, it then follows that (a_i) is not peculiar for all odd s for which $r \geq 2$.

So let us suppose that (10.27) holds, $p \geq 3$, and $(p, s) = 1$. We establish that (10.28) is valid for $c \geq 1$. Let

$$q = Q(s) - 1.$$

Then

$$T(q) = \frac{q^2 + q}{2} \geq s. \tag{10.30}$$

Moreover, by (10.24),

$$T(Q(p^c)q)) \geq T\left(\frac{p^{c-1}(p-1) + 2}{2}q\right)$$

$$= \frac{1}{8}(q(p^{c-1}(p-1) + 2))(q(p^{c-1}(p-1) + 2) + 2). \tag{10.31}$$

(1) Suppose first that $c = 1$. Then by (10.30) and (10.31),

$$T(Q(p^c)q) \geq \frac{q}{8}(p+1)(q(p+1) + 2) = \frac{p+1}{2}\frac{q^2 + q}{2} + \left(\frac{(p+1)^2}{8} - \frac{p+1}{4}\right)q^2$$

$$\geq \frac{p+1}{2}s + \frac{p+1}{4}\frac{p-1}{2}q^2 \geq \frac{p+1}{2}s + \frac{p-1}{2}q^2$$

$$\geq \frac{p+1}{2}s + \frac{p-1}{2}\frac{q^2 + q}{2} \geq \left(\frac{p+1}{2} + \frac{p-1}{2}\right)s = ps,$$

as desired.

(2) Now suppose that $c \geq 2$ and $p \geq 5$. Then by (10.31),

$$T(Q(p^c)q) > \frac{(p^{c-1}(p-1)q)^2}{8} \geq 2p^{2c-2}q^2 \geq 2p^c q^2$$

$$\geq 2p^c \frac{q^2+q}{2} \geq 2p^c s > p^c s,$$

as desired. Hence, (10.28) is valid. We have now established that (a_i) is not peculiar if s is odd and $s > 9$.

(3) Next suppose that $p = 3$ and $c = 2$. Then by (10.31),

$$T(Q(p^c)q) \geq \frac{1}{8}(q(p^{c-1}(p-1)+2))(q(p^{c-1}(p-1)+2)+2)$$

$$= \frac{8q(8q+2)}{8} = 8q^2 + 2q = 4\frac{q^2+q}{2} + 6q^2$$

$$\geq 4s + 6\frac{q^2+q}{2} \geq 10s > 3^2 s,$$

as desired.

Finally, suppose that $p = 3$ and $c \geq 3$. Then

$$T(Q(p^c)q) \geq \frac{1}{8}(q(p^{c-1}(p-1)+2))(q(p^{c-1}(p-1)+2)+2)$$

$$> \frac{(p^{c-1}(p-1)q)^2}{8} = \frac{4 \cdot 3^{2c-2}q^2}{8} \geq \frac{4 \cdot 3^{c+1}q^2}{8}$$

$$> 3^c q^2 \geq 3^c \frac{q^2+q}{2} \geq 3^c s,$$

as desired in (10.28).

□

10.7 Astronomical Dial

The astronomical dial of the Prague horologe represents a geocentric model of the Universe with a motionless Earth in the center. It shows the current position of the Sun and Moon on the ecliptic, the Moon's phases, culminations and settings of the Sun, Moon, and the zodiac signs (see Fig. 10.1). The astronomical clock is actually one of the first analog computers, since it demonstrates the movements of celestial bodies. Sometimes we talk about "high technology" of the 15th century.

The gilded solar hand shows the Central European Time (CET) in the ring of Roman numerals. Note that the difference between CET and the original Prague local mean time is only 138 s. In the 16th century, Jan Táborský from Klokotská Hora took care of the astronomical clock. He is also the author of the oldest known description of the Prague astronomical clock from 1570 (see Táborský [392]).

Fig. 10.7 Stereographic
projection of the point A
lying on the circle k

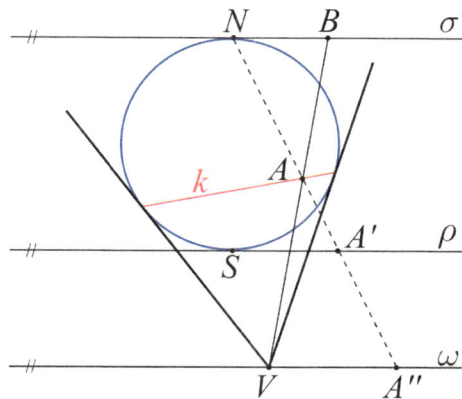

The astronomical dial was obtained by the standard stereographic projection (see e.g. [179, 334]) of the celestial sphere from its North Pole N onto the tangent plane passing through the South Pole S which is at the center of the dial. In the case of the Prague Astronomical Clock, the celestial sphere was represented by a spherical surface with a radius of about 40 cm (cf. Fig. 10.9 below). In the projection plane of the dial, the smallest interior circle around the South Pole illustrates the Tropic of Capricorn, while the exterior circle illustrates the Tropic of Cancer (see Figs. 10.1 and 10.9). The concentric circle between them corresponds to the equator of the celestial sphere. The black circular area at the at the bottom of the dial-plate corresponds to astronomical night, when the Sun is lower than 18° below Prague's horizon.

The following important theorem on the stereographic projection was already known to Claudius Ptolemy (born $\approx$ 100 AD).

Theorem 10.13 (Ptolemy) *Any circle on the sphere which does not pass through the North Pole is mapped onto a circle as well.*

Proof Let k be an arbitrary circle on the sphere that does not pass through the center N of the stereographic projection. First assume that k is not a great circle, i.e., its center differs from the center of the sphere. Consider the tangent cone to the sphere which touches it at k (see Fig. 10.7). Denote by V the vertex of this cone. Let ω be the plane passing through V which is parallel to the projection plane ρ. First we shall project onto the plane ω instead of ρ. Choose an arbitrary point A on k and construct its stereographic projection $A'' \in \omega$.

Let the surface line of the cone passing through the points V and A intersect the plane σ at point B. Obviously $|AB| = |NB|$, since all tangents drawn from the point B to the spherical surface have the same length. Moreover, we see that the triangles ABN and AVA'' are similar. From this we also get that $|AV| = |A''V|$. Since the point V has the same distance from all points of the circle k, the segment $A''V$ also has this constant length (independently of the choice of $A \in k$). Hence, the stereographic projection to the plane ω maps the circle k onto a circle with center V and radius $|A''V|$.

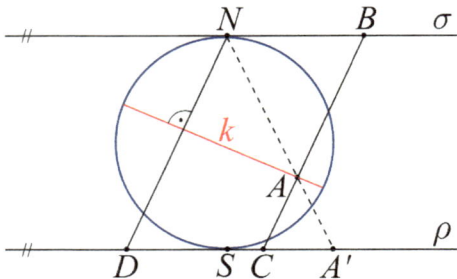

Since the planes ρ and ω are parallel, the stereographic projection of k to the plane ρ will also be a circle with center $\rho \cap NV$ and radius $c|A''V|$, where $c = |A'S|/|A''S|$ is the coefficient of similarity and A' is the stereographic projection of the point A to the plane ρ. The coefficient c is independent of the choice of A, since it is equal to the ratio of the distances of N from the planes ρ and ω.

Second, assume that k is the great circle (i.e., the tangent cone in this case degenerates into a cylinder and therefore the previous method cannot be used). Let $A \in k$ be an arbitrary point. Let σ denote the plane tangent to the spherical surface at the point N. Further, consider the perpendicular from the center of the projection N to the plane of the circle k. Its intersection with the plane ρ is denoted by D (see Fig. 10.8).

Now we prove that the circle k is mapped by the stereographic projection onto the circle $k' \subset \rho$ with center D and radius $|ND|$. Consider the tangent from the point A to the spherical surface parallel to ND. Denote by $B \in \sigma$ and $C \in \rho$ its intersections with the planes σ and ρ. Since all tangents from the point B to the spherical surface have the same length a since $NBCD$ is a parallelogram, we have

$$|AB| = |NB| = |CD|. \tag{10.32}$$

From this and the similarity of the triangles ABN and ACA' we find that

$$|AC| = |A'C|, \tag{10.33}$$

where A' is the stereographic projection of the point A. From (10.32) and (10.33) we can express the distance

$$|A'D| = |A'C| + |CD| = |AC| + |AB| = |BC| = |ND| \tag{10.34}$$

independently of the choice of A. (Notice that all segments in (10.34) lie in the plane of the parallelogram $NBCD$.) Hence, the great circle k is projected to the plane ρ onto the circle k' with center D and radius $|A'D| = |ND|$.   □

Remark There are six important circles on the celestial sphere. Three of them are great circles: equator, ecliptic, Prague's horizon, and the other three are Tropic of Cancer, Tropic of Capricorn, and the circle of astronomical night. By Ptolemy's Theorem 10.13, all these six circles are mapped again onto circles of the astronomical dial (see Fig. 10.9). Precisely measured values of all radii of these circles are presented in [194]. The ecliptic is projected onto a circle, which is represented by the gilded ring with zodiac signs along the ecliptic. However, its center is not in the South Pole S, but the ring with the Sun's and Moon's pointers eccentrically rotates around this pole.

Theorem 10.14 *Any circle on the sphere which passes through the North Pole is mapped onto a straight line.*

Proof Let k be an arbitrary circle on the sphere that passes thought the center N of the stereographic projection. The circle k thus determines the plane containing also the stereographic projection of k. The image of this projection must also lie in the projection plane ρ. Since these two planes are not parallel, their intersection is the sought line. Each of its points A' is a stereographic projection of exactly one point $A \in k$ (different from N).   □

Remark By Theorem 10.14, the meridians of the celestial sphere corresponding to $0°$, $90°$, $180°$, and $270°$ are projected onto the cross whose center is at the South Pole S, see Fig. 10.1.

Remark Another important property of the stereographic projection is that it is a conformal mapping, i.e., all angles are preserved. For instance, the angle $23.5°$ between the equator and ecliptic on the celestial sphere is the same as that on the astronomical dial (see Fig. 10.1).

Twelve black Arabic numerals on the astronomical dial denote planetary hours of the Babylonian time measured from sunrise to sunset (see Figs. 10.1 and 10.9). This time interval is divided into 12 equal parts each day. The number 12 corresponds to the Prague horizon. During the summer solstice the length of 1 planetary hour in Prague is about 80 min while during the winter solstice it is only about 40 min. Karel Sandler found that the arcs corresponding to planetary hours are not parts of circles, assuming that the Sun's pointer moves each day along circular arcs, see Sandler [348].

Theorem 10.15 (Sandler) *If the arcs of the planetary hours were circular, then the trisection of any angle would be possible by ruler and compass.*

Proof On the right part of the horizon consider three points P_1, P_2, P_3 such that the angle between the line segment SP_i and the axis y is $72°$, $90°$, and $108°$, respectively (see Fig. 10.9). These angles can be trisected by ruler and compass, since the regular pentagon has a Euclidean construction. Thus, one planetary hour on the circle with radius SP_i, $i = 1, 2, 3$, corresponds to the constructable angle $12°$, $15°$, and $18°$, respectively.

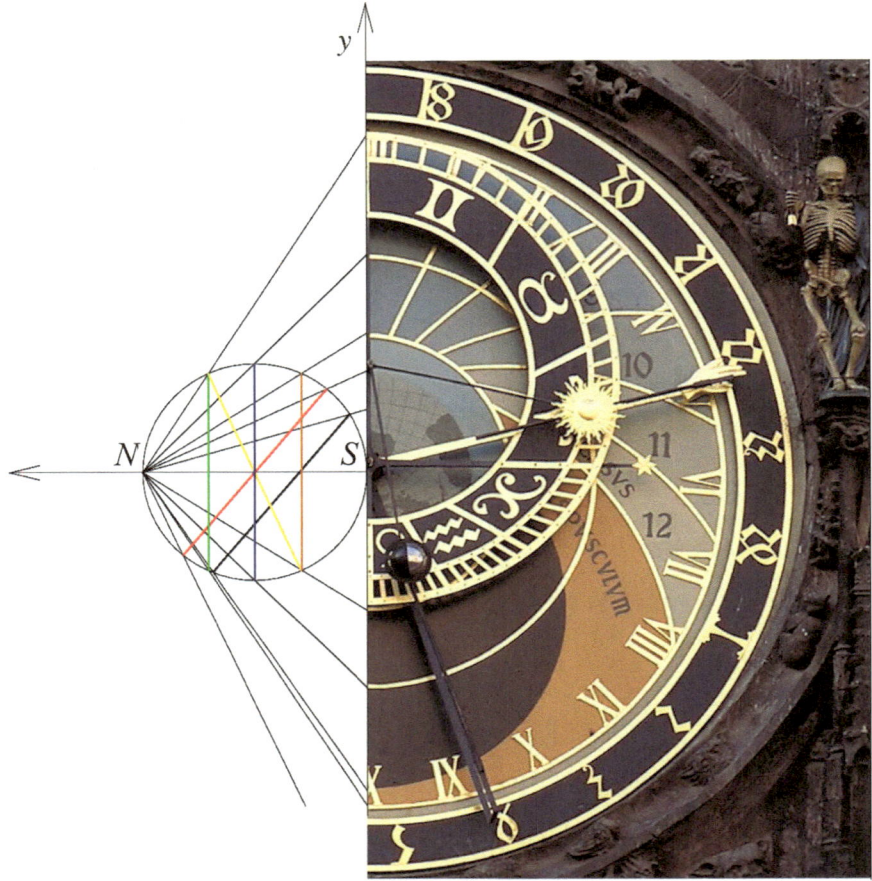

Fig. 10.9 Stereographic projection of the celestial sphere onto the astronomical dial of the Prague horologe. The six most important circles are colored as follows: equator (blue), ecliptic (yellow), Prague's horizon (red), Tropic of Cancer (green), Tropic of Capricorn (brown), and the circle of astronomical night (black)

The planetary arc corresponding to the noon is a straight line segment on the axis y. Let us assume that all other arcs of planetary hours are circles. They are uniquely determined by three points lying on three circles with radii SP_i, $i = 1, 2, 3$. Now we take an arbitrary point P on the horizon circle and consider the angle between the line segment SP and the axis y. Since we assume that the arcs of planetary hours are parts of circles, this angle could be trisected by these arcs which is generally impossible. $\qquad\square$

10.8   What Mathematics Is Hidden Behind the Main Clock?

In the main clockwork of the Prague horologe, there are three large gears of the same diameter 116 cm (see Fig. 10.10) which were originally driven on one axis by three pinions, each with 24 teeth. The first gear has 365 teeth and turns around the zodiac ring once per sidereal day (23 h 56 min 4 s). The second gear, which has 366 teeth, leads the solar pointer and turns around once per mean solar day (24 h). The third gear, which has 379 teeth, leads the Moon's hand and rotates according to the mean apparent motion of the Moon. The number of these teeth were precisely calculated from a deep knowledge of the motion of the Sun and Moon on the rotating celestial sphere.

Let us ask the question, how could clockmakers in the beginning of the 15th century make such large gears with a relatively high accuracy. As a consequence of Gauss's Theorem 4.12 we get:

Theorem 10.16 *Let p be an odd prime. Then there exists a Euclidean construction by ruler and compass of the regular p-gon if and only if p is of the form $p = 2^n + 1$.*

We see that the number 379 of teeth on the third gear is not an odd prime of the form $2^n + 1$. Therefore, the regular 379-gon cannot be constructed by the ruler (straight-edge) and compass. By Theorem 4.12, it is also not possible to find a Euclidean construction of the regular 365-gon and 366-gon, since their prime factorizations

$$365 = 5 \cdot 73 \quad \text{and} \quad 366 = 2 \cdot 3 \cdot 61$$

contain primes which are not of the form $2^n + 1$.

Suppose for a moment that the clockmakers calculated the length of each tooth with a very high precision of 0.01 mm and that they were able to measure this length subsequently around the circumference of the wheel. But then the total error would be more than 3.5 mm which would be not sufficient for the precise construction of each wheel. Zdeněk Horský [148, p. 64] believed that the clockmakers proceeded by way of trial and error. Below we propose another hypothesis that could be used to mark off 379 teeth around a given wheel with sufficient precision.

The clockmakers could take a rope (or a wire) and indicate on it the circumference of the wheel by two marks A and B. The distance AB on the stretched rope can be subdivided into 379 segments of equal length by means of similarity of triangles. First, we lay out 379 segments of equal length one after another onto a straight line passing through the point A. In this way we obtain a point C such that AC is uniformly divided into 379 segments. Then parallels with BC will determine a uniform subdivision of AB into 379 segments. Then it is enough to wind the rope around the wheel and transfer the marks on the circumference.

Fig. 10.10 The main clock contains three large gears controlling the movement of the Sun, the Moon and the ecliptic with its zodiac signs

Chapter 11
Application of Primes

11.1 The Prime 11 in Coding

Humankind has been concerned with the study of prime numbers for several millennia. According to some scientists, this is proved by archaeological findings from the African country Zaire, where a bone was dug up with 11, 13, 17, and 19 notches (see Fig. 11.1). Using the radiocarbon method (based on carbon chronometry of the isotope ratio of ^{12}C and ^{14}C), it was found to be 8000 years old. The number 11, 13, 17, and 19 probably seemed a bit strange to somebody, so he/she recorded them on the bone. Of course, this does not mean that people were seriously thinking about properties of primes. However, already in ancient Greece primes were surely studied. For instance, Euclid (4th–3rd century BC) could prove there are infinitely many of them (see Theorem 2.3). Also Eratosthenes of Cyrene (3rd century BC) became famous by his prime sieve (see (2.4)). But it was not evident until the 20th century that primes may also have a number of interesting technical applications. In this section we will show the practical use of the prime number 11 in *error-detecting codes*.

In the seventies of the last century, the correctness of data on eight-track punched tapes was controlled by so-called parity, i.e., the eighth track was supplemented so that the number of holes in each row was even. In this way, data punching errors were detected. Check-sums for verification of the correctness of various data files (e.g. column or row sums in some table) have a similar function. These are simple examples of error-detecting codes.

Before we show how the prime number 11 is used, we derive a simple criterion for divisibility by 11. In fact, it is enough to alternately subtract and add individual digits of a given number (cf. (11.6)), and then verify whether the total sum is divisible by 11.

Theorem 11.1 *Let $k \in \{0, 1, 2, \dots\}$ and*

$$m = \sum_{n=0}^{k} c_n 10^n \quad for \quad c_n \in \{0, 1, \dots, 9\}, \quad c_k \neq 0,$$

© The Author(s), under exclusive license to Springer Nature Switzerland AG 2021
M. Křížek et al., *From Great Discoveries in Number Theory to Applications*,
https://doi.org/10.1007/978-3-030-83899-7_11

Fig. 11.1 Representation of prime numbers on a bone 8 thousand years old

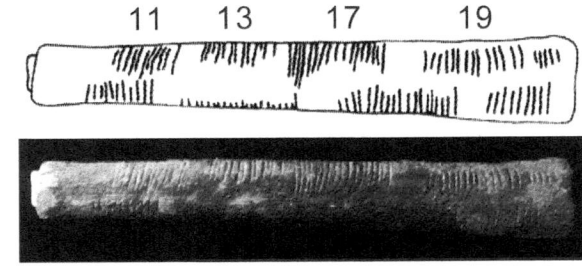

i.e., $c_k, \ldots, c_0$ *are digits of a natural number m in the decimal system. Then*

$$11 \mid m \iff 11 \mid \left(\sum_{n=0}^{k}(-1)^n c_n\right). \tag{11.1}$$

Proof We have

$$10^n - (-1)^n = (10+1)\left[10^{n-1} + 10^{n-2}(-1) + \cdots + 10(-1)^{n-2} + (-1)^{n-1}\right], \tag{11.2}$$

where the brackets contain exactly n terms. The difference $10^n - (-1)^n$ is thus divisible by 11.

Let m be divisible by 11, that is

$$11 \mid (c_k 10^k + c_{k-1}10^{k-1} + \cdots + c_1 10 + c_0). \tag{11.3}$$

Using now relation (11.2) to every term in (11.3) except for the last one, we obtain

$$11 \mid ((-1)^k c_k + (-1)^{k-1}c_{k-1} + \cdots - c_1 + c_0). \tag{11.4}$$

Conversely, we similarly find that if (11.4) is valid, then (11.3) is valid as well. $\square$

In the Czech Republic, all newborns get a so-called *birth number* ($\approx$ social security number) issued by birth registry offices. Since 1986, these numbers are divisible by the prime number 11. The last four digits are chosen so that the whole ten-digit birth number (apart from the slash) is divisible by 11. Try, for example, that the birth number

$$975811/0428 \tag{11.5}$$

(corresponding to the birth of a girl on August 11, 1997) is divisible by 11. According to the above criterion (11.1), we have

$$-9 + 7 - 5 + 8 - 1 + 1 - 0 + 4 - 2 + 8 = 11, \tag{11.6}$$

which is clearly divisible by 11.

What is the advantage of having birth numbers chosen in this way? This is that the computer immediately detects an error as soon as you type a birth number with exactly one wrong digit. Then the difference between the correct and incorrectly entered birth number is $\pm c \cdot 10^n$, where $c \in \{1, 2, \ldots, 9\}$. This is never divisible by 11, but it can be divisible by composite numbers 12, 14, 15, 16, ... If we write e.g. by mistake 975811/0728 instead of the number in (11.5), then the computer would not detect an error when dividing by 12, because both the numbers are divisible by 12. Composite numbers are therefore not suitable for this purpose. Since the prime 11 allows us to detect an error, this code is called an *error-detecting code*.

If making a mistake in more than one digit, the computer also detects an error with a high probability $\approx \frac{10}{11}$. Quality software is, however, also able to find other inconsistencies. For instance, it must be able to check the number of digits entered or discard an artificial birth number 830229/0425, which is divisible by 11, but which corresponds to the nonexisting February 29, 1983. If somebody exchanges two consecutive digits, this code will also detect it.

Eleven is the smallest two-digit prime number. Note that single-digit prime numbers are not suitable for error detection. It would generally not be possible to detect an error when entering one incorrect digit. On the other hand, we could use larger prime numbers. However, then we would have less choices of birth numbers. The reason is that there are more ten-digit numbers that are divisible by 11 than ten-digit numbers that are divisible by e.g. 13. Therefore, the prime 11 is for the decimal system "optimal" in the above sense.

The ordinary Czech double crown can be easily used in connection with criterion (11.1). Its shape is the regular 11-gon with slightly rounded corners. So you can try to check the equivalence (11.1) e.g. for the number (11.5). Initially orient the coin as in Fig. 11.2. Then turn it alternately clockwise and counterclockwise by as many vertices as the corresponding digit. After that the coin must be in the initial position again due to (11.6).

The ISBN numbers of books are protected against possible errors similarly as birth numbers. We describe the ten-digit ISBN code, which was used worldwide until 2007. The ten digits $x_1 x_2 \ldots x_{10}$ are divided into four parts, between which are 3 hyphens. The first three parts have a variable length. This code has the form:

ISBN country-publishing house-identification number-check digit

Fig. 11.2 The Czech double crown has a shape the regular 11-gon

For instance, the book [199] has

$$\text{ISBN } 0\text{-}387\text{-}95332\text{-}9, \tag{11.7}$$

where ISBN is an abbreviation of the *International Standard Book Number,* the first number 0 corresponds to the country or language (Anglo-Saxon countries have 0 or 1, Francophone countries 2, German speaking countries 3, ..., Czech and Slovak Republic 80, etc.), 387 is the code of the publishing house Springer, 95332 is the identification number of the book, and last check digit $x_{10} = 9$ is chosen so that the number

$$x_1 + 2x_2 + 3x_3 + \cdots + 10x_{10} \tag{11.8}$$

is divisible by 11, i.e.,

$$x_{10} \equiv \sum_{k=1}^{9} k x_k \quad (\text{mod } 11).$$

If $x_{10} = 10$, then instead of x_{10} the Roman number ten X is used. Further details can be found at [432]. Substituting the code (11.7) into (11.8), we get

$$2 \cdot 3 + 3 \cdot 8 + 4 \cdot 7 + 5 \cdot 9 + 6 \cdot 5 + 7 \cdot 3 + 8 \cdot 3 + 9 \cdot 2 + 10 \cdot 9 = 286,$$

which after division by 11 gives 26. If we wrongly type one digit or mistakenly swap two unequal digits, then the divisibility by 11 will be violated and we will easily reveal that an error occurred somewhere. As of 2011, all books are provided by a different 13-digit ISBN code starting with EAN prefix 987 (EAN = European Article Number).

ISSN codes (*International Standard Serial Number*) serve to identify various periodicals (newspapers, journal, magazines). Until 2007 they were written as two four-digit numbers separated by a hyphen: $y_1 y_2 y_3 y_4$-$y_5 y_6 y_7 y_8$. The last check digit y_8 was chosen so that the number

$$8y_1 + 7y_2 + \cdots + 2y_7 + y_8$$

is divisible by 11 (if $y_8 = 10$, then the Roman character X is again used). For instance, for the international journal Applications of Mathematics with ISSN 0862-7940 we obtain $\frac{165}{11} = 15$. As of 2007 a new ISSN standard with EAN Prefix 977 was defined. For more details see [433].

Account numbers of the Czech Commercial Bank consist of two parts:

$$\text{account number} : \ b_5 b_4 b_3 b_2 b_1 b_0 \text{-} a_9 a_8 a_7 a_6 a_5 a_4 a_3 a_2 a_1 a_0$$

where the digits b_i and a_i are chosen so that

Fig. 11.3 One-dimensional and two-dimensional barcodes

$$11 \mid \left(\sum_{i=0}^{5} b_i 2^i \right), \quad 11 \mid \left(\sum_{i=0}^{9} a_i 2^i \right). \tag{11.9}$$

Other financial institutions have account numbers formed in another way.

A great variety of error-detecting codes arose by aesthetic feelings of their creators rather than their substantial theoretical advantages. Most of these codes are designed similarly, for example, ISMN codes (*International Standard Music Number*), identification numbers of organizations in the Czech Republic, codes containing biometric data, codes on credit cards, telephone cards or codes on cell phones, even though the number 11 is not always used. For instance, the largest two-digit prime 97 is used to protect IBAN (*International Bank Account Number*).

When using error-detecting codes, we can find out that an error has occurred somewhere, but in general we do not know in which digit. This disadvantage can be eliminated by using error-correcting codes (see Sect. 12.1). They allow us to use some redundant information contained in codewords to determine in which bit (character) an error has occurred and correct it.

New generations of two-dimensional codes with very high information capacity (over 1 kB) and the ability to detect and correct errors have been constructed, which greatly improve previous codes. For instance, the two-dimensional code given in Fig. 11.3 (right) is based on the divisibility by the prime 17. Such codes can be easily and cheaply printed on paper or sent to a smart phone screen. Another of their advantages consists in the possibility of data transfer without the need to insert them from the keyboard, where typos are often made. Since 1994, the two-dimensional QR code (an abbreviation of Quick Response) is widely used. This code can encode a much larger amount of data than the classic one-dimensional EAN barcode, see Fig. 11.3 (left).

11.2   Encryption of Secret Messages by Large Prime Numbers

When transmitting secret military messages, bank data, strategic data, and other confidential and sensitive information, maximum caution is required. For example, business transactions today commonly conducted through the internet can be read by someone who is unauthorized and then misused. Therefore, such data must be encrypted during transmission.

There is a large difference between the terms *encryption* and *coding*. For encryption some secret information (key) is used without whose knowledge it is practically impossible to retrieve the contents of an encrypted message (see Mlýnek [274]). On the other hand, coding is a transformation of one form of information into another form which is for some reason more advantageous for a given situation (e.g. ASCII code, Morse code, genetic code, or codes from Sects. 12.1 and 12.2).

Cryptography is a branch of science which designs suitable encryption methods and which deals with problems of secure data transmission, their protection and storage. For encryption, the so-called secret key was used in the past. Its disadvantage was that all communicators had to know it and thus it could easily have been revealed. Once the secret key is known, the transmission of messages is no longer secret. Nice historical overviews of encryption methods are given in the books Grošek and Porubský [127], and Trappe and Washington [402].

A method which uses a public key for encryption was introduced in the article Rivest, Shamir, and Adleman [329]. Now it is called the *RSA method* by the initials of the authors' surnames. In this method, anyone can encrypt messages, but they can only be decrypted by someone who also knows the secret decryption key. At present this is undoubtedly one of the most secure encryption techniques. Its main idea is based on the fact that if we can encrypt a message, it does not mean that we can decrypt it. In other words, a big advantage the RSA method is that from the knowledge of an encryption key, it is practically impossible to derive a decryption key. This follows from properties of one-way functions whose values can be easily calculated, but it is almost impossible to invert these functions. A secret message is first converted to a string of digits (i.e. a natural number). The main trick of the RSA method is that multiplying two prime numbers that have, say, about a hundred digits takes a small fraction of a second on a usual personal computer. On the other hand, to decompose this product back into its two prime factors would take by the best known methods on the most modern supercomputers a much longer time than is the age of the universe. Therefore, considerable attention is paid to factorization methods and during the last forty years great progress has been made, see e.g. [82, 227, 276, 307, 402, 408]. In the following paragraphs we will get acquainted with only a basic idea of the RSA method (not with its practical implementation).

Recall that the relation

$$x \equiv y \pmod{n}$$

means that the difference $x - y$ is divisible by n. This congruence can be multiplied by any integer c and raised to the power of any natural number k (see Sect. 1.8), i.e.,

$$cx \equiv cy \pmod{n} \quad \text{and} \quad x^k \equiv y^k \pmod{n}.$$

We first convert a secret message to a natural number x. For example, the well-known ASCII code ($A = 65, B = 66, \ldots, Z = 90, \ldots$), can be used for this purpose. There are, of course, other more effective possibilities. Further, we shall assume that

$$x < n,$$

where n is the product of two different primes, which are not publicly known and have more than 100 digits. If the message is longer than n, we divide it into several shorter parts so that the previous inequality holds for each message.

The encryption algorithm itself is relatively a simple operation in which x is raised to a natural exponent e modulo n. Let us denote the encrypted message by the symbol x^*. It is that natural number x^* which is uniquely determined by the inequality $x^* < n$ and the congruence

$$x^* \equiv x^e \pmod{n} \quad \text{(encrypted message)}, \tag{11.10}$$

where e is called the *encryption exponent*. Both the numbers e and n are publicly known and only these two numbers are enough to encrypt. Therefore, anyone can encrypt.

Decryption is performed analogously. We determine the number $(x^*)^\wedge \in \mathbb{N}$ satisfying the inequality $(x^*)^\wedge < n$ so that

$$(x^*)^\wedge \equiv (x^*)^d \pmod{n} \quad \text{(decrypted message)}. \tag{11.11}$$

However, the *decryption exponent* d is not publicly known.

Below we show how to choose the exponents e and d so that $(x^*)^\wedge = x$, i.e., the encrypted message is identical after decryption with the original message x. Before that we recall some properties of the Euler totient function from Sect. 2.8. For any $n \in \mathbb{N}$ its value $\phi(n)$ is defined as the number of natural numbers not exceeding n that are coprime with n, i.e.,

$$\phi(n) = |\{m \in \mathbb{N}; \ 1 \leq m \leq n, \ (m, n) = 1\}|, \tag{11.12}$$

where the symbol $| \cdot |$ stands for the number of elements. We can easily find that

$$\phi(p) = p - 1,$$

if p is a prime. Moreover, for any ith power one can prove that

$$\phi(p^i) = (p - 1)p^{i-1}, \quad i \in \mathbb{N}. \tag{11.13}$$

However, we will be mainly interested in the value of Euler's function for

$$n = pq, \tag{11.14}$$

where p and q are different primes. The number n is thus divisible by p and q and the number of all natural numbers less than n is clearly $pq - 1$. Such a set includes $p - 1$ multiples of q and $q - 1$ multiples of p, which are mutually different and divide n. Therefore,

$$\phi(n) = (pq - 1) - (p - 1) - (q - 1) = pq - p - q + 1 = (p - 1)(q - 1).$$
(11.15)

An important property of Euler's function is the implication

$$(m, n) = 1 \quad \Longrightarrow \quad \phi(mn) = \phi(m)\phi(n),$$
(11.16)

which is actually a generalization of the fact that (11.14) implies (11.15).

As mentioned in Sect. 2.8, using the function ϕ, Leonhard Euler generalized Fermat's Little Theorem without knowing that it would have large applications in a few centuries. The RSA method (based on the Euler–Fermat Theorem 2.21) is used, for example, by the US army or in banks for secure transmission of secret or confidential data. For completeness, let us recall this theorem.

Theorem 11.2 (Euler–Fermat) *For* $x, n \in \mathbb{N}$ *we have*

$$(x, n) = 1$$
(11.17)

if and only if

$$x^{\phi(n)} \equiv 1 \quad (\text{mod } n).$$
(11.18)

Now we prove the following theorem that establishes a sufficient condition for the existence of the inverse element modulo $\phi(n)$.

Theorem 11.3 *If* $e \in \mathbb{N}$ *satisfies the equality*

$$(e, \phi(n)) = 1,$$
(11.19)

then there exists exactly one $d \in \mathbb{N}$ *less than* $\phi(n)$ *such that*

$$ed \equiv 1 \quad (\text{mod } \phi(n)).$$
(11.20)

Proof For $k = 1, \ldots, \phi(n) - 1$ define the residue $z_k \in \{1, \ldots, \phi(n) - 1\}$ by the congruence

$$ek \equiv z_k \quad (\text{mod } \phi(n)).$$

If two residues are equal, i.e. $z_{k_1} = z_{k_2}$, then

$$e(k_1 - k_2) \equiv 0 \quad (\text{mod } \phi(n)).$$

In a similar way as in the proof of Euclid's Theorem 2.1 we can first prove that $k_1 = k_2$. By the assumption (11.19) and Theorem 1.3 there exist integers x and y such that $ex + \phi(n)y = 1$, i.e.,

$$e(k_1 - k_2)x + \phi(n)(k_1 - k_2)y = k_1 - k_2.$$

From this it follows that $k_1 - k_2 \equiv 0 \pmod{\phi(n)}$ which gives $k_1 = k_2$. Hence, we see that all z_k are mutually different. Therefore, there exists exactly one d corresponding to the remainder 1 which fulfills (11.20). $\qquad \square$

Further, we prove that the encrypted message x^* is the same after decryption with the original message x.

Theorem 11.4 *If (11.19) holds, then*

$$(x^*)\hat{} = x. \tag{11.21}$$

Proof According to the congruence (11.20), there exists r such that

$$ed = 1 + r\phi(n). \tag{11.22}$$

Let us distinguish two cases:

(1) Let the assumption (11.17) hold. Then taking the rth power of the quantities in Euler's relation (11.18) and multiplying them by x, we obtain

$$x^{1+r\phi(n)} \equiv x \pmod{n}. \tag{11.23}$$

Now using successively (11.11), (11.10), (11.22), and (11.23), we find that

$$(x^*)\hat{} \equiv (x^*)^d \equiv x^{ed} \equiv x^{1+r\phi(n)} \equiv x \pmod{n}. \tag{11.24}$$

Thus the relation (11.21) is valid, since both the natural numbers x and $(x^*)\hat{}$ are less than n.

(2) Suppose to the contrary that the assumption (11.17) does not hold. Then by (11.14) we have either $p \mid x$, or $q \mid x$. Assume, for instance, that $q \mid x$ (the case $x = p$ could be treated analogously). Since $(q, p) = 1$ and $x < n$, we can raise the Fermat relation $x^{p-1} \equiv 1 \pmod{p}$ to the power of $r(q-1)$, i.e.,

$$x^{r(p-1)(q-1)} \equiv 1 \pmod{p}.$$

By (11.15) we get $\phi(n) = (p-1)(q-1)$ and therefore, we easily find that

$$x^{1+r\phi(n)} \equiv x \pmod{px}.$$

However, this is again the relation (11.23), since $n \mid px$. The proof of the equality (11.21) is then the same as in (11.24). $\qquad \square$

The encryption exponent e is chosen so that $3 \le e < \phi(n)$ and that (11.19) holds. This can be, of course, easily satisfied due to (11.15). Moreover, e has to be chosen so that $e^m \not\equiv 1 \pmod{\phi(n)}$ for "not large" m. Otherwise, due to the relation (11.20), anyone could decrypt messages for $d = e^{m-1}$.

The primes p and q do not appear in congruences (11.10) and (11.11). Hence, if we do not know them, it is practically impossible to determine the value of the decryption exponent d. However, from Theorem 11.3 we know that there exists exactly one natural number $d < \phi(n)$ satisfying the congruence (11.20). However, how do we determine its value in a specific case, when we know p and q?

Finding the prime factorization of the numbers $p - 1$ and $q - 1$ is much easier than factoring n, because they have only about half the number of digits than n has and they are even. If we can factor $\phi(n)$ from (11.15) into prime numbers, then using (11.13) and (11.16), we can easily calculate the value $\phi(\phi(n))$, see e.g. (11.26). The exponent d can then be determined relatively easily. From the Euler–Fermat Theorem 11.2, where instead of x we write e and instead of n we write $\phi(n)$, we obtain

$$(e, \phi(n)) = 1 \implies e^{\phi(\phi(n))} \equiv 1 \pmod{\phi(n)}.$$

Multiplying the above congruence by d and using (11.20), we get an explicit expression for the decryption exponent $d < \phi(n)$,

$$d \equiv de^{\phi(\phi(n))} \equiv ede^{\phi(\phi(n))-1} \equiv e^{\phi(\phi(n))-1} \pmod{\phi(n)}. \tag{11.25}$$

If we cannot factor $\phi(n)$ into primes, we can calculate d directly from the congruence (11.20), e.g., by the Euclidean algorithm from Sect. 1.6, or we choose just another p or q.

Example Let $p = 491$ and $q = 701$ be primes from the factorization (11.14). Then $n = 344191$ and by (11.15) we have $\phi(n) = 343000$. Choosing the encryption exponent $e = 3$, the equality (11.19) obviously holds, since 343000 is not divisible by three.

Assume that the "secret message", which we want to send, is: SOS (see [184]). Setting, e.g., A = 01, B = 02, C = 03, ..., O = 15, ..., S = 19, ..., Z = 26, the message is converted to the number $x = 191519$. Now we will encrypt the secret message x by the public key (11.10). We obtain $x^* = 224717$, since

$$191519^3 \equiv 224717 \pmod{344191}.$$

Next, let us deal with the calculation of d. Using repeatedly relations (11.16) and (11.13), we get

$$\phi(\phi(n)) = \phi(343000) = \phi(2^3 5^3 7^3) = \phi(2^3)\phi(5^3)\phi(7^3)$$
$$= 4 \cdot 100 \cdot 294 = 117600. \tag{11.26}$$

Using a computer, we can verify that

$$3^{117599} \equiv 228667 \pmod{343000},$$

and thus, by (11.25) we get $d = 228667$. To perform a fast calculation of the previous congruence, the exponent should be decomposed into powers of two, i.e.,

$$117599 = 2^0 + 2^1 + 2^2 + 2^3 + 2^4 + 2^6 + 2^8 + 2^9 + 2^{11} + 2^{14} + 2^{15} + 2^{16}.$$

Then the powers 3^{2^k} can be expressed by means of already calculated lower powers $3^{2^{k-1}}$ modulo $\phi(n)$, and then multiplying the appropriate powers again modulo $\phi(n)$.

Finally from (11.11) and (11.21) (or again on computer) we get that

$$224717^{228667} \equiv 191519 \pmod{344191},$$

i.e., $(x^*)\hat{} = 191519$, which means SOS.

A mathematical description of the RSA method was actually invented by Diffie and Hellman as early as 1976 by introducing the concept of one-way functions (see the next Sect. 11.3). Although Diffie and Hellman were the first to present the relationship (11.10) for encryption (see Diffie and Hellman [92]), they did not provide any example as did Rivest, Shamir and Adleman. These three authors then started to be famous namely by publishing a fast algorithm to calculate the congruence (11.10). They clearly described in detail a number of advantages of the RSA method and its features, they suggested how to choose primes p and q, etc. An analysis of the RSA method is also given in Grošek and Porubský [127, p. 205]. The RSA method is currently considered to be one of the most secure tools and there are companies selling 20-decimal primes to banks for RSA purposes.

Finally, let us repeat the meaning of some of the symbols used:

p, q—two different prime numbers that are not publicly known,
n—the product of p and q, which is publicly known (it is a part of the public key),
e—the encryption exponent that is publicly known (it is a part of the public key),
d—the decryption exponent that is not publicly known (it is the so-called private key),
x—a message converted to a natural number less than n,
x^*—a message encrypted with exponent e.

11.3 Digital Signature

Suppose that the assumption (11.19) is satisfied. Since e and d are by (11.20) a pair of mutually inverse elements modulo $\phi(n)$, we can easily verify that (cf. (11.21))

$$(x\hat{})^* = x \tag{11.27}$$

holds as well. This equality is used for the famous *digital signature*, which guarantees the authenticity of the message.

For example, the commander of a military unit will first use his private key (11.11) for some important (but not necessarily secret) message. Only he knows this key. Due to the validity of relationship (11.27), then soldiers can decrypt this message using the public key (11.10). In this way, they are sure that the whole message was actually sent by their commander and not by someone else.

Since a digital signature cannot be falsified, it is much more reliable than the real signature or fingerprint. Moreover, the undersigned cannot later deny that it is his signature. Another possibility will be treated in the next section.

Digital signatures also play an important role in data protection, in designing computer systems against unwanted inputs and manipulations, against "unauthorized readers", falsification or destruction of various files, etc.

The RSA method has found wide applications especially in the military and in banking transactions. It is also used to transfer secret keys (see Sect. 12.2), PINs, and wherever it is necessary to ensure high confidentiality.

Next, we will briefly describe the concept of one-way functions. Let A and B be some subsets of integer numbers. Then a function $f : A \rightarrow B$ is called *one-way*, if for any $x \in A$ it is "easy" to calculate the function value $f(x)$, but it is "very difficult" (technically impossible) to find $x \in A$ for an arbitrary value y from the range

$$f(A) = \{y \in B; \ \exists x \in A : y = f(x)\}$$

so that

$$y = f(x).$$

At the same time, for a given $x_1 \in A$ it is not possible in real time to determine $x_2 \in A$, $x_2 \neq x_1$ such that

$$f(x_1) = f(x_2). \tag{11.28}$$

An example of a one-way function is given by a relation

$$f(x) \equiv g^x \pmod{p}, \tag{11.29}$$

where p is a large prime, g is a primitive root modulo p, and $0 \leq f(x) < p$ is the remainder. Even for a large natural number x, it is relatively easy to calculate the value $f(x)$. However, the backward establishing of the value x from $f(x)$ (i.e., calculation of a discrete logarithm) is not possible in real time, even though p and g are publicly known. One-way functions are also used to create digital signatures, see e.g. [39, 275].

11.4 Hashing Functions

Hashing functions are special cases of one-way functions $f : \{0, 1\}^N \rightarrow \{0, 1\}^n$ for $N \gg n$. This means that they assign to data of an arbitrary length a string (called *hash*) of bits of a fixed length, for example, 128 bits = 16 bytes. At present also much longer hashes are used. Various constructions of some frequently used hashing functions are described e.g. in [21, 333].

A hashing function is designed so that they are easily computed with $\mathcal{O}(N)$ operations and it is not possible to find two different input data in real time, which would yield the same value of the hashing function (see (11.28)). Hashing functions are used in the implementation of digital signatures, verifying passwords, cryptographic check-sums in data transmission, etc. For example, when checking the identity of two databases, we do not mutually compare all files, but instead of complicated data transfers only their hashes $f(x_1)$ and $f(x_2)$ are compared. If $f(x_1) = f(x_2)$, then with a very high probability almost equal to 1 both the databases are identical.

The hashing function can also be used for later demonstration of the authenticity of any document x that was not officially published. It is sufficient if only its value $f(x)$ is published. At any time it is then possible to additionally verify that the hash $f(x)$ corresponds to the proper file x. For example, the author of the last will written in electronic form can inform his relatives only about its hash. After his death and the announcement of his last will, anyone can verify that it is not a post-modified electronic document, i.e., that nothing has been added to or removed from the last will.

Let us give an example on computational complexity for obtaining a file x whose hash is equal to a given value y, i.e. $f(x) = y$. Assume, for simplicity, that we only select x from 16-byte files ($= 128$ bits). Consider the hashing function whose output length is also 16 bytes and suppose that about 2^{127} attempts would be necessary to obtain the searched file x. Performing 10^9 attempts per second, we would need about $5.4 \cdot 10^{21}$ years, which is much more than the present age of the universe $\approx 13.7 \cdot 10^9$ years. In the book [199, p. 174], we show how to use the Euler–Fermat Theorem 11.2 and Fermat primes for constructions of hashing functions for efficiently saving data files and finding necessary information in these data.

11.5 Generators of Pseudorandom Numbers

Primes can also be used to design generators of pseudorandom numbers. They have, for instance, this form

$$r_i \equiv kr_{i-1} \pmod{m}, \quad i = 1, 2, \ldots,$$

where the modulus m is a prime or a power of a prime, r_0 and k are suitably selected constants and $0 \leq r_i < m$. To guarantee that such a generator does not always create the same sequence in every run, it must be restarted from a different initial value r_0, in general. This is usually achieved by hardware.

Pseudorandom number generators are used to obtain prime factorizations, when testing prime numbers, in simulations of some physical processes, in the solution of partial differential equations by the Monte Carlo method (in particular, in a higher dimensional space), in cryptography for generating pseudorandom sequences of bits (see Sect. 12.2), but also in all computer games where some randomness is needed, etc.

A sequence of pseudorandom numbers $x_i \in [0, 1)$ can be defined, for instance, as follows

$$x_i = \frac{r_i - 1}{F_4 - 1} \quad \text{for } i = 1, 2, \ldots,$$

where $F_4 = 2^{2^4} + 1 = 65537$ is the Fermat prime and $r_i \in \{1, \ldots, F_4 - 1\}$ is the remainder such that

$$r_i \equiv 75r_{i-1} \pmod{F_4} \quad \text{and} \quad r_0 = 1. \tag{11.30}$$

The length of this sequence is equal to $2^{16} = 65\,536$. This is the maximum possible length for the prime modulus F_4.

The sequence of powers $75, 75^2, 75^3, \ldots$ is first divided by the Fermat number $F_4 = 2^{16} + 1$ and the corresponding remainders are then normalized such that they are divided by $F_4 - 1$. The periodic sequence x_i of pseudorandom numbers thus attains values between 0 and 1 (including 0 but excluding 1). The constant preceding r_{i-1} in relation (11.30) was chosen to be 75, since this is a primitive root modulo F_4. To verify this, we will first use Theorem 2.26 (ii), (iii), (v), Theorem 2.27, and relation (4.10),

$$\left(\frac{75}{F_4}\right) = \left(\frac{3}{F_4}\right)\left(\frac{5^2}{F_4}\right) = \left(\frac{3}{F_4}\right) = \left(\frac{F_4}{3}\right) = \left(\frac{2}{3}\right) = -1.$$

From the definition of the Legendre symbol we now see that 75 is a quadratic non-residue modulo F_4. By Carmichael's Theorem 2.24 we further get that 75 is also a primitive root modulo F_4, i.e., $\mathrm{ord}_{F_4} 75 = F_4 - 1$. (Theorems 2.29 and 4.16 can also be applied.)

In the 1980s, the pseudorandom number generator (11.30) was used in some home computers, e.g. ZX Spectrum. However, later certain weaknesses of this simple generator were found, see e.g. the monograph Ripley [328, p. 40] (cf. also Knuth [171]). On the other hand, the following generator has better properties and much longer period

$$s_i \equiv 279470273s_{i-1} \pmod{p}, \quad s_0 = 1,$$

where $p = F_5 - 6$ is a prime with 10 digits and $F_5 = 2^{2^5} + 1$ is the Fermat number. Although its results are fully deterministic, it seemingly behaves very randomly.

Some disadvantages of linear pseudorandom number generators of the form

$$y_{i+1} \equiv by_i + c \pmod{m}, \quad i = 0, 1, 2, \ldots,$$

can be avoided by using quadratic generators (see e.g. Strauch and Porubský [387, Sect. 2.25.5]). They define a sequence of pseudorandom numbers x_i by

$$x_i = \frac{y_i}{m},$$

where

$$y_{i+1} \equiv ay_i^2 + by_i + c \pmod{m}, \quad i = 0, 1, 2, \dots,$$

where the modulus m is a prime or a power of a prime, $0 \le y_i < m$, and y_0, a, b, c are suitably selected integer constants. In Knuth [170, p. 34], necessary and sufficient conditions on these constants are given so that the sequence has the maximum possible length m.

Remark The sequence of pseudorandom numbers $(x_i)_{i=1}^{\infty}$ defined by the above generators attains values between 0 and 1 after normalization. Let $F = F(x)$ corresponds to some probability distribution, i.e., such a distribution function F fulfills $F(x) \to 0$ for $x \to -\infty$ and $F(x) \to 1$ for $x \to \infty$. If F is increasing, then we can consider the inverse function F^{-1} (e.g. for the normal Gaussian distribution). Then $(F^{-1}(x_i))_{i=1}^{\infty}$ is a sequence of pseudorandom numbers corresponding to the considered probability distribution.

11.6 A Message to Extraterrestrial Civilizations

In 1974, a message to extraterrestrial civilizations was sent from the world's largest radio telescope Arecibo with diameter 305 m in Puerto Rico to the globular star cluster M13 in the constellation Hercules. This message was written in the binary system (instead of zeros and ones, short signals were transmitted on two different frequencies). It contained exactly 1679 bits. If there is a sufficiently advanced civilization on some planet in this huge system of stars, then it is assumed that they know the Fundamental Theorem of Arithmetic 2.2. If so, then they factor the number 1679 into the product of two prime numbers

$$1679 = 73 \times 23.$$

This allows one to create a rectangular image composed of squares (see Fig. 11.4), which has a very special arrangement, i.e., it has a low entropy. White squares correspond to 0 and black squares to 1. Assembling individual bits into the longitudinal rectangle 23×73, we get an image (see [273, p. 335]) with high entropy, which resembles scattered tea leaves. Therefore, a message whose length is the square of a prime number was considered as well, because then the order of factors does not matter.

When assembling individual bits into the 73×23 rectangle, we can proceed in several ways. One of them is drawn in Fig. 11.4, where the message has to be read from right to left. The bits can also be rearranged in a mirror image, and then the message can be read as usual from left to right.

Fig. 11.4 A message to extraterrestrial civilizations from 1974. It is admirable how much information its authors managed to put into just 1679 bits $\approx$ 210 bytes. For instance, reading from the right above line 15 the sugar deoxyribose H_7C_5O, adenine $H_4C_5N_5$, thymine $H_5C_5N_2O_2$, and sugar deoxyribose are coded. Below this line there are two phosphate ions PO_4. Above line 25 the sugar deoxyribose, cytosine $H_4C_4N_3O$, guanine $H_4C_5N_5O$, and sugar deoxyribose are coded. Below this line there are again two phosphate ions PO_4. This is a detail of a DNA double helix that is unfolded and depicted in the middle of this message

The numbers that are written on the first three lines of Fig. 11.4 are 1, 2, 3, 4, 5, 6, 7, 8, 9, 10 in the binary system.

The position of the black squares in the fourth line under each number seems striking, but it indicates a kind of label (e.g. the decimal point), where the number is to be read from. The authors (Frank Drake and Carl Sagan) named this upper part of the message: an introductory lesson.

Another group of characters on lines 6–10 thus determines five numbers

$$1, 6, 7, 8, 15,$$

which corresponds to the atomic numbers of the biogenic elements

H, C, N, O, P.

The following lines encode components of the DNA acid. The structure of the DNA double helix is illustrated in the middle of the figure. Between the two strands of DNA there is information about the total number of nucleotides. Below it we find the silhouette of a human, and then the Sun with nine planets. To highlight the Earth, it is moved one line closer to the human. The giant planets Jupiter and Saturn are marked with three black squares, Uranus and Neptune with two squares, and Pluto was at that time a planet.

The radio telescope, from which the message was sent, is at the bottom of this figure. Rays emanating from the focus are parallel after reflection from the antenna mirror. The last two lines encode the approximate diameter of the mirror, expressed in the wavelength of 12.6 cm of the transmitted signal.

A similar message containing $31 \times 41 = 1271$ bits is given in the book [85, p. 106] by Delahaye. Many more complicated messages to extraterrestrial civilizations have been sent since 1974. They are always based on number theory. We cannot communicate with them in English or Czech, but we can send them pictures.

11.7   Further Applications of Primes

Consider the set

$$\mathbb{Z}_p = \{0, 1, \ldots, p-1\}$$

for $p > 1$ with the addition and multiplication operations modulo p. If p is a prime, then by Theorem 3.4 for any nonzero element of the set $\mathbb{Z}_p$ there exists exactly one inverse element. Such an algebraic structure is called a *finite field*. If p is not a prime, then the inverse element cannot be defined. For instance, for $\mathbb{Z}_6$ we get nontrivial zero divisors since $2 \cdot 3 \equiv 0 \pmod{6}$. Algebraic finite fields have a number of practical applications in quantum logic, combinatorics, theory of rings, etc. We also should not forget that computer architecture since about 1940 is built precisely on arithmetic modulo the prime number $p = 2$, i.e., using the field $\mathbb{Z}_2$.

In arithmetic modulo 2^n or $2^n \pm 1$ is very easy to define the so-called fast multiplication. For example, multiplication by two can be converted to a shift of one bit left (see [199, p. 172]). In 1971, Schönhage and Strassen[1] presented a fast algorithm to multiply two large numbers of N digits. It requires only $\mathcal{O}(N \log N \log \log N)$ arithmetic operations (see [349]). Note that the standard algorithm for multiplying two numbers, that children use at school, requires $\mathcal{O}(N^2)$ operations. For example, if one multiplies two numbers with a thousand digits, the advantages of the fast algorithm immediately appear.

Biologists describe an interesting species of cicada (genus Magicicada) whose pupae live underground for 17 years (see [82, Sect. 8.6], [355], and [416, p. 27]). Then they hatch, in a few weeks they mature, lay eggs, and die. A certain parasite

---

[1] In 1969, V. Strassen became famous by his surprising result [386] that the number of arithmetic operations of the Gaussian elimination is not asymptotically optimal.

that endangers adults, has a two-year or three-year life cycle. In the first case, it encounters a cicada every 34 years and in the second case every 51 years, because these numbers are coprime. With its unusual life cycle, a cicada protects against parasites. Pupae of other types of cicadas live underground 7 or 13 years.

For another application of primes and Fermat's Christmas Theorem 3.12 in biology we refer to [421]. One can assign to any amino acid from the standard DNA genetic code a prime of the form $4k + 1 = a^2 + b^2$, where $a \leq b$ are from the set $\{0, 1, \ldots, 7\}$. For instance, leucine $17 = 1^2 + 4^2$, serine $29 = 2^2 + 5^2$, valine $41 = 4^2 + 5^2$, alanine $53 = 2^2 + 7^2$, glycine $61 = 5^2 + 6^2$. Stop codons are represented by the identity $0 = 0^2 + 0^2$. In this way, one can save computer memory for a large amount of genetic data describing coding of proteins.

Prime numbers and coprime numbers play an important role in designing gear wheels for various devices such as bicycles, gearboxes, mechanical clocks, factorization machines, see e.g. [85, p. 112], [141].

Another interesting application of primes is flipping coins electronically. Its main idea was first introduced by Manuel Blum [33] in 1982. Assume that Alice and Bob want to flip a coin, for instance, by telephone. Alice chooses two large primes p and q with $p \equiv q \equiv 3 \pmod 4$ and sends their product $n = pq$ to Bob. Bob randomly picks a natural number $x < n$ and sends to Alice the integer $a \in \{1, 2, \ldots, n - 1\}$ for which

$$x^2 \equiv a \pmod n. \tag{11.31}$$

Then Alice finds the four solutions $x, y, n - x, n - y$ of the congruence (11.31). After that Alice sends one of these solutions to Bob. Since $x + y \not\equiv 0 \pmod p$ and $x + y \equiv 0 \pmod q$, we find that $(x + y, n) = q$ and analogously $(x + (n - y), n) = p$. Consequently, if Bob receives either y or $n - y$, then he can easily factor n by the Euclidean algorithm from Sect. 1.6. On the other hand, if Bob receives either x or $n - x$, then he has no way to factor n in a reasonable length of time. Therefore, Bob wins the coin flip if he can factor n, while Alice wins if Bob cannot factor n. From the above exposition it follows that there is an equal chance for Bob to receive a solution of (11.31) that helps him rapidly factor n, or a solution of (11.31) that does not help him factor n. Thus, the coin flip is fair.

Prime numbers and coprime numbers are used in speech recognition or when using X-rays in astronomy or when designing directional antennas or concert halls—see the book by Schroeder [350], which contains many other practical applications. Fermat and Mersenne primes are employed in the fast Fermat and Mersenne transforms [4, 93, 123] in investigating fractals and chaos, in the Euclidean construction of the regular polygons, etc., see the monograph by Křížek, Luca, and Somer [199].

Chapter 12
Further Applications of Number Theory

12.1 Error-Correcting Codes

Number theory has a lot of applications in various areas of our lives, e.g., in error-detecting and error-correction codes, in encoding and decoding a television signal, in astronomy, in solving finite difference equations and differential equations, in creating JPG or PNG formats, in group and lattice theory, in computer science, in optics, in nuclear physics, in spectroscopy, in the study of atom structure or resonances in oscillation theory (see [37, 316, 318, 350, 424], etc.). For instance, the wavelength λ of the absorbed/emitted light of hydrogen satisfies

$$\frac{1}{\lambda} = R_{\mathrm{H}}\left(\frac{1}{n^2} - \frac{1}{m^2}\right),$$

where R_{H} is the Rydberg constant and $n < m$ are positive integers associated with levels of electron orbits of hydrogen. The case $n = 1$ (or $n = 2$) corresponds to the Lyman (or Balmer) series of hydrogen lines.

Number theory can also be used for entertainment as we shall see in Sect. 12.6. In Sect. 11.1, we dealt with application of primes in error-detecting codes. Using them, we can find out that there is an error in a given data set, but we generally do not know which bit was transmitted or displayed incorrectly. However, we can again use number theory to eliminate this drawback. The battle against noise is a subject of *error-correcting codes*. Their main idea was invented in 1947 by Richard Wesley Hamming (1915–1998) who had to remove a considerable unreliability in relay computers of that time.

To explain the main idea of error-correcting codes, we first introduce the *Hamming distance* $d(u, v)$ between two vectors $u = (u_1, u_2, \ldots, u_k)$ and $v = (v_1, v_2, \ldots, v_k)$, $u_i, v_i \in \{0, 1\}$, as the number of positions where u differs from v.

M. Křížek et al., *From Great Discoveries in Number Theory to Applications*, https://doi.org/10.1007/978-3-030-83899-7_12

Fig. 12.1 The first three
components of the
codewords are the
coordinates of the vertices of
the cube. The Hamming
distance any two codewords
is 4

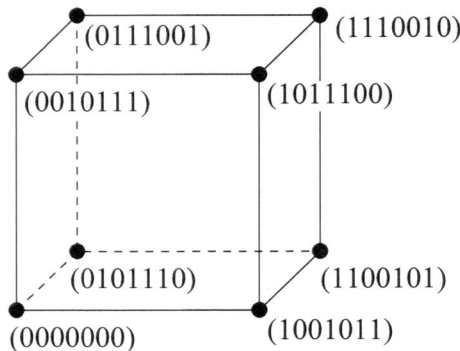

For instance, if $k = 3$, then $d((000), (011)) = 2$ and $d((101), (010)) = 3$, where
commas between the components of the vectors are omitted for simplicity. A simple
error-correcting code uses the following eight *codewords*, which are vectors of length
$k = 7$,

$$(0000000)$$
$$(0010111)$$
$$(1001011)$$
$$(1100101)$$
$$(1110010)$$
$$(0111001)$$
$$(1011100)$$
$$(0101110) \tag{12.1}$$

We observe that the nonzero elements in (12.1) cyclically shift one bit to the right.
Further notice that the first 3 bits (digits) in each line of (12.1) are different from
each other and that no other such triple of bits exists, because $2^3 = 8$. Thus, the first
three bits represent coordinates of the vertices of a unit cube (see Fig. 12.1). The
remaining 4 bits are chosen so that the Hamming distance between any two lines in
(12.1) is just 4.

Denote by G the set of all eight codewords from (12.1) with the operation

$$u \oplus v = (u_1 \oplus v_1, u_2 \oplus v_2, \ldots, u_7 \oplus v_7),$$

where the individual components are added according to the following rules:

$$0 \oplus 0 = 0, \quad 1 \oplus 0 = 1, \quad 0 \oplus 1 = 1, \quad 1 \oplus 1 = 0.$$

This operation is clearly associative. Since the Hamming distance of every two different elements of G is 4, by summing them, we obtain a codeword composed of four ones and three zeros. We can easily verify that this codeword again belongs to G. For example,

$$(0101110) = (1001011) \oplus (1100101) = (1011100) \oplus (1110010)$$
$$= (0010111) \oplus (0111001) = (0101110) \oplus (0000000).$$

Note that the set G along with the operation addition $\oplus$ forms a group (for details see [312]). Its neutral element is the first codeword in (12.1) composed only of zeros alone and each element is inverse to itself.

Now we will briefly describe how error-correcting codes can be applied to eliminate accidental or random errors, for example, when transferring data from interplanetary probes. The need to employ error-correcting codes arises if there is a certain (not too big) probability of an error in the transmission of information, e.g., when radio waves pass through an interplanetary cloud of dust or an asteroid belt.

Example Data transmission in the case of code words (12.1) proceeds as follows. The first 3 bits of transmitted information are uniquely complemented by another 4 bits according to (12.1) and the corresponding codeword is sent. Then the other three bits of the transmitted information are taken. They are again extended to seven bits according to (12.1) and sent, etc. If one bit changes in a codeword during its transfer, then the corresponding Hamming distance from the original codeword will be 1, while it will be 3 or 5 from the other codewords. Thus, the reception equipment will immediately find which bit was transferred incorrectly and corrects it. When two of the seven bits are lost, then the reception device only recognizes that an error has occurred and the entire codeword must be sent again. After reception, only the first 3 bits of each codeword are used which carry the original information. The remaining 4 bits are redundant and thus are omitted.

If data transfer were even more unreliable, then another error-correcting code with longer codewords has to be applied (see Thompson [399]). One could also use a larger antenna, etc. For example, Mars' first orbiting satellite Mariner 9 was already equipped in 1971 with an error-correcting code having the possibility of up to seven corrections per 32 bits long codewords. To each point of the transmitted photo of Mars' surface, one of $2^6 = 64$ degrees of brightness was assigned. The corresponding 6 bits were supplemented by 26 bits of redundant information to correct any error caused by noise during data transmission through interplanetary space. For more information about this code see [2, p. 90].

Remark In the densest arrangement (packing) of equally large circles in the plane, each circle touches its six neighbors. Consider now a uniform packing of congruent spheres in d-dimensional Euclidean space. Denote by $K(d)$ the maximum number of points of contact of a given central ball with its neighbors, which is called the *kissing number*. Then we have $K(1) = 2$, $K(2) = 6$, $K(3) = 12$, $K(4) = 24$, and

$K(8) = 240$, see [212]. For the other d we only know some very rough lower and upper estimates of $K(d)$, except for $d = 24$ when the lower estimate is equal to the upper estimate. In this case $K(24) = 196560$. This property is used in a very efficient error-correcting code, see e.g. [77, 298, 399].

Error-correcting codes are also used to ensure the proper functioning of some computer memories. The well-known compact discs are similarly "protected" against mechanical damage. If the scratching is limited, it still works.

Finally note that error-correcting codes and self-correcting processes were already discovered by nature several billion years ago. For example, if one strand of DNA acid is damaged, the lost information will be replenished according to the complementary strand (see Fig. 1.3). Minor damage of a living organism will also "self-repair". For instance, when you scratch yourself, you do not have to worry about anything, and it will heal automatically. The standard genetic code is also error-detecting, since frame-shifted errors quickly terminate synthesis of RNA or proteins [193].

12.2   Coding By a Symmetric Key

Cryptography is a part of a broader scientific field called cryptology, which also includes the field of cryptanalysis, i.e. analysis of security algorithms. Cryptography deals with protection of data transmission and data storage, primarily by ensuring their confidentiality, data integrity, authenticity of information, and indisputability of their origin. The rapid development of cryptography can be linked to the beginning of the 20th century, when the telegraph and later radio became a widespread manner of information transfer. For example, the German army used the electromechanical encryption machine Enigma. Its code was deciphered in 1932 by the mathematicians Marian Rejewski, Jerzy Rozycki, and Hendryk Zygalski working for the Polish secret service. At the end of the Second World War, an enhanced code was decrypted by the brilliant mathematician Alan Turing (1912–1954). This achievement later helped to shorten the war and save many people lives.

In this section we will introduce a very simple symmetric key encryption method (called the Vernam cipher). Symmetric ciphers with secret keys are used for the encryption of large data files, because they are significantly faster than asymmetric encryption methods (such as the RSA method, see Sect. 11.2). To send a secret message, we need an encryption key, without whose knowledge the message cannot be deciphered. We generate this key as a sufficiently long sequence of completely randomly selected zeros and ones (e.g., by tossing coins or using generators of pseudorandom numbers, see Sect. 11.5). Below we show that the secret key selected in this way happens to be an encryption and decryption key at the same time. Therefore, it is called a *symmetric key*.

On the set $G = \{0, 1\}$ let us define the addition operation $\oplus$ as in the previous section,

$$0 \oplus 0 = 0, \quad 1 \oplus 0 = 1, \quad 0 \oplus 1 = 1, \quad 1 \oplus 1 = 0. \tag{12.2}$$

We will now apply these arithmetic rules for a very simple encryption and decryption of secret messages as follows:

We first recode the text of the message into a sequence of zeros and ones. The difference between coding and encryption is often confused. When encoding, one form of writing is converted to another, however, it is not kept secret. The conversion algorithm is known (e.g., bar code on goods, ASCII code, radiotelegraphic code Q, etc.). Therefore, coding only serves to facilitate the transfer of information. On the other hand, encryption is used to keep the message secret (see also Sect. 11.2). The message in the binary system is then corrected with the encryption key so that we gradually use relations (12.2) on the individual digits—see the first three rows of the following example:

$$100001110\ldots \text{ secret message,}$$
$$\underline{000101101}\ldots \text{ encryption key,}$$
$$100100011\ldots \text{ transmitted message encrypted,}$$
$$\underline{000101101}\ldots \text{ decryption key,}$$
$$100001110\ldots \text{ decrypted message.}$$

The receiver adds the decryption key to the encrypted message—see the last three lines of this example. Notice also that the sum of the encryption and decryption keys is zero, i.e., the encryption (and decryption) key is inverse to itself. If the encryption key was chosen to be a truly random sequence of zeros and ones (see Balakrishnan and Koutras [19]), then the encrypted message itself is again a completely random sequence (which does not carry any information).

Theorem 12.1 *The encrypted message is the same after decryption with the original message.*

Proof From (12.2) it follows that

$$u \oplus v \oplus v = u \oplus 0 = u \quad \text{for all } u, v \in G,$$

and therefore, after a double use of the symmetric key a decrypted message has to be identical to the original message. □

This encryption method is used, among others, in banks in combination with the RSA method having a public encryption key (see Sect. 11.2). However, a big advantage of the symmetric key is that it is computationally extremely cheap while the RSA method is very time consuming.

Let us describe a simple way how to use a combination of the RSA method and the symmetric key for sending a longer secret message:

A symmetric key of a certain length (e.g. 128 bytes = 1024 bits) is randomly generated. The secret message is then divided into eight-byte blocks and each block is encrypted with the symmetric key. The symmetric key itself is encrypted by the public RSA key of the recipient and is sent together with the encrypted message. In this way, the recipient will get the symmetric key without revealing it, and can decrypt the secret message.

12.3   Kepler's Semiregular Tilings

The German mathematician and astronomer Johannes Kepler in his pivotal work *Harmonices mundi* (1619) addressed the question: which tilings (i.e. mosaics, parquet, tessellations) can be created from regular n-gons so that adjacent n-gons always share the whole side. In addition, he required that each vertex is of the same type $(n_1, n_2, \ldots, n_k)$, i.e., it is surrounded in each case by the regular n_1-gon, n_2-gon, etc. We assume that the k-tuple $(n_1, n_2, \ldots, n_k)$ is equivalent to $(n_k, \ldots, n_2, n_1)$. Hence, we do not distinguish if the vertices of n-gons around a given vertex are numbered in the clockwise or counterclockwise direction. Also the k-tuples $(n_1, n_2, \ldots, n_k)$ and $(n_2, \ldots, n_k, n_1)$ are considered to be equivalent, i.e., it does not matter from where we start to number the n-gons. Such a tiling is called *semiregular*. In particular, if $n_1 = n_2 = \cdots = n_k$, then it is called *regular*. Moreover, two tilings will be considered to be the same if we get one from the other by rotation or dilation.

Theorem 12.2 (Kepler) *There exist exactly* 12 *different semiregular tilings of the plane,* 3 *of them are regular.*

Proof The interior angle of a regular n_i-gon is equal to $(n_i - 2)180°/n_i$. Therefore, for a vertex of type $(n_1, n_2, \ldots, n_k)$, the following necessary (but not sufficient) condition for the existence of a semiregular tiling holds:

$$\frac{n_1 - 2}{n_1}180 + \frac{n_2 - 2}{n_2}180 + \cdots + \frac{n_k - 2}{n_k}180 = 360.$$

From this we get by simple manipulations the Diophantine equation

$$\frac{1}{n_1} + \frac{1}{n_2} + \cdots + \frac{1}{n_k} = \frac{k - 2}{2}. \tag{12.3}$$

The right-hand side of (12.3) has to be positive, and thus $k \geq 3$. Since a point can be surrounded by 6 equilateral triangles and all other n-gons have larger interior angles, we obtain another necessary condition $k \leq 6$. If we order, for clarity, all the components of the resulting k-tuples by size, we get the following 17 solutions of equation (12.3).
Triples:

$$(3, 7, 42), \quad (3, 8, 24), \quad (3, 9, 18), \quad (3, 10, 15), \quad (3, 12, 12),$$
$$(4, 5, 20), \quad (4, 6, 12), \quad (4, 8, 8), \quad (5, 5, 10), \quad (6, 6, 6);$$

quadruples:

$$(3, 3, 4, 12), \quad (3, 3, 6, 6), \quad (3, 4, 4, 6), \quad (4, 4, 4, 4);$$

pentuples:

$$(3, 3, 3, 3, 6), \quad (3, 3, 3, 4, 4);$$

hextuples:

$$(3, 3, 3, 3, 3, 3). \tag{12.4}$$

However, not all of them correspond to semiregular tilings of the entire plane. For example, a point can be surrounded by two pentagons and one decagon, but we can easily verify that this is not enough to tile the whole plane. In addition, the components of the above seventeen k-tuples were ordered by size, which we will not require anymore.

By inspection we obtain only the following 12 solutions (see Fig. 12.2):

$$(3, 3, 3, 3, 3, 3), \quad (4, 4, 4, 4), \quad (6, 6, 6),$$
$$(3, 12, 12), \quad (4, 8, 8), \quad (4, 6, 12),$$
$$(3, 6, 3, 6), \quad (3, 4, 6, 4), \quad (3, 3, 4, 3, 4),$$
$$(3, 3, 3, 4, 4), \quad (3, 3, 3, 3, 6), \quad (3, 3, 3, 3, 6).$$

The last two solutions are numerically the same. However, from Fig. 12.2 we see that the last tiling is a mirror image of the previous one. The other 10 tilings have an axis of symmetry.   □

Kepler's semiregular tilings are used for the decorative paving of some sidewalks, for parquets, and artistic mosaics, such as patterns for wallpapers, fabrics, etc., see Grünbaum and Shephard [128].

12.4 Platonic Solids

The Greek mathematician Plato (427–347 BC) tried to interpret cosmology on purely geometric considerations. In doing so, he discovered that in the three-dimensional space there are exactly 5 regular bodies (cf. Fig. 12.3). Recall that a *regular polyhedron* (also called a *Platonic solid*) is a convex polyhedron all of whose faces are congruent regular polygons and for which the same number of faces meet at each vertex.

Theorem 12.3 *There exist exactly 5 Platonic solids.*

Proof We show that there are at most 5 Platonic solids, i.e., that there is no other than in Fig. 12.3. The interior angle in the regular n-gon is equal to $(n-2)180°/n$. Since

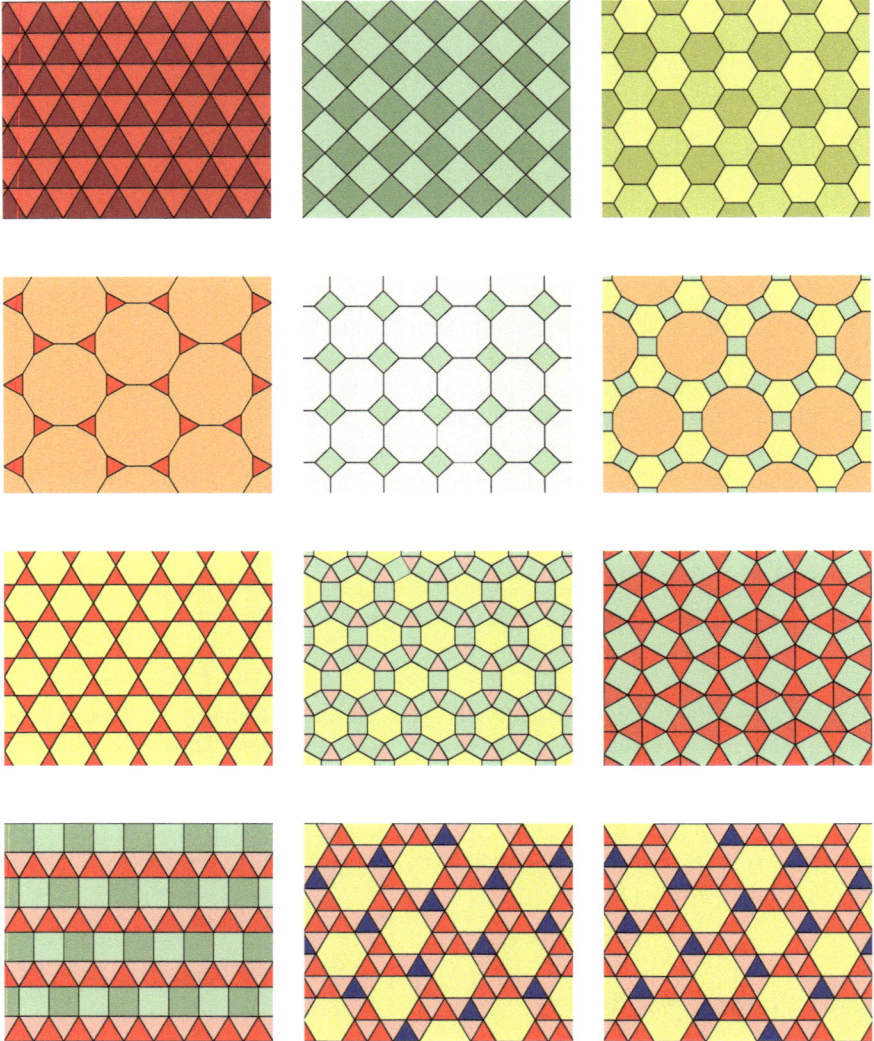

Fig. 12.2 Kepler's regular and semiregular tilings of the plane by regular polygons. Since the last tiling is a mirror image of the previous one, sometimes only 11 of Kepler's tilings are presented, see [206]

each Platonic solid is a convex polyhedron, the sum all interior angles around a given vertex is less than 360°. Denoting by k the number of regular polygons surrounding an arbitrary vertex, we get the following Diophantine inequality

$$k\frac{n-2}{n}180 < 360,$$

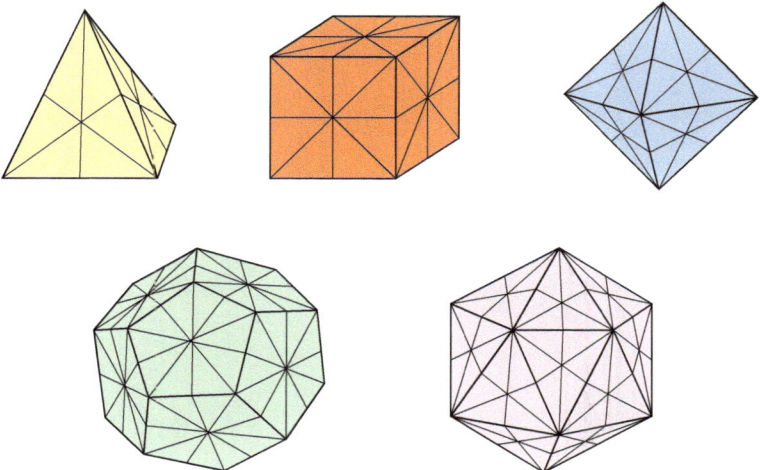

Fig. 12.3 Platonic bodies and their planes of symmetries

where clearly $k \geq 3$. From this we obtain

$$3(n - 2) \leq k(n - 2) < 2n, \qquad (12.5)$$

and thus

$$n < 6.$$

If $n = 3$ then the right inequality in (12.5) has exactly three solutions $k \in \{3, 4, 5\}$ which correspond to the regular tetrahedron, octahedron, and icosahedron. For $n = 4$ the inequality in (12.5) has the only solution $k = 3$ which leads to a cube. Also for $n = 5$ we get only one solution $k = 3$ corresponding to the regular dodecahedron. Each of these five solutions is therefore achievable. □

A cube has 8 vertices and 6 faces. It is dual to an octahedron, since that has 6 vertices and 8 faces. Similarly, an icosahedron has 20 vertices and it is dual to a dodecahedron which has 12 vertices. A tetrahedron is dual to itself. For any convex polyhedron the well-known *Euler formula* holds

$$v + f = e + 2,$$

where v is the number of vertices, e is the number of edges, and f is the number of faces. This relation was known to René Descartes already in the first half of the 17th century.

Just as we investigated regular and semiregular planar tilings by regular polygons, we can similarly study semiregular polyhedra. A *semiregular polyhedron* is a convex polyhedron all of whose faces are regular polygons and all its spatial (solid)

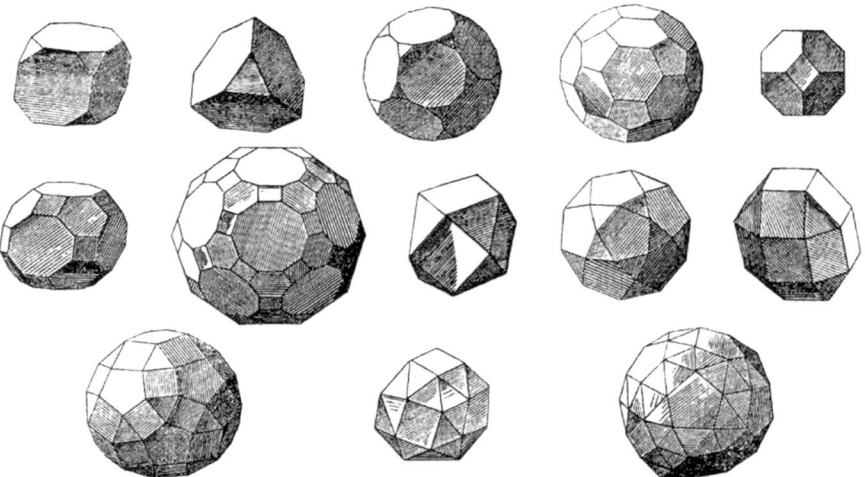

Fig. 12.4 Archimedean solids by Johannes Kepler

angles at the vertices are directly or indirectly identical.[1] A special case of semiregular bodies are Platonic solids whose surface is formed by regular polygons of one type. Semiregular polyhedra whose surface is formed by regular polygons of two or more types, are divided into Archimedean solids (see Fig. 12.4), regular prisms, and regular antiprisms. Their existence can be investigated similarly as in Theorems 12.2 and 12.3. Regular prisms (or regular antiprisms) have two opposite faces formed by the same regular n-gons and the other faces are squares (or equilateral triangles). A special case of a regular prism (or a regular antiprism) is the cube (or the regular octahedron). An overview of the thirteen Archimedean solids was given by Johannes Kepler in the second chapter of his famous book *Harmonices Mundi*. These bodies are named after the ancient mathematician, physicist, and inventor Archimedes of Syracuse (287–212 BC). In the 20th century, the fourteenth Archimedean solid was discovered. Its body arises by rotating the "top layer" about 45° of the tenth solid from Fig. 12.4.

Regular and semiregular solid have many applications in crystallography and the theory of point groups. They are also used for decorative purposes. Also a soccer ball resembles an Archimedean solid with 12 pentagonal and 20 hexagonal faces (see the fourth body in Fig. 12.4). This polyhedron has 60 vertices and has also found applications in chemistry. It turned out that there exists a stable carbon molecule, called fullerene C_{60}, which has 60 atoms located just at the vertices of this semiregular body. Regular triangulations of the triangular faces of the regular icosahedron are used together with central projection to construct triangulations of the surface of a sphere. The regular icosahedron (cf. Fig. 12.3) can also be used for partitioning of the whole three-dimensional Euclidean space into acute tetrahedra (see [41, p. 54]

---

[1] A spatial angle is indirectly identical with its mirror image.

Table 12.1 Regular polytopes in $\mathbb{R}^4$. The last column shows the volume provided that each edge has length a (see Conway and Sloane [77, pp. 452–453]), where $\alpha = \frac{1}{2}(1 + \sqrt{5})$ is the golden ratio

Name	c	f	e	v	Duality	Volume
4-simplex	5 tetrahedra	10	10	5	Self-dual	$\sqrt{10}a^4/192$
4-cube	8 cubes	24	32	16	Dual to 4-orthoplex	a^4
4-orthoplex	16 tetrahedra	32	24	8	Dual to 4-cube	$a^4/6$
24-cell	24 octahedra	96	96	24	Self-dual	$2a^4$
120-cell	120 dodeca-hedra	720	1200	600	Dual to 600-cell	$15\sqrt{5}a^4\alpha^8/2$
600-cell	600 tetrahedra	1200	720	120	Dual to 120-cell	$25a^4\alpha^3/4$

for details). The only Platonic body that tiles $\mathbb{R}^3$ is a cube. This tiling has a variety of practical applications. Three-dimensional space can also be alternately tessellated by regular tetrahedra and regular octahedra. This semiregular tilling forms the lattice of a diamond.

A *regular polytope* in $\mathbb{R}^d$ can be defined by induction: All its $(d - 1)$-dimensional facets are congruent regular polytopes and all its vertices have the same vertex degree (i.e., the number of edges coming together at a given vertex).

For $d = 4$ there are exactly six regular polytopes: 4-simplex, 4-cube, 4-orthoplex, 24-cell, 120-cell, and 600-cell, see Table 12.1. They were discovered by Ludwig Schläfli (1814–1895). The above Euler formula can be then generalized as follows

$$v + f = e + c \quad \text{(Euler–Poincaré formula)},$$

where the letter c stands for the number of three-dimensional convex polyhedral cells on their surfaces and the other symbols have the same meaning as for $d = 3$. The space $\mathbb{R}^4$ can be tessellated by congruent copies of 4-cubes or 4-orthoplexes or 24-cells.

The following theorem is well known (Stillwell [384]).

Theorem 12.4 *For any $d \geq 5$ there are only three regular polytopes in $\mathbb{R}^d$: the d-simplex, the d-cube, and the d-orthoplex.*

The only regular polytope that tiles $\mathbb{R}^d$ for $d \geq 5$ is a d-cube. We may also consider regular and semiregular tilings of hyperspheres or hyperbolic pseudospheres by curved regular polytopes, see e.g. [41, p. 167], [78].

12.5   Tetrahedral Space-Fillers

In 1900, the famous German mathematician David Hilbert (1862–1943) asked the question: *What are the polyhedra whose congruent copies can be used to fill all of three-dimensional space without gaps?* This question has not yet been fully resolved. It is closely related to a part of Hilbert's eighteenth problem which asks whether there exists a polyhedron which tiles three-dimensional Euclidean space but does not admit a tile-transitive tiling. A large number of partial solutions were discovered by crystallographers studying the structure of mono-crystal lattices. In this section, we will deal with the simplest case, where a polyhedron is a tetrahedron. A tetrahedron whose congruent copies fill the whole space without gaps and overlaps is called a *space-filler.*

The Greek scholar Aristotle inspired by the thoughts of his teacher Plato in his treatise *On the Heavens* (350 BC) mistakenly thought that space can be filled without gaps by regular tetrahedra of the same size (see [12, Vol. 3, Chap. 8]). However, this would require that the angle between two faces of a regular tetrahedron be 72°. If so, we could divide the regular icosahedron into 20 regular tetrahedra whose common point is at the center of the icosahedron. Because Aristotle was a respected person, nobody doubted his claim. It was not proved false until the Arab mathematician Averroës (1126–1198) discovered that each edge of the regular icosahedron (see Fig. 12.3) inscribed into a sphere of radius 1 measures

$$\frac{1}{5}\sqrt{10(5 - \sqrt{5})} \approx 1.05.$$

This is not equal to 1 as would follow from the Aristotle conjecture. So Aristotle was wrong. Each dihedral angle of the regular tetrahedron equals $\arccos \frac{1}{3}$, which rounded to the nearest degree is 71° (see Fig. 12.5).

Now let us ask a question (see [181]), whether there exists a tetrahedral space-filler all of whose edges have integer lengths. The answer will be based on a construction which was proposed by Michael Goldberg in 1974. In the article Goldberg [119] (see also Sommerville [381]), he divided space into infinitely long triangular prisms whose cross-section is an equilateral triangle whose sides have length e. On three parallel edges of one fixed prism (see Fig. 12.6), we place points A, B, C, D, E, and F whose coordinates in the direction of the vertical axis are successively: 0, a, $2a$, $3a$, $4a$, and $5a$, where $a > 0$ is an arbitrary real number. It is easy to see that the obliquely cut triangular prism $ABCDEF$ can fill the whole space, since its faces ABC and DEF are parallel.

From Fig. 12.6 it is further obvious that the prism $ABCDEF$ can be divided into three congruent tetrahedra $ABCD$, $BCDE$, and $CDEF$. Denoting by b and c the length of the edge AB and AC, respectively, we get

$$b^2 = e^2 + a^2, \quad c^2 = e^2 + 4a^2. \tag{12.6}$$

Fig. 12.5 Three-dimensional space cannot be filled with congruent copies of regular tetrahedra. To a given face of a regular tetrahedron it is possible to connect another regular tetrahedron of the same size only in a unique manner. If we "arrange" 5 regular tetrahedra around one common edge, then a small gap appears

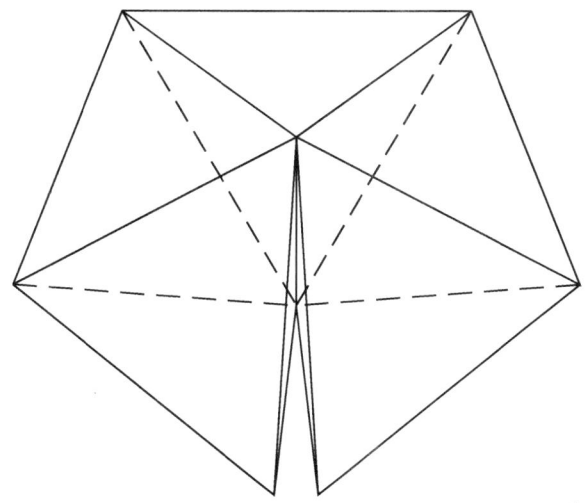

We observe that edges BC, CD, DE, and EF have length b, and edges BD, DF, and CE have length c. Since the ratio a/e can be an arbitrary positive number, there exist infinitely many tetrahedral space-fillers.

Theorem 12.5 *There exist infinitely many tetrahedral space fillers whose edge lengths are integer numbers.*

Proof Eliminating e^2 from Eq. (12.6), we obtain the Diophantine equation

$$3a^2 + b^2 = c^2. \tag{12.7}$$

As in Theorem 2.5, we can find that Eq. (12.7) has positive integer solutions of the form

$$a = 2kmn, \quad b = k(im^2 - jn^2), \quad c = k(im^2 + jn^2),$$

where i, j, k, m, n are arbitrary natural numbers such that

$$im^2 > jn^2$$

and $ij = 3$. $\qquad\square$

For example, if $n = 1$ and $m = 2, 3, 4, 5$ we get the following triples $\langle 3a, b, c\rangle$:

$$\langle 12, 11, 13\rangle, \quad \langle 9, 13, 14\rangle, \quad \langle 24, 13, 19\rangle \quad \text{and} \quad \langle 15, 11, 14\rangle.$$

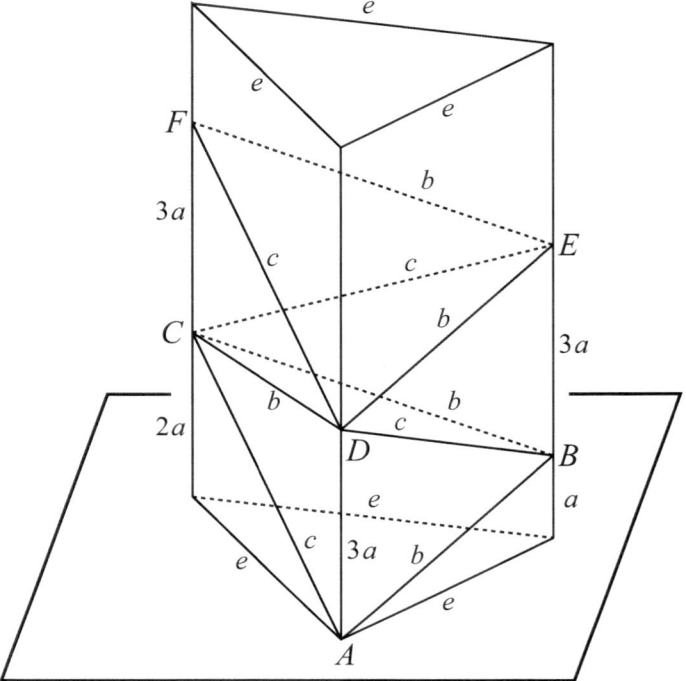

Fig. 12.6 Decomposition of a triangular prism into congruent tetrahedra

12.6   Tricks with Numbers

Number theory is also largely used in recreational mathematics (see e.g. [1, 70, 86, 160, 300, 301]). In this section we will demonstrate this by a few examples.

Why do so many people say that they hate mathematics? The truth is that they often do not even know what mathematics is. The British mathematician David Acheson believes that mathematicians could change this situation, for example, by mediating some of their ideas to the general public and showing that they have joy from their profession.

One way to do this is to use the element of surprise, which often accompanies mathematical problems, and everyone likes pleasant surprises. Acheson himself in [1] describes how he experienced his first mathematical surprise when he was 10 years old. At that time he liked various magic tricks. One day he came across the following "trick with numbers":

Choose any three-digit number so that the first digit and the last digits differ by at least two. Now create a number whose digits are in reverse order, and subtract the smaller of these numbers from the larger one. (e.g. $782 - 287 = 495$). Now add the result to the number whose digits are again in reverse order (in our case $495 + 594 = 1089$).

It is remarkable that the result will always be 1089 under the above assumptions. Acheson says: "If you see this trick for the first time when you are ten years old, it will literally enchant you."

Theorem 12.6 *The previous trick with three-digit numbers always yields* 1089.

Proof It is enough to consider three digits a, b, c such that $a \geq c + 2$. Then

$$100a + 10b + c - 100c - 10b - a = 100(a - c) - a + c$$
$$= 100(a - c - 1) + 90 + (10 - a + c).$$

Adding to this result the number $100(10 - a + c) + 90 + (a - c - 1)$, we get $900 + 180 + 9 = 1089$. $\qquad\square$

$$\bullet \quad \bullet \quad \bullet$$

Further, we present another trick with the number 153. Consider a natural number

$$n = c_k 10^k + \cdots + c_1 10 + c_0, \tag{12.8}$$

where $c_k, \ldots, c_1, c_0$ are its digits from the set $\{0, 1, 2, \ldots, 9\}$ and $c_k > 0$. Assume that n is divisible by 3 and let us form the number

$$m = c_k^3 + \cdots + c_1^3 + c_0^3. \tag{12.9}$$

Now set $n := m$ and repeat the previous procedure. Then after a finite number of steps we get the number 153.

Example We have

$$3 \;\mapsto\; 3^3 = 27 \;\mapsto\; 2^3 + 7^3 = 351 \;\mapsto\; 3^3 + 5^3 + 1^3 = 153,$$
$$18 \;\mapsto\; 1^3 + 8^3 = 513 \;\mapsto\; 5^3 + 1^3 + 3^3 = 153,$$
$$153 \;\mapsto\; 1^3 + 5^3 + 3^3 = 153,$$
$$9876 \;\mapsto\; 9^3 + 8^3 + 7^3 + 6^3 = 1800 \;\mapsto\; 1^3 + 8^3 = 513 \;\mapsto\; 5^3 + 1^3 + 3^3 = 153.$$

Theorem 12.7 *When we repeatedly sum the cubes of individual digits of any natural number divisible by three, then after a finite number of steps we get* 153.

Proof First we prove that if $3 \mid n$, then $3 \mid m$. By Theorem 1.1(b) and (12.8) we have $n \equiv \sum_{i=0}^{k} c_i \pmod 3$ and by Fermat's Little Theorem 2.17 we obtain $c_i^3 \equiv c_i \pmod 3$. From this and (12.9) it follows that

$$m = \sum_{i=0}^{k} c_i^3 \equiv \sum_{i=0}^{k} c_i \equiv n \pmod 3.$$

Next we show that for $n \geq 10^4$ we have $n > m$. Since $c_k > 0$ and $k \geq 4$, by (12.8) and (12.9) we get

$$n = \sum_{i=0}^{k} c_i 10^i \geq c_k 10^k \geq 10^k > (k+1)9^3 \geq \sum_{i=0}^{k} c_i^3 = m,$$

i.e., for $n \geq 10^4$ the number m is less than n. We repeat this procedure as long as the result is less than 10^4. Then we have only a finite number of possibilities and we can verify on computer in integer arithmetic that all possibilities lead to the number 153. $\square$

Remark The assumption of divisibility by three in Theorem 12.7 is substantial. There are other natural numbers that are equal to the sum of the cubes of their digits: $1 = 1^3$, $370 = 3^3 + 7^3 + 0^3$, $371 = 3^3 + 7^3 + 1^3$, and $407 = 4^3 + 0^3 + 7^3$. However, among all natural numbers divisible by three, the number 153 is the only one with this property. Note also that $153 = 1! + 2! + 3! + 4! + 5!$.

$$\bullet \quad \bullet \quad \bullet$$

The Indian mathematician D. R. Kaprekar discovered that the number 6174 has the following interesting feature (see [160]). Choose an arbitrary four-digit number n_1 whose digits are not all the same. Now consider 2 four-digit numbers whose digits are ordered by size from the largest to the smallest digit and from the smallest to the largest digit. Let n_2 be the difference of these two numbers. Repeating this process, we always get after at most 7 steps the number 6174.

Example Choose $n_1 = 1470$. Then

$$n_2 = 7410 - 0147 = 7263,$$
$$n_3 = 7632 - 2367 = 5265,$$
$$n_4 = 6552 - 2556 = 3996,$$
$$n_5 = 9963 - 3699 = 6264,$$
$$n_6 = 6642 - 2466 = 4176,$$
$$n_7 = 7641 - 1467 = 6174.$$

Fixed points of the *Kaprekar mapping* $f(n) = n' - n''$, where the digits of n are not all the same and in n' these digits are arranged in descending, in n'' in ascending order, are called *Kaprekar constants*. Hence, 6174 is the Kaprekar constant.

For three-digit numbers the Kaprekar constant is equal to 495. The proof of these remarkable properties can be easily verified by computer in integer arithmetic. However, for two-digit numbers such a constant does not exist, since $54 - 45 = 9$. Also for five-digit numbers there is no such constant.

• • •

In Hančl and Rucki [135], we can find the following statement.

Theorem 12.8 *If $x + \dfrac{1}{x}$ is an integer number for some real number $x \neq 0$, then $x^n + \dfrac{1}{x^n}$ is also an integer number for any $n \in \mathbb{N}$.*

Proof We will proceed by induction. The statement obviously holds for $n = 1$. So let it hold also for an arbitrary $n \in \mathbb{N}$. Then

$$x^{n+1} + \frac{1}{x^{n+1}} = \left(x^n + \frac{1}{x^n}\right)\left(x + \frac{1}{x}\right) - \left(x^{n-1} + \frac{1}{x^{n-1}}\right),$$

where the number on the right-hand side is an integer by the induction assumption and the fact that $x^{n-1} + 1/x^{n-1} = 2$ if $n = 1$. □

• • •

The following, which is likely the most beautiful mathematical formula, is attributed to Euler

$$e^{i\pi} + 1 = 0$$

which connects the five most used mathematical constants: 0, 1, the imaginary unit i, the Euler number e and Ludolph's number π. From this we have

$$e^{i\pi} = -1 = (-1)^3 = \left(e^{i\pi}\right)^3 = e^{3i\pi}.$$

Since the left-hand and right-hand sides are equal, we can evaluate their logarithms

$$\log e^{i\pi} = i\pi \log e = \log e^{3i\pi} = 3i\pi \log e$$

and from the equality $\log e = 1$ we get $i\pi = 3i\pi$. Dividing both sides by the nonzero number $i\pi$, we get

$$1 = 3. \tag{12.10}$$

The paradoxical equality (12.10) is, of course, wrong. The reason is that the logarithm in the complex plane is not a single-valued function, and therefore taking unique logarithms in the complex plane is generally not possible, contrary to the real logarithm.

• • •

We close this section with several simple examples.

Example Check that the first six multiples of the number 142857 have the same digits that cyclically repeat in the same order and start with a different digit each time.

Example Choose an arbitrary two-digit number whose last digit is 5. To get its square the first digit a has to be multiplied by $a + 1$ and after that we write 25. Clearly,

$$(10a + 5)^2 = 100a^2 + 100a + 25 = 100a(a + 1) + 25 \qquad (12.11)$$

which proves the statement. For instance, by (12.11) we have $65^2 = 6 \cdot 7 \cdot 100 + 25 = 4225$.

Example Is the number $\log_2 3$ irrational? Suppose to the contrary that $\log_2 3 = \frac{m}{n}$ for some $m, n \in \mathbb{N}$. Then $2^{m/n} = 3$. This implies that $2^m = 3^n$, which is a contradiction, since 2^m is even, while 3^n is odd.

Example Is the number $\sqrt{2}$ irrational? Assume to the contrary that it is rational of the form $\sqrt{2} = m/n$, where $m, n \in \mathbb{N}$ and $(m, n) = 1$. Then $2n^2 = m^2$ and by Euclid's Theorem 2.1 we find that $2 \mid m$. So let $m = 2k$ for some $k \in \mathbb{N}$. Hence, $n^2 = 2k^2$ which again means that $2 \mid n$, and thus m and n are coprime. This contradiction implies that $\sqrt{2}$ is irrational.

Now we shall present another elegant proof of this statement. Suppose again that $\sqrt{2}$ is rational. Since $\mathbb{N}$ is well ordered, we can define the number $k = \min\{m \in \mathbb{N}; \; m\sqrt{2} \in \mathbb{N}\}$. However, then $k(\sqrt{2} - 1) \in \mathbb{N}$, since $k, k\sqrt{2} \in \mathbb{N}$ and $\sqrt{2} > 1$. We see that $k(\sqrt{2} - 1)\sqrt{2} = 2k - k\sqrt{2} \in \mathbb{N}$ and $k(\sqrt{2} - 1) < k$, which is a contradiction, i.e., $\sqrt{2}$ is irrational.

Example We show that there are irrational numbers a, b such that a^b is rational. If $\sqrt{2}^{\sqrt{6}}$ is rational, we are done. If it is irrational, then

$$\left(\sqrt{2}^{\sqrt{6}}\right)^{\sqrt{6}} = \sqrt{2}^6 = 2^3 = 8$$

and we can choose $a = \sqrt{2}^{\sqrt{6}}$ and $b = \sqrt{6}$.

12.7 Application of Congruences

A well-known classical puzzle (see Landau [223]) is the problem of placing n queens Q on an $n \times n$ chessboard so that no two of those queens attack each other, that is, no pair of the queens lie on the same row, column, or diagonal of the chessboard. By inspection, one immediately sees that there is no solution if $n \in \{2, 3\}$. If $p > 3$ is a prime, one can easily obtain a solution to this problems for the $p \times p$ chessboard by means of congruences modulo p (see Koshy [176, pp. 267–270]).

Fig. 12.7 Solution of the
nonattacking queens puzzle
on a 5 × 5 chessboard

$i \backslash j$	1	2	3	4	5
1	·	·	Q	·	·
2	Q	·	·	·	·
3	·	·	·	Q	·
4	·	Q	·	·	·
5	·	·	·	·	Q

Let $f(i)$ be the column in which we wish to place the queen in row i, where $1 \le i \le p$ and $1 \le f(i) \le p$. Setting

$$f(i) \equiv \frac{p+1}{2} i \pmod{p},$$

we obtain a solution, see Fig. 12.7 for $p = 5$.

• • •

One can also use congruences modulo n to design round-robin tournaments between n teams when $n \ge 3$ is odd (see Koshy [176, pp. 271–275]). In a round-robin tournament, each team plays with every other team exactly once. Since the number n of teams participating in the tournament is odd, at least one team does not play in each round of the tournament. We say that a team receives a *bye* in a particular round if the team does not compete in that round. Since each team competes with $n - 1$ other teams and there is at least one bye in each round, we see that one needs to schedule at least n rounds in the tournament. We show how to schedule a round-robin tournament containing n teams with exactly n rounds and exactly one bye in each round.

Let the teams be numbered $1, 2, \ldots, n$. Let $g(i, j)$ denote the team that team j plays in round i, where $i, j \in \{1, 2, \ldots, n\}$ and

$$1 \le g(i, j) \le n.$$

If $g(i, j) = j$, we say that team j receives a bye in round i, since no team can compete with itself. We set

$$g(i, j) \equiv i - j \pmod{n}, \tag{12.12}$$

where $1 \le g(i, j) \le n$. Team j receives a bye in round i if and only if

$$g(i, j) \equiv i - j \equiv j \pmod{n} \iff 2j \equiv i \pmod{n}. \tag{12.13}$$

Since $(2, n) = 1$, we see that 2 has the unique multiplicative inverse $(n + 1)/2$ modulo n, and thus by (12.13), team j gets a bye in round i if and only if

Fig. 12.8 Positions of byes
are given by congruence
(12.13), cf. Fig. 12.7. Teams
denoted by × (or by +) play
together in each round, i.e.,
by (12.14) the teams j and k
play together in round i

$i \backslash j$	1	2	3	4	5
1	+	×	bye	×	+
2	bye	×	+	+	×
3	×	×	+	bye	+
4	×	bye	×	+	+
5	+	×	×	+	bye

$$j \equiv \frac{n+1}{2} i \pmod{n},$$

where $1 \leq j \leq n$. Therefore, each round of the tournament has exactly one bye. Let the teams j and k be fixed, where $j \neq k$. Then by the definition of $g(i, j)$,

$$g(i, j) = k \iff k \equiv i - j \pmod{n} \iff i \equiv j + k \pmod{n},$$
$$(12.14)$$

where $1 \leq i \leq n$. Thus, i is uniquely determined. Therefore, the function g gives a schedule for a round-robin tournament of n teams with exactly n rounds and exactly one bye in each round.

Example According to (12.12) for $n = 5$, we get

$$g(i, j) = \begin{bmatrix} 5 & 4 & 3 & 2 & 1 \\ 1 & 5 & 4 & 3 & 2 \\ 2 & 1 & 5 & 4 & 3 \\ 3 & 2 & 1 & 5 & 4 \\ 4 & 3 & 2 & 1 & 5 \end{bmatrix}.$$

The team j receives a bye in round i if (12.13) holds. Hence, by (12.14) we obtain the following solution marked in Fig. 12.8.

•   •   •

Lawther, [224] in 1935 presented a nice application of congruences and the Carmichael lambda function (see also Ore [293, pp. 302–310]). Consider a telephone cable used for long-distance communication that is produced in uniform sections, say 300 m long, with each section consisting of $m \geq 3$ concentric insulated wires or conductors. The sections of the cable are spliced end to end. At the splices the order of the wires should be scrambled substantially so as to minimize interference. It is desirable to avoid having wires that are adjacent in one section also be adjacent in sections following closely afterward. To accomplish this goal of separating adjacent wires, we adopt the following four rules:

1. The same splicing instructions should be used at each intersection.
2. The wires in one concentric layer are spliced to those in the corresponding layer of the next section.

Fig. 12.9 The splicing table for the case in which $m = 11$ and $s = 2$ or $s = 3$

$s = 2$	$s = 3$
$1 \to 1$	$1 \to 1$
$2 \to 3$	$2 \to 4$
$3 \to 5$	$3 \to 7$
$4 \to 7$	$4 \to 10$
$5 \to 9$	$5 \to 2$
$6 \to 11$	$6 \to 5$
$7 \to 2$	$7 \to 8$
$8 \to 4$	$8 \to 11$
$9 \to 6$	$9 \to 3$
$10 \to 8$	$10 \to 6$
$11 \to 10$	$11 \to 9$

3. When some wire in one section S_1 is spliced to a wire in the next section S_2, the adjacent wire in S_1 should be spliced to to the wire in S_2 that is obtained from the one last spliced by counting forward a fixed number s. The number s is called the *spread* of the splicing rule.
4. The number s in rule 3 should be chosen so that wires that are adjacent in one section are not adjacent in succeeding sections for as long as possible.

Figure 12.9 gives the splicing for two values of the spread s when the number of wires m in each section of the cable is equal to 11.

In general, for a given spread s, the splicing table will give

$$i \to 1 + (i - 1)s \quad (\mathrm{mod}\ m), \tag{12.15}$$

where $1 \le i \le m$, $2 \le s \le m$, and the number on the right-hand side of (12.15) is reduced to the least positive integer modulo m. We of course require that any two wires in section S_1 are spliced to distinct wires in section S_2. Thus, from (12.15), if $1 \le i < j \le m$, then the congruence

$$1 + (i - 1)s \equiv 1 + (j - 1)s \quad (\mathrm{mod}\ s)$$

or

$$(j - i)s \equiv 0 \quad (\mathrm{mod}\ m) \tag{12.16}$$

will not be possible. If $(s, m) = d > 1$, then the congruence (12.16) is satisfied for $i = 1$ and $j = 1 + m/d$. Hence, we see that we can only consider spreads s for which $2 \le s \le m$ and $(s, m) = 1$.

Let us now repeat the splicing procedure. During the first operation, the ith wire in section S_1 was joined to the wire with number

$$i_2 \equiv 1 + (i - 1)s \quad (\text{mod } m)$$

in section S_2. By the second splice, this wire leads to wire number

$$i_3 \equiv 1 + (i_2 - 1)s \equiv 1 + (i - 1)s^2 \quad (\text{mod } m)$$

in section S_3. Continuing this process, we find that after n splices, the ith wire in the first section S_1 will be connected to wire number

$$i_{n+1} \equiv (i - 1)s^n \quad (\text{mod } m) \tag{12.17}$$

in section S_{n+1}.

The objective of our splicing arrangement was to generate a distribution of the connections so that two wires that were adjacent in some section would stay separated by a greater amount in the subsequent sections for as long as possible. Let us observe how our scheme behaves in this regard. Two adjacent wires in some section may be numbered i and $i + 1$. By (12.17), after n splices, they become connected with the wires numbered

$$1 + (i - 1)s^n, \quad 1 + is^n \quad (\text{mod } m),$$

respectively. These two wires are adjacent only if their difference is congruent to ± 1 modulo m, that is, when

$$1 + is^n - (1 + (i - 1)s^n) \equiv \pm 1 \quad (\text{mod } m)$$

or

$$s^n \equiv \pm 1 \quad (\text{mod } m).$$

We have the following theorem.

Theorem 12.9 *Let $m \geq 3$ and let $\lambda_0(m)$ be the least positive integer t such that*

$$r^t \equiv \pm 1 \quad (\text{mod } m)$$

for all r such that $(r, t) = 1$ and $1 \leq r \leq m$. Then

$$\lambda_0(m) = \frac{1}{2}\lambda(m) \tag{12.18}$$

if and only if there exists a primitive root modulo m, and otherwise

$$\lambda_0(m) = \lambda(m),$$

where $\lambda(m)$ is the Carmichael lambda function. In particular, (12.18) is satisfied when m is of the form 4, p, p^k, $2p^k$, where p is an odd prime and $k \in \mathbb{N}$. Furthermore, *if r is any integer such that* $(r, m) = 1$, $1 \leq r \leq m$, *and* $r^t \not\equiv \pm 1$ (mod m) *for any* $t \in \{1, 2, \ldots, \lambda_0(m) - 1\}$, *then* $\text{ord}_m r = \lambda(m)$.

The proof of Theorem 12.9 follows from Theorem 2.19 and [293, pp. 302–305].

Remark Consider the splicing scheme given above for a cable with $m \geq 3$ wires, sections $S_1, S_2, \ldots$, and spread $s \in \{2, \ldots, m\}$. If $(s, m) = 1$ and s is the smallest integer for which $\text{ord}_m s = \lambda(m)$, then by Theorem 12.9, for the sections $S_1, S_2, \ldots$, $S_{\lambda_0(m)}$, any two wires that are adjacent in one of these sections are not adjacent in any of the other given sections. By the properties of the Carmichael lambda function λ,

$$\lambda_0(m) \leq \frac{m-1}{2}.$$

Moreover, $\lambda_0(m) = \frac{1}{2}(m - 1)$ for $m \geq 3$ if and only if m is a prime.

Additionally, let us make the observation that when $m \geq 3$ is a prime, the result obtained above is the best possible by any method. To see this, notice that when one starts with any wire in section S_1, it can only remain separated from adjacent ones as long as it is possible to find a new pair of wires between which it can be placed at each splice, and altogether there are only $\frac{1}{2}(m - 1)$ such different pairs.

Below is Table 12.2 giving the values of $\lambda_0(m)$ for all m up to 20 as well as a suitable spread s.

•   •   •

Our final application of congruences involves showing the existence of Latin squares of order n for all n and the existence of a complete set of $n - 1$ mutually orthogonal Latin squares of order n when n is an odd prime, see [133, Chap. 13], [232, pp. 513–514], [345, Chap. 7]. We recall the definition of a Latin square and orthogonal Latin squares given in Sect. 9.5.

Theorem 12.10 *There exists a Latin square of order n for each positive integer n.*

Table 12.2 Values of $\lambda_0(m)$ for $m \in \{3, \ldots, 20\}$ with the corresponding values of s

m	$\lambda_0(m)$	s	m	$\lambda_0(m)$	s
3	1	2	12	2	5
4	1	3	13	6	2
5	2	2	14	3	3
6	1	5	15	4	2
7	3	3	16	4	3
8	2	3	17	8	3
9	3	2	18	3	5
10	2	3	19	9	2
11	5	2	20	4	3

Proof Consider the $n \times n$ matrix $L = (a_{ij})$, where $1 \le i, j \le n$ and $0 \le a_{ij} \le n - 1$. Let $a_{ij} \equiv i + j \pmod{n}$. If $a_{ij} \equiv a_{im} \pmod{n}$, then

$$i + j \equiv i + m \pmod{n},$$

which implies that $j = m$, since $j, m \in \{1, \ldots, n\}$. Similarly, if $a_{ij} \equiv a_{kj} \pmod{n}$, then $i = k$. Hence, each row and column of L consists of n different numbers and L is a Latin square of order n. $\qquad\square$

Theorem 12.11 *Let p be an odd prime. Then there exists a complete set of $p - 1$ mutually orthogonal Latin squares of order p.*

Proof Let $k \in \{1, \ldots, p - 1\}$. Define the $p \times p$ matrix $L_k = (a_{ij}^{(k)}), i, j = 1, \ldots p,$ by

$$a_{ij}^{(k)} \equiv k(i - 1) + j - 1 \pmod{p},$$

where $0 \le a_{ij}^{(k)} \le p - 1$. We show that L_k is a Latin square for $k \in \{1, \ldots, p - 1\}$ and that L_k is orthogonal to L_m if $m \ne k$ and $m \in \{1, \ldots, p - 1\}$.

Consider the matrix L_k. Suppose that $a_{ij}^{(k)} \equiv a_{im}^{(k)} \pmod{p}$, where $i, j, m \in \{1, \ldots, p\}$. Then

$$k(i - 1) + j - 1 \equiv k(i - 1) + m - 1 \pmod{p},$$

which implies that $j = m$. Moreover, if $a_{ij}^{(k)} \equiv a_{mj}^{(k)} \pmod{p}$, then

$$k(i - 1) + j - 1 \equiv k(m - 1) + j - 1 \pmod{p}$$

or

$$k(m - i) \equiv 0 \pmod{p}.$$

Since k and p are coprime, it follows from Theorem 2.1 that $i = m$. Hence, L_k is a Latin square for $k \in \{1, \ldots, p - 1\}$.

We now demonstrate that L_k and L_m are orthogonal for $m \ne k$ and $k, m \in \{1, \ldots, p - 1\}$. We need to show that the ordered pair

$$\langle a_{ij}^{(k)}, a_{ij}^{(m)} \rangle \ne \langle a_{gh}^{(k)}, a_{gh}^{(m)} \rangle$$

for $\langle i, j \rangle \ne \langle g, h \rangle$, where $i, j, g, h \in \{1, \ldots, p\}$. Suppose that

$$\langle a_{ij}^{(k)}, a_{ij}^{(m)} \rangle \equiv \langle k(i - 1) + j - 1, m(i - 1) + j - 1 \rangle$$
$$\equiv \langle a_{gh}^{(k)}, a_{gh}^{(m)} \rangle \equiv \langle k(g - 1) + h - 1, m(g - 1) + h - 1 \rangle \pmod{p}$$

This implies that

$$k(i - g) \equiv h - j, \quad m(i - g) \equiv h - j \quad (\text{mod } p), \tag{12.19}$$

which yields that

$$(k - m)(i - g) \equiv 0 \quad (\text{mod } p).$$

Since $k \neq m$, it follows from (12.19) that $i \equiv g$ (mod p), so that $i = g$. We now see from Theorem 2.1 that $j \equiv h$ (mod p), from which it follows that $j = h$. Hence, $\langle i, j \rangle = \langle g, h \rangle$ and consequently, L_k and L_m are orthogonal. $\qquad \square$

Example Suppose that $p = 5$. Below we present a complete set of four mutually orthogonal Latin squares of order 5 given by the construction in Theorem 12.11:

$$L_1 = \begin{bmatrix} 0 & 1 & 2 & 3 & 4 \\ 1 & 2 & 3 & 4 & 0 \\ 2 & 3 & 4 & 0 & 1 \\ 3 & 4 & 0 & 1 & 2 \\ 4 & 0 & 1 & 2 & 3 \end{bmatrix}, \quad L_2 = \begin{bmatrix} 0 & 1 & 2 & 3 & 4 \\ 2 & 3 & 4 & 0 & 1 \\ 4 & 0 & 1 & 2 & 3 \\ 1 & 2 & 3 & 4 & 0 \\ 3 & 4 & 0 & 1 & 2 \end{bmatrix},$$

$$L_3 = \begin{bmatrix} 0 & 1 & 2 & 3 & 4 \\ 3 & 4 & 0 & 1 & 2 \\ 1 & 2 & 3 & 4 & 0 \\ 4 & 0 & 1 & 2 & 3 \\ 2 & 3 & 4 & 0 & 1 \end{bmatrix}, \quad L_4 = \begin{bmatrix} 0 & 1 & 2 & 3 & 4 \\ 4 & 0 & 1 & 2 & 3 \\ 3 & 4 & 0 & 1 & 2 \\ 2 & 3 & 4 & 0 & 1 \\ 1 & 2 & 3 & 4 & 0 \end{bmatrix}.$$

12.8 Paradoxes in Numerical Computations

In preforming any real-life calculations in finite noninteger computer arithmetic, we should keep in mind that rounding errors are produced. Hence, we should very carefully and critically evaluate the numerical output and should not blindly trust computer results as it is often done. Let us present several cautionary and shocking examples which one should have in mind before using the computer [42].

Example In 1966, Ivo Babuška, Milan Práger, and Emil Vitásek investigated numerical instability of successive evaluations of the expression

$$\ldots(((((1 : 2) \cdot 2) : 3) \cdot 3) : 4) \cdot 4 \ldots \tag{12.20}$$

They obtained various results for this expression on different computers involving thousands of divisions and multiplications, see [15, p. 6].

Remark When subtracting two numbers of almost the same size, we usually do not produce rounding errors. However, the mantissa of the difference contains only a few significant digits which leads to the loss of accuracy. For example, suppose that the mantissa has 5 digits and let us evaluate the difference $3.1416 \cdot 10^0 - 3.1415 \cdot 10^0$. Then the computer saves the result $1.0000 \cdot 10^{-4}$, where the four zeros in the mantissa are not significant digits. This means the the error of the difference moves from the originally last digit of the mantissa to the second digit.

Example Can a decreasing sequence be increasing in computer arithmetic? Let $a_0 = 1$, $a_1 = \frac{1}{11}$, and consider the second-order linear recurrence

$$a_{n+2} = \frac{34}{11}a_{n+1} - \frac{3}{11}a_n \quad \text{for } n = 0, 1, 2, \ldots \tag{12.21}$$

The exact solution

$$a_n = \frac{1}{11^n}$$

is clearly a decreasing sequence. However, calculating (12.21) with MATLAB (Matrix Laboratory) on a standard PC computer, we observe that the associated sequence decreases until $\tilde{a}_{12} = 0.2068 \cdot 10^{-11}$, then it increases, reaches the value $\tilde{a}_{40} = 40.44$, and grows quite rapidly. This example demonstrates that subtraction of two numbers of almost the same size is not a good practice in scientific computing. Below we list some data for (12.21) with different precision arithmetic,

$$\underline{a}_{50} \approx 2 \cdot 10^{11} \quad \text{in single-precision arithmetic (4 bytes),}$$
$$\tilde{a}_{50} \approx 10^{10} \quad \text{in double-precision arithmetic (8 bytes),}$$
$$\overline{a}_{50} \approx 10^3 \quad \text{in extended-precision arithmetic (10 bytes).}$$

Moreover, the corresponding sequences $(\underline{a}_n)$, $(\tilde{a}_n)$, and $(\overline{a}_n)$ are increasing from $n = 8, 9$, and 14, respectively. Rounding to two significant digits only yields an increasing sequence from $n = 2$.

Example (*Kahan*). Set $a_0 = 2$, $a_1 = -4$, and consider the recurrence (see Muller et al. [280])

$$a_n = 111 - \frac{1130}{a_{n-1}} + \frac{3000}{a_{n-1}a_{n-2}} \quad \text{for } n = 2, 3, \ldots \tag{12.22}$$

Its general solution has the form

$$a_n = \frac{\alpha 100^{n+1} + \beta 6^{n+1} + \gamma 5^{n+1}}{\alpha 100^n + \beta 6^n + \gamma 5^n},$$

where α, β, and γ depend on the initial values a_0 and a_1.

Thus, if $\alpha \neq 0$ then the limit of the sequence (a_n) is 100. Nevertheless, we observe that if $\alpha = 0$ and $\beta \neq 0$, then the limit is 6. For the initial values $a_0 = 2$ and $a_1 = -4$ we get $\alpha = 0$, $\beta = -3$ and $\gamma = 4$. From this we find that

$$a_{20} \doteq 6.034, \quad a_{30} \doteq 6.006.$$

However, calculating (12.22) in double precision, we get

$$\tilde{a}_{20} \doteq 98.351, \quad \tilde{a}_{30} \doteq 99.999$$

due to rounding errors.

Example (Muller). Let $a_1 = e - 1 = 1.718281828\ldots$ and consider the thoroughly innocent-looking sequence (see Muller et al. [280])

$$a_n = n(a_{n-1} - 1) \quad \text{for } n = 2, 3, \ldots \tag{12.23}$$

This first-order recurrence is only a slight modification of an example from Babuška, Práger, and Vitásek [14]. By induction we can easily prove that

$$a_n = n!\left(\frac{1}{n!} + \frac{1}{(n+1)!} + \frac{1}{(n+2)!} + \cdots\right). \tag{12.24}$$

Indeed, for $n = 1$ we obtain $a_1 = e - 1$ and assuming the validity of (12.24) for $n - 1 \geq 0$, we get by (12.23)

$$a_n = n(a_{n-1} - 1) = n\left((n-1)!\left(\frac{1}{(n-1)!} + \frac{1}{n!} + \frac{1}{(n+1)!} + \cdots\right) - \frac{(n-1)!}{(n-1)!}\right)$$
$$= n!\left(\frac{1}{n!} + \frac{1}{(n+1)!} + \frac{1}{(n+2)!} + \cdots\right).$$

From (12.24) we immediately see that the sequence (a_n) is decreasing, all a_n have a very reasonable size in the interval $(1, 2)$, and

$$\lim_{n \to \infty} a_n = 1.$$

However, if we run the iteration in double precision on MATLAB, we observe that the sequence decreases until $\tilde{a}_{15} = 1.0668$, and then a strange thing happens: $\tilde{a}_{16} = 1.0685$, $\tilde{a}_{17} = 1.1652$, $\tilde{a}_{18} = 2.9731$, $\tilde{a}_{19} = 37.4882$, $\tilde{a}_{20} = 729.7637, \ldots$, and the sequence grows out of control.

Performing only 24 subtractions and 24 multiplications in (12.23) with different numbers of significant digits, we obtain:

$\underline{a}_{25} \approx -6.204484 \cdot 10^{23}$ in single-precision arithmetic (4 bytes),

$\tilde{a}_{25} \approx 1.201807248 \cdot 10^{9}$ in double-precision arithmetic (8 bytes),

$\overline{a}_{25} \approx -7.3557319606 \cdot 10^{6}$ in extended-precision arithmetic (10 bytes).

However, by (12.22) we have

$$1.038 < 1 + \frac{1}{26} < a_{25} < 1 + \frac{1}{26} + \frac{1}{26^2} + \cdots = \frac{26}{25} = 1.04,$$

where the standard formula for the sum of a geometric sequence was applied. Now, let us extend the number of significant digits. In arithmetic with D decimal digits, the first twelve significant digits are:

D	$a_{25}(D)$
20	615990.413139,
21	$-\,4457.98859386$,
22	$-\,4457.98859386$,
23	195.374419140,
24	40.2623187072,
25	$-\,6.27131142281$,
26	1.48429359885.

The values corresponding to $D = 21$ and $D = 22$ are the same, since the 21st decimal digit of the Euler number $e = 2.718281828459045235360287445\ldots$ is equal to zero. Only for $D > 25$ do the numerical results start to resemble the exact value $a_{25} = 1.03993872967\ldots$ For instance, if $D = 30$ we get $a_{25}(30) = 1.039897\ldots$

Now let us take a closer look at the main reason of this strange behavior. Denote by ε_i the rounding error at the ith step. Then we have

$$\tilde{a}_1 = e - 1 + \varepsilon_1 = a_1 + \varepsilon_1,$$
$$\tilde{a}_2 = 2(\tilde{a}_1 - 1) + \varepsilon_2 = 2(a_1 + \varepsilon_1 - 1) + \varepsilon_2 = a_2 + 2!\varepsilon_1 + \varepsilon_2,$$
$$\tilde{a}_3 = 3(\tilde{a}_2 - 1) + \varepsilon_3 = 3(a_2 + 2\varepsilon_1 + \varepsilon_2 - 1) + \varepsilon_3 = a_3 + 3!\varepsilon_1 + 3\varepsilon_2 + \varepsilon_3,$$

$$\vdots$$

$$\tilde{a}_{25} = 25(\tilde{a}_{24} - 1) + \varepsilon_{25} = a_{25} + 25!\varepsilon_1 + \frac{25!}{2!}\varepsilon_2 + \frac{25!}{3!}\varepsilon_3 + \cdots + \varepsilon_{25}.$$

Therefore, the total rounding error is

$$a_{25} - \tilde{a}_{25} = -25! \left(\varepsilon_1 + \frac{1}{2!}\varepsilon_2 + \frac{1}{3!}\varepsilon_3 + \cdots + \frac{1}{25!}\varepsilon_{25} \right) \qquad (12.25)$$

and its size depends particularly on the several initial rounding errors $\varepsilon_1, \varepsilon_2, \ldots$

This sophisticated example shows why it is necessary to try to avoid the subtraction of two numbers that are almost equal. We observe from (12.25) why the above recurrence (12.23) is so sensitive to round-off errors. Computer arithmetic with variable length does not improve this numerical effect for large n. An interesting application of the sequence (12.23) in banking can be found in Muller et al. [280].

Example (*Rump*). The use of a much smaller number of subtractions than in the previous example may also lead to absolutely catastrophic numerical results. Evaluate the function

$$u(x, y) = 333.75y^6 + x^2(11x^2y^2 - y^6 - 121y^4 - 2) + 5.5y^8 + \frac{x}{2y} \qquad (12.26)$$

at $x = 77617.0$ and $y = 33096.0$. Note that this is a polynomial of degree eight plus a simple rational function $x/(2y)$. We observe that no recurrence relation as in (12.23) is evaluated and we perform only three subtractions and a few other arithmetic operations. Contrary to the previous example, we get almost the same numbers:

$\underline{u}(x, y) = 1.172603$ in single-precision arithmetic (4 bytes),
$\tilde{u}(x, y) = 1.1726039400531$ in double-precision arithmetic (8 bytes),
$\overline{u}(x, y) = 1.172603940053178$ in extended-precision arithmetic (10 bytes)

on an outmoded IMB 370 computer (see Rump [340]). The programming package MAPLE, with

$$D = 7, 8, 9, 10, 12, 18, 26, 27$$

decimal digits produces very similar results. However, we should not rejoice over the above results, since the exact value is

$$u(x, y) = -0.827396 \ldots \text{ NEGATIVE!}$$

As discussed in [84], for $z = 333.75y^6 + x^2(11x^2y^2 - y^6 - 121y^4 - 2)$ and $w = 5.5y^8$ we can evaluate exactly that

$$z = -7917111340668961361101134701524942850,$$
$$w = 7917111340668961361101134701524942848,$$

which have 35 out of 37 digits in common. Thus we are dealing with huge numbers with insufficient floating points. The exact value is

$$u(x, y) = z + w + \frac{x}{2y} = -2 + \frac{x}{2y} = -\frac{54767}{66192} = -0.827396\ldots$$

Numerical results by MAPLE yield:

$$\begin{aligned}
D = 20, \quad & u_D(x, y) = 1073741825.1726039401, \\
D = 21, \quad & u_D(x, y) = 1.17260394005317863186, \\
D = 28, \quad & u_D(x, y) = 1.1726039400531786318588349045, \\
D = 29, \quad & u_D(x, y) = -0.82739605994682136814116509548, \\
D = 37, \quad & u_D(x, y) = -0.8273960599468213681411650954798162920,
\end{aligned}$$

where the subscript D stands for the use of computer arithmetic with D decimal digits. We see that when $D = 21$ to 28, the result is close to Rump's 1988 data, and the "correct" answer appears starting from $D = 29$. This example shows that the arithmetic of large numbers should be performed very cautiously in scientific computing. We should keep in mind that a very small number of subtractions and roundings (see (12.23) and (12.26)) may completely destroy the exact solution [341].

Example Numerical differentiation represents another very instable operation. To illustrate it, consider the rational function (see [220])

$$g(x) = \frac{4970x - 4923}{4970x^2 - 9799x + 4830}, \quad x \geq 0.99, \tag{12.27}$$

and calculate its second derivative at the point 1 using the second central differences, i.e.

$$\delta_h^2 g(x) = \frac{g(1 + h) - 2g(1) + g(1 - h)}{h^2} \tag{12.28}$$

which tends to $\ddot{g}(1)$ as $h \to 0$. However, from the second row of Table 12.3 we can only barely deduce which value of h is the best approximation of $\ddot{g}(1)$ without knowing that the exact value is $\ddot{g}(1) = 94$. For "large" h the function $g(1 - h)$ quickly changes. On the other hand, for "small" h total loss of accuracy arises, since we subtract almost the same numbers in the numerator of (12.28).

The remedy is to rearrange formula (12.28) by means of (12.27) as follows:

$$\delta_h^2 g(x) = \frac{94(1 - 70^2 71^2 h^2)}{(1 - 71^2 h^2)(1 - 70^2 h^2)} \tag{12.29}$$

which produces relatively good numerical results—see the last row of Table 12.3.

Table 12.3 Numerical values of $\ddot{g}(1)$ calculated by (12.28) and (12.29), respectively

$\delta_h^2 g(x)$	$h = 10^{-2}$	$h = 10^{-3}$	$h = 10^{-4}$	$h = 10^{-5}$	$h = 10^{-6}$	$h = 10^{-7}$
(12.28)	−91769.95	−2250.2	70.94	93.71	116.42	−151345
(12.29)	−91769.95	−2250.2	70.79	93.77	94.00	94.00

Remark In computer arithmetic, only a finite number of significant digits is used. This fact may lead to a catastrophic loss of accuracy, in particular, when numerical schemes are not stable. The aim of this section was to present several thoroughly innocent-looking examples that produce completely nonsensical results. They are of several types "in limit":

(1) the expression in (12.20) is of type 1^∞,
(2) the product in (12.23) is of type $0 \cdot \infty$,
(3) the right-hand side of (12.26) is of type $\infty - \infty$,
(4) the fraction in (12.29) is of type $0/0$.

Chapter 13
Tables

Table 13.1 The first one thousand prime numbers

2	3	5	7	11	13	17	19	23	29
31	37	41	43	47	53	59	61	67	71
73	79	83	89	97	101	103	107	109	113
127	131	137	139	149	151	157	163	167	173
179	181	191	193	197	199	211	223	227	229
233	239	241	251	257	263	269	271	277	281
283	293	307	311	313	317	331	337	347	349
353	359	367	373	379	383	389	397	401	409
419	421	431	433	439	443	449	457	461	463
467	479	487	491	499	503	509	521	523	541
547	557	563	569	571	577	587	593	599	601
607	613	617	619	631	641	643	647	653	659
661	673	677	683	691	701	709	719	727	733
739	743	751	757	761	769	773	787	797	809
811	821	823	827	829	839	853	857	859	863
877	881	883	887	907	911	919	929	937	941
947	953	967	971	977	983	991	997	1009	1013
1019	1021	1031	1033	1039	1049	1051	1061	1063	1069
1087	1091	1093	1097	1103	1109	1117	1123	1129	1151
1153	1163	1171	1181	1187	1193	1201	1213	1217	1223
1229	1231	1237	1249	1259	1277	1279	1283	1289	1291
1297	1301	1303	1307	1319	1321	1327	1361	1367	1373
1381	1399	1409	1423	1427	1429	1433	1439	1447	1451
1453	1459	1471	1481	1483	1487	1489	1493	1499	1511

(continued)

© The Author(s), under exclusive license to Springer Nature Switzerland AG 2021   303
M. Křížek et al., *From Great Discoveries in Number Theory to Applications*,
https://doi.org/10.1007/978-3-030-83899-7_13

Table 13.1 (continued)

1523	1531	1543	1549	1553	1559	1567	1571	1579	1583
1597	1601	1607	1609	1613	1619	1621	1627	1637	1657
1663	1667	1669	1693	1697	1699	1709	1721	1723	1733
1741	1747	1753	1759	1777	1783	1787	1789	1801	1811
1823	1831	1847	1861	1867	1871	1873	1877	1879	1889
1901	1907	1913	1931	1933	1949	1951	1973	1979	1987
1993	1997	1999	2003	2011	2017	2027	2029	2039	2053
2063	2069	2081	2083	2087	2089	2099	2111	2113	2129
2131	2137	2141	2143	2153	2161	2179	2203	2207	2213
2221	2237	2239	2243	2251	2267	2269	2273	2281	2287
2293	2297	2309	2311	2333	2339	2341	2347	2351	2357
2371	2377	2381	2383	2389	2393	2399	2411	2417	2423
2437	2441	2447	2459	2467	2473	2477	2503	2521	2531
2539	2543	2549	2551	2557	2579	2591	2593	2609	2617
2621	2633	2647	2657	2659	2663	2671	2677	2683	2687
2689	2693	2699	2707	2711	2713	2719	2729	2731	2741
2749	2753	2767	2777	2789	2791	2797	2801	2803	2819
2833	2837	2843	2851	2857	2861	2879	2887	2897	2903
2909	2917	2927	2939	2953	2957	2963	2969	2971	2999
3001	3011	3019	3023	3037	3041	3049	3061	3067	3079
3083	3089	3109	3119	3121	3137	3163	3167	3169	3181
3187	3191	3203	3209	3217	3221	3229	3251	3253	3257
3259	3271	3299	3301	3307	3313	3319	3323	3329	3331
3343	3347	3359	3361	3371	3373	3389	3391	3407	3413
3433	3449	3457	3461	3463	3467	3469	3491	3499	3511
3517	3527	3529	3533	3539	3541	3547	3557	3559	3571
3581	3583	3593	3607	3613	3617	3623	3631	3637	3643
3659	3671	3673	3677	3691	3697	3701	3709	3719	3727
3733	3739	3761	3767	3769	3779	3793	3797	3803	3821
3823	3833	3847	3851	3853	3863	3877	3881	3889	3907
3911	3917	3919	3923	3929	3931	3943	3947	3967	3989
4001	4003	4007	4013	4019	4021	4027	4049	4051	4057
4073	4079	4091	4093	4099	4111	4127	4129	4133	4139
4153	4157	4159	4177	4201	4211	4217	4219	4229	4231
4241	4243	4253	4259	4261	4271	4273	4283	4289	4297
4327	4337	4339	4349	4357	4363	4373	4391	4397	4409
4421	4423	4441	4447	4451	4457	4463	4481	4483	4493

(continued)

Table 13.1 (continued)

4507	4513	4517	4519	4523	4547	4549	4561	4567	4583
4591	4597	4603	4621	4637	4639	4643	4649	4651	4657
4663	4673	4679	4691	4703	4721	4723	4729	4733	4751
4759	4783	4787	4789	4793	4799	4801	4813	4817	4831
4861	4871	4877	4889	4903	4909	4919	4931	4933	4937
4943	4951	4957	4967	4969	4973	4987	4993	4999	5003
5009	5011	5021	5023	5039	5051	5059	5077	5081	5087
5099	5101	5107	5113	5119	5147	5153	5167	5171	5179
5189	5197	5209	5227	5231	5233	5237	5261	5273	5279
5281	5297	5303	5309	5323	5333	5347	5351	5381	5387
5393	5399	5407	5413	5417	5419	5431	5437	5441	5443
5449	5471	5477	5479	5483	5501	5503	5507	5519	5521
5527	5531	5557	5563	5569	5573	5581	5591	5623	5639
5641	5647	5651	5653	5657	5659	5669	5683	5689	5693
5701	5711	5717	5737	5741	5743	5749	5779	5783	5791
5801	5807	5813	5821	5827	5839	5843	5849	5851	5857
5861	5867	5869	5879	5881	5897	5903	5923	5927	5939
5953	5981	5987	6007	6011	6029	6037	6043	6047	6053
6067	6073	6079	6089	6091	6101	6113	6121	6131	6133
6143	6151	6163	6173	6197	6199	6203	6211	6217	6221
6229	6247	6257	6263	6269	6271	6277	6287	6299	6301
6311	6317	6323	6329	6337	6343	6353	6359	6361	6367
6373	6379	6389	6397	6421	6427	6449	6451	6469	6473
6481	6491	6521	6529	6547	6551	6553	6563	6569	6571
6577	6581	6599	6607	6619	6637	6653	6659	6661	6673
6679	6689	6691	6701	6703	6709	6719	6733	6737	6761
6763	6779	6781	6791	6793	6803	6823	6827	6829	6833
6841	6857	6863	6869	6871	6883	6899	6907	6911	6917
6947	6949	6959	6961	6967	6971	6977	6983	6991	6997
7001	7013	7019	7027	7039	7043	7057	7069	7079	7103
7109	7121	7127	7129	7151	7159	7177	7187	7193	7207
7211	7213	7219	7229	7237	7243	7247	7253	7283	7297
7307	7309	7321	7331	7333	7349	7351	7369	7393	7411
7417	7433	7451	7457	7459	7477	7481	7487	7489	7499
7507	7517	7523	7529	7537	7541	7547	7549	7559	7561
7573	7577	7583	7589	7591	7603	7607	7621	7639	7643
7649	7669	7673	7681	7687	7691	7699	7703	7717	7723
7727	7741	7753	7757	7759	7789	7793	7817	7823	7829
7841	7853	7867	7873	7877	7879	7883	7901	7907	7919

Table 13.2 Fermat numbers $F_m = 2^{2^m} + 1$

$F_0 = 3$
$F_1 = 5$
$F_2 = 17$
$F_3 = 257$
$F_4 = 65537$
$F_5 = 4294967297$
$F_6 = 18446744073709551617$
$F_7 = 340282366920938463463374607431768211457$
$F_8 = 115792089237316195423570985008687907853$
$\quad\quad 2699846656405640394575840079131296399937$
$F_9 = 134078079299425970995740249982058461274$
$\quad\quad 7936582059239337772356144372176403007354$
$\quad\quad 6976801874298166903427690031858186486054$
$\quad\quad 5085375388281194656994643364900608409765$
$F_{10} = 179769313486231590772930519078902473361$
$\quad\quad 7976978942306572734300811577326758055000$
$\quad\quad 9631327084773224075360211201138798713933$
$\quad\quad 5765878976881441662249284743063947412400$
$\quad\quad 3777678934248654852763022196012460941190$
$\quad\quad 4530829520850057688381506823424628814733$
$\quad\quad 9131105408272371633505106845862982399472$
$\quad\quad 4593847971630483535632962422413721721700$

Table 13.3 Values of some arithmetic functions

n	$\tau(n)$	$\sigma(n)$	$\phi(n)$	$\lambda(n)$	$\omega(n)$	$\varepsilon(n)$
1	1	1	1	1	0	0
2	2	3	1	1	1	−1
3	2	4	2	2	1	0
4	3	7	2	2	1	0
5	2	6	4	4	1	0
6	4	12	2	2	2	−1
7	2	8	6	6	1	0
8	4	15	4	2	1	1
9	3	13	6	6	1	0
10	4	18	4	4	2	−1
11	2	12	10	10	1	0
12	6	28	4	2	2	0

(continued)

Table 13.3 (continued)

n	$\tau(n)$	$\sigma(n)$	$\phi(n)$	$\lambda(n)$	$\omega(n)$	$\varepsilon(n)$
13	2	14	12	12	1	0
14	4	24	6	6	2	-1
15	4	24	8	4	2	0
16	5	31	8	4	1	1
17	2	18	16	16	1	0
18	6	39	6	6	2	-1
19	2	20	18	18	1	0
20	6	42	8	4	2	0
21	4	32	12	6	2	0
22	4	36	10	10	2	-1
23	2	24	22	22	1	0
24	8	60	8	2	2	1
25	3	31	20	20	1	0
26	4	42	12	12	2	-1
27	4	40	18	18	1	0
28	6	56	12	6	2	0
29	2	30	28	28	1	0
30	8	72	8	4	3	-1
31	2	32	30	30	1	0
32	6	63	16	8	1	1

Notation $\tau(n)$ number of all positive divisors of n, $\sigma(n)$ sum of all positive divisors of n, $\phi(n)$ Euler totient function, $\lambda(n)$ Carmichael lambda function, $\omega(n)$ number of different prime divisors of n, and $\varepsilon(n)$ is given by (5.5)

Table 13.4 Primitive roots of odd primes less than 50

3	5	7	11	13	17	19	23	29	31	37	41	43	47
2	2	3	2	2	3	2	5	2	3	2	6	3	5
	3	5	6	6	5	3	7	3	11	5	7	5	10
			7	7	6	10	10	8	12	13	11	12	11
			8	11	7	13	11	10	13	15	12	18	13
					10	14	14	11	17	17	13	19	15
					11	15	15	14	21	18	15	20	19
					12		17	15	22	19	17	26	20
					14		19	18	24	20	19	28	22
							20	19		22	22	29	23
							21	21		24	24	30	26
								26		32	26	33	29

(continued)

Table 13.4 (continued)

3	5	7	11	13	17	19	23	29	31	37	41	43	47
								27		35	28	34	30
											29		31
											30		33
											34		35
											35		38
													39
													40
													41
													43
													44
													45

Table 13.5 Various types of primes

Name	Primes
Anti-elite	2, 13, 17, 97, 193, 241, 257, 641, 673, 769, 2689, 5953, 8929,...
Baťa	19, 29, 59, 79, 89, 199, 499, 599, 1999, 2999, 4999, 8999,...
Cullen	3, $141 \cdot 2^{141} + 1$, $4713 \cdot 2^{4713} + 1$, $5795 \cdot 2^{5795} + 1$,...
Cyclic	3, 5, 7, 11, 13, 17, 37, 79, 113, 197, 199, 337, 1193, 11939,...
Eisenstein real	2, 5, 11, 17, 23, 29, 41, 53, 59, 71, 83, 89, 89, 101, 107,...
Elite	3, 5, 7, 41, 15361, 23041, 26881, 61441, 87041, 163841,...
Euclidean	2, 3, 5, 7, 11, 31, 379, 1019, 1021, 2657, 3229, 4547, 4787,...
Factorial	2, 3, 5, 7, 23, 719, 5039, 39916801, 479001599, 87178291199,...
Fermat	3, 5, 17, 257, 65537
Fibonacci	2, 3, 5, 13, 89, 233, 1597, 28657, 514229, 433494437,...
Gaussian real	3, 7, 11, 19, 23, 31, 43, 47, 59, 67, 71, 79, 83, 103, 107,...
Irregular	37, 59, 67, 101, 103, 131, 149, 157, 233, 257, 263, 271, 283,...
Lucas	2, 3, 7, 11, 29, 47, 199, 521, 2207, 3571, 9349, 3010349,...
Mersenne	3, 7, 31, 127, 8191, 131071, 524287, 2147483647, ...
Palindromic	3, 5, 7, 11, 101, 131, 151, 181, 191, 313, 353, 373, 383,...
Permutation	2, 3, 5, 7, 11, 13, 17, 37, 79, 113, 199, 337, $(10^{19} - 1)/9$,...
Primorial	3, 5, 11, 13, 41, 89, 317, 337, 991, 1873, 2053, 2377, 4093,...
Regular	3, 5, 7, 11, 13, 17, 19, 23, 29, 31, 41, 43, 47, 53, 61, 71, 73,...
Repunit	11, 1111111111111111111, 11111111111111111111111,...
Safe	5, 7, 11, 23, 47, 59, 83, 107, 167, 179, 227, 263, 347, 359,...
Siamese	7, 11, 79, 83, 223, 227, 439, 443, 1087, 1091, 13687, 13691,...
Sophie Germain	2, 3, 5, 11, 23, 29, 41, 53, 83, 89, 113, 131, 173, 179, 191,...
Thabit	2, 5, 11, 23, 47, 191, 383, 6143, 786431, 51539607551,...

(continued)

Table 13.5 (continued)

Name	Primes
Unique	3, 11, 37, 101, 9091, 9901, 333667, 909091, 99990001,…
Wieferich	1093, 3511
Wilson	5, 13, 563
Woodall	7, 23, 383, 32212254719, 2833419889721787128217599,…

Table 13.6 Prime factors of the first twenty Mersenne numbers $M_p = 2^p - 1$

$M_2 = 3$
$M_3 = 7$
$M_5 = 31$
$M_7 = 127$
$M_{11} = 2047 = 23 \cdot 89$
$M_{13} = 8191$
$M_{17} = 131071$
$M_{19} = 524287$
$M_{23} = 8388607 = 47 \cdot 178481$
$M_{29} = 536870911 = 233 \cdot 1103 \cdot 2089$
$M_{31} = 2147483647$
$M_{37} = 137438953471 = 223 \cdot 616318177$
$M_{41} = 2199023255551 = 13367 \cdot 164511353$
$M_{43} = 8796093022207 = 431 \cdot 9719 \cdot 2099863$
$M_{47} = 140737488355327 = 2351 \cdot 4513 \cdot 13264529$
$M_{53} = 9007199254740991 = 6361 \cdot 69431 \cdot 20394401$
$M_{59} = 576460752303423487 = 179951 \cdot 3203431780337$
$M_{61} = 2305843009213693951$
$M_{67} = 147573952589676412927 = 193707721 \cdot 761838257287$
$M_{71} = 2361183241434822606847 = 228479 \cdot 48544121 \cdot 212885833$

References

1. Acheson, D.: 1089 and all that. The element of surprise in mathematics. EMS Newslett. **49**, 9–11 (2003)
2. Adámek, J.: Coding (in Czech). SNTL, Prague (1982)
3. Adleman, L.M., Pomerance, C., Rumely, R.S.: On distinguishing prime numbers from composite numbers. Ann. Math. **117**, 173–206 (1983)
4. Agrawal, M., Kayal, N., Saxena, N.: Primes is in P. Ann. Math. (2), 160, 781–793 (2004)
5. Aigner, A.: On prime numbers for which (almost) all Fermat numbers are quadratic non-residues (in German). Monatsh. Math. **101**, 85–93 (1986)
6. Aigner, M., Ziegler, G.M.: Proofs from The Book, 3rd edn. Springer, Berlin (2004)
7. Alford, W.R., Granville, A., Pomerance, C.: There are infinitely many Carmichael numbers. Ann. Math. **140**, 703–722 (1994)
8. André-Jeannin, R., Irrationalité de la somme des inverses de certaines suites récurrentes, C. R. Acad. Sci. Paris Sér. I Math. **308**, 539–541 (1989)
9. Andrews, W.S.: Magic Squares and Cubes. Dover Publications, New York (1960)
10. Anglin, W.S.: The square pyramid puzzle. Am. Math. Monthly **97**, 120–124 (1990)
11. Antonyuk, P. N., Stanyukovich, K. P., Periodic solutions of the logistic difference equation (in Russian), Dokl. Akad. Nauk SSSR 313: 1033–1036. English Transl. Soviet Math. Dokl. **42**(1991), 116–119 (1990)
12. Aristote, Du.: ciel, text établi et traduit par Paul Moraux. Les Belles Lettres, Paris (1965)
13. Avila, B., Chen, Y.: On moduli for which the Fibonacci numbers contain a complete residue system. Fibonacci Quart. **51**, 151–152 (2013)
14. Babuška, I., Práger, M., Vitásek, E.: Numerical stability of computational processes. Wiss. Z. Tech. Univ. Dresden **12**, 101–110 (1963)
15. Babuška, I., Práger, M., Vitásek, E.: Numerical Processes in Differential Equations. Willey, London, New York, Sydney (1966)
16. Badea, C.: The irrationality of certain infinite series. Glasgow Math. J. **29**, 221–228 (1987)
17. Baillie, R., Wagstaff, S.S.: Lucas pseudoprimes. Math. Comp. **35**, 1391–1417 (1980)
18. Baker, R.C., Irving, A.J.: Bounded intervals containing many primes. Math. Z. **286**, 821–841 (2017)
19. Balakrishnan, N., Koutras, M.V.: Runs and Scans with Applications, Wiley Series in Probability and Statistics. Wiley, New York (2002)

M. Křížek et al., *From Great Discoveries in Number Theory to Applications*,
https://doi.org/10.1007/978-3-030-83899-7

20. Balasubramanian, R., Deshoullers, J.-M., Dress, F.: Waring's problem for biquadrates, Part I, II (in French), C. R. Acad. Sci. Paris Sér. I Math. **303**, 85–88, 161–163 (1986)
21. Balková, Ľ., Legerský, J.: Hash functions and combinatorics on words (in Czech), Pokroky Mat. Fyz. Astronom. **58**, 274–284 (2013)
22. Bammel, S., Rothstein, J.: The number of 9×9 Latin squares. Discrete Math. **11**, 93–95 (1975)
23. Banachiewicz, T.: O związku pomiędzy pewnym twierdzeniem matematyków chińskich a formą Fermata na liczby pierwsze, Spraw. Tow. Nauk, Warszawa **2**, 7–11 (1909)
24. Barton, O., Sudbery, J.: Magic squares and matrix models of Lie algebras. Adv. Math. **180**, 596–647 (2003)
25. Batenburg, K.J., Sijbers, J.: Automatic multiple threshold scheme for segmentation of tomograms. Proc. SPIE 2007, Article 65123D
26. Beauregard, R.A., Suryanarayan, E.R.: Square-plus-two primes. Math. Gazette **85**, 90–91 (2001)
27. Beeger, N.G.W.H.: On even numbers m dividing $2^m - 2$. Am. Math. Monthly **58**, 553–555 (1951)
28. Beiler, A.H.: Recreations in the Theory of Numbers: The Queen of Mathematics Entertains, 2nd edn. Dover, New York (1964)
29. Benham, C.J., Harvey, S., Olson, W.K., Sumners, D.W.L., Swigon, D.: Mathematics of DNA Structure, Function and Interactions, IMA Volumes in Mathematics and Applications, vol. 150. Springer, New York (2009)
30. Beran, L.: Fermat's factorization method in your hands (in Czech). Rozhledy Mat.-Fyz. **72**, 277–283 (1995)
31. Bezuszka, S.J., Kenney, M.J.: Even perfect numbers: (update)^2. Math. Teacher **90**, 628–633 (1997)
32. Bilu, Y., Hanrot, G., Voutier, P.M.: Existence of primitive divisors of Lucas and Lehmer numbers. J. Reine Angew. Math. **539**, 75–122 (2001)
33. Blum, M.: Coin-Flipping by Telephone: A Protocol for Solving Impossible Problems, pp. 133–137. Compcon Spring, IEEE Proc. (1982)
34. Boase, M.: A result about the primes dividing Fibonacci numbers. Fibonacci Quart. **39**, 386–391 (2001)
35. Bonse, H.: Über eine bekannte Eigenschaft der Zahl 30 und ihre Verallsemeinerung. Ach. Math. Phys. **3**, 292–295 (1907)
36. Borho, W.: On Thabit ibn Kurrah's formula for amicable numbers. Math. Comp. **26**, 571–578 (1972)
37. Borho, W.: Living numbers. Math. Miniaturen 1, Birkhäuser, Basel, 5–38 (1981)
38. Borning, A.: Some results for $k! \pm 1$ and $2 \cdot 3 \cdot 5 \cdots p \pm 1$. Math. Comp. **26**, 567–570 (1972)
39. Bosáková, D., et al.: Electronic Signature (in Czech). Grada, Prague (2002)
40. Bose, R.C., Parker, E.T., Shrikhande, S.S.: Further results on the construction of mutually orthogonal Latin squares and the falsity of Euler's conjecture. Canad. Math. J. **12**, 189–203 (1960)
41. Brandts, J., Korotov, S., Křížek, M.: Simplicial Partitions with Applications to the Finite Element Method. Springer International Publishing, Cham, Nature Switzerland AG (2020)
42. Brandts, J., Křížek, M., Zhang, Z.: Paradoxes in numerical calculations. Neural Netw. World **26**, 317–330 (2016)
43. Bressoud, D., Wagon, S.: Computational Number Theory. Key College Publ, Springer, New York (2000)
44. Breuil, C., Conrad, B., Diamond, F., Taylor, R.: On the modularity of elliptic curves over Q: wild 3-adic exercises. J. Am. Math. Soc. **14**, 843–939 (2001)
45. Brillhart, J., Lehmer, D.H., Selfridge, J.L., Tuckerman, B., Wagstaff, S.S.: Factorization of $b^n \pm 1$, $b = 2, 3, 5, 6, 7, 10, 11, 12$ up to high powers. Contemporary Math. vol. 22, 2nd edn., Amer. Math. Soc., Providence, (1988)
46. Brown, J.L.: Zeckendorf's theorem and some applications. Fibonacci Quart. **2**, 163–168 (1964)

47. Bruck, R.H., Ryser, H.J.: On the non-existence of certain finite projective planes. Canad. Math. J. **1**, 88–93 (1949)
48. Bryant, S.: Groups, graphs, and Fermat's last theorem. Am. Math. Monthly **74**, 152–156 (1967)
49. Bugeaud, Y., Mignotte, M.: Catalan's conjecture proved (in Czech). Pokroky Mat. Fyz. Astronom. **50**, 280–285 (2005)
50. Bugeaud, Y., Mignotte, M., Siksek, S.: Sur les nombres de Fibonacci de la forme $q^k y^p$. C. R. Math. Acad. Sci. Paris **339**, 327–330 (2004)
51. Bugeaud, Y., Mignotte, M., Siksek, S.: Classical and modular approaches to exponential Diophantine equations. I. Fibonacci and Lucas perfect powers. Ann. Math. **163**, 969–1018 (2006)
52. Buhler, J., Crandall, R., Ernvall, R., Metsänkylä, T.: Irregular primes and cyclotomic invariants to four million. Math. Comp. **61**, 151–153 (1993)
53. Buhler, J.P., Harvey, D.: Irregular primes to 163 million. Math. Comp. **80**, 2435–2444 (2011)
54. Bukovský, L., Kluvánek, I.: Dirichlet's Principle (in Slovak). ÚV MO, Mladá fronta, Prague (1970)
55. Burr, S.A.: On moduli for which the Fibonacci sequence contains a complete system of residues. Fibonacci Quart. **9**, 497–504 (1971)
56. Burton, D.M.: Elementary Number Theory, 4th edn. McGraw-Hill, New York (1998)
57. Cahill, N.D., D'Errico, J.R., Spence, J.P.: Complex factorizations of the Fibonacci and Lucas numbers. Fibonacci Quart. **41**, 13–19 (2003)
58. Calda, E.: Fibonacci numbers and the Pascal triangle (in Czech). Rozhledy Mat.-fyz. **71**, 15–19 (1993)
59. Caldwell, C., Dubner, H.: Primorial, factorial and multifactorial primes. Math. Spectrum **26**, 1–7 (1993/1994)
60. Caldwell, C., Gallot, Y.: On the primality of $n! \pm 1$ and $2 \times 3 \times 5 \times \cdots \times p \pm 1$. Math. Comp. **71**, 441–448 (2002)
61. Carlip, W., Mincheva, M.: Symmetry of iteration digraphs. Czechoslovak Math. J. **58**, 131–145 (2008)
62. Carmichael, R.D.: Note on a new number theory function. Bull. Am. Math. Soc. **16**, 232–238 (1910)
63. Carmichael, R.D.: On composite numbers P which satisfy the Fermat congruence $a^{P-1} \equiv 1 \pmod{P}$. Am. Math. Monthly **19**, 22–27 (1912)
64. Carmichael, R.D.: On the numerical factors of the arithmetic forms $\alpha^n \pm \beta^n$. Ann. Math. **15**(2), 30–70 (1913)
65. Chartrand, G., Lesniak, L.: Graphs and Digraphs, 3rd edn. Chapman & Hall, London (1996)
66. Chassé, G.: Applications d'un corps fini dans lui-même, Dissertation. Univ. de Rennes I, (1984)
67. Chassé, G.: Combinatorial cycles of a polynomial map over a commutative field. Discrete Math. **61**, 21–26 (1986)
68. Chaumont, A., Müller, T.: All elite primes up to 250 billion. J. Integer Seq. **9**, Article 06.3.8, 1–5 (2005)
69. Chen, J.R.: On the representation of larger even integer as the sum of a prime and the product of at most two primes. Sci. Sinica **16**, 157–176 (1973)
70. Cihlář, J., Vopravil, V.: Games and Numbers (in Czech). Pedagogical Faculty, Ústí nad Labem, (1983), (1995)
71. Cipolla, M.: Sui numeri composti P, che verificano la congruenza di Fermat $a^{P-1} \equiv 1 \pmod{P}$. Annali di Matematica (3) **9**, 139–160 (1904)
72. Cipra, B.: How number theory got the best of the Pentium chip. Science **267**, 175 (1995)
73. Cochran, W., Crick, F.H.C., Vand, V.: The structure of synthetic polypeptides. I. Transformation of atoms on a helix. Acta Cryst. **5**, 581–586 (1952)
74. Cohn, J.H.E.: Square Fibonacci numbers. J. Lond. Math. Soc. **39**, 537–540 (1964)
75. Cohn, J.H.E.: Square Fibonacci numbers, etc. Fibonacci Quart. **2**, 109–113 (1964)

76. Conlon, D., Fox, J., Zhao, Y.: The Green-Tao Theorem: An Exposition, pp. 1–26 (2014). arXiv: 1403.2957

77. Conway, J.H., Sloane, N.J.A.: Sphere Packing, Lattices and Groups. Springer, New York (1988)

78. Coxeter, H.S.M.: Regular honeycombs in hyperbolic space. In: Proceedings International Congress Math. vol. 3, pp. 155–169. Amsterdam (1954)

79. Crandall, R., Dilcher, K., Pomerance, C.: A search for Wieferich and Wilson primes. Math. Comp. **66**, 433–449 (1997)

80. Crandall, R., Fagin, B.: Discrete weighted transforms and large-integer arithmetic. Math. Comp. **62**, 305–324 (1994)

81. Crandall, R.E., Mayer, E., Papadopoulos, J.: The twenty-fourth Fermat number is composite. Math. Comp. **72**, 1555–1572 (2003)

82. Crandall, R.E., Pomerance, C.: Prime Numbers. A computational perspective, Springer, New York (2001), (2005)

83. Cullen, J.: Question 15897. Math. Quest. Educ. Times **9**, 534 (1905)

84. Cuyt, A., Verdonk, B., Becuwe, S., Kuterna, P.: A remarkable example of catastrophic cancellation unraveled. Computing **66**, 309–320 (2011)

85. Delahaye, J.P.: Merveilleux nombres premiers. Voyage au cœur de l'arithmétique (in French), Belin—Pour la Science, Paris (2000)

86. Delahaye, J.-P.: The science behind SUDOKU. Sci. Am. **294**, 71–77 (2006)

87. Dénes, J., Keedwell, A.D.: Latin Squares and Their Applications. Budapest; Academic Press, New York, Akadémiai Kiadó (1974)

88. Deng, G., Yuan, P.: Symmetric digraphs from powers modulo n. Open J. Discrete Math. **1**, 103–107 (2011)

89. Deng, G., Yuan, P.: On the symmetric digraphs from powers modulo n. Discrete Math. **312**, 720–728 (2012)

90. Devaney, R.L.: The Mandelbrot set, the Farey tree, and the Fibonacci sequence. Am. Math. Monthly **106**, 289–302 (1999)

91. Dickson, L.E.: History of the Theory of Numbers, vol. I. Carnegie Inst., Washington, Divisibility and primality (1919)

92. Diffie, W., Hellman, M.E.: New directions in cryptography. IEEE Trans. Inform. Theory **22**, 644–654 (1976)

93. Dimitrov, V.S., Cooklev, T.V., Donevsky, B.D.: Generalized Fermat–Mersenne number theoretic transform. IEEE Trans. Circuits Syst. II, Analog Digit. Signal Process **41**, 133–139 (1994)

94. Dirichlet, P.G.L., Beweis des Satzes dass jede unbegrenzte arithmetische Progression, deren erstes Glied und Differenz ganze Zahlen ohne gemeinschaftlichen Factor sind, unendlich viele Primzahlen enthält, Abh. d. Königl. Akad. d. Wiss. (1837), 45–81; reprinted in Werke, vol. 1, 315–350, G. Reimer, Berlin (1889)

95. Dorais, F.G., Klyve, D.: A Wieferich prime search up to 6.7×10^{15}. J. Integer Seq. **14**, Article 11.9.2, 1–14 (2011)

96. Drobot, V.: On primes in the Fibonacci sequence. Fibonacci Quart. **38**, 71–72 (2000)

97. Dujella, A.: A proof of the Hoggatt-Bergum conjecture. Proc. Am. Math. Soc. **127**, 1999–2005 (1999)

98. Duparc, H.J.A.: On Carmichael numbers, Poulet numbers, Mersenne numbers and Fermat numbers, Rapport ZW 1953-004, Math. Centrum Amsterdam, 1–7 (1953)

99. Duverney, D.: Irrationalité de la somme des inverses de la suite de Fibonacci. Elem. Math. **52**, 31–36 (1997)

100. Dvořáková, L'., Kruml, S., Ryzák, D.: Antipalindromic numbers. Acta Polytech. **61**, 428–434 (2021)

101. Edwards, H.M.: The background of Kummer's proof of Fermat's last theorem for regular primes. Arch. History Exact Sci. **14**, 219–236 (1975)

102. Elliott, D.F., Rao, K.R.: Fast transforms. Algorithms, Analyses, Applications. Academic Press, London (1982)

103. Erdős, P.: On almost primes. Am. Math. Monthly **57**, 404–407 (1950)
104. Erdős, P., Graham, R.L.: Old and New Problems and Results in Combinatorial Number Theory, Monographie 28 de L'Enseign. Math, Imprimerie Kundig, Genéve (1980)
105. Erdős, P., Selfridge, J.L.: The product of consecutive integers is never a power. Illinois J. Math. **19**, 292–301 (1975)
106. Euler, L.: Observationes de theoremate quodam Fermatiano aliisque ad numeros primos spectantibus, Comment. Acad. Sci. Petropol. 6, ad annos 1732–33, 103–107 (1738)
107. Euler, L.: Recherches sur une novelle espèce de quarrés magiques, Verh. Zeeuwsch. Genootsch. Wetensch. Vlissengen **9**, 85–239 (1782)
108. Euler, L.: De numeris amicabilibus. Leonahardi Euleri Opera Omnia, Teubner, Leipzig and Berlin, Ser. I., **2**, 63–162 (1915)
109. Faltings, G.: Finiteness theorems for abelian varieties over number fields (in German). Invent. Math. **73**, 349–366 (1983)
110. Fatou, P.: Sur les solutions uniformes de certaines équations fonctionnelles (in French). C. R. Acad. Sci. Paris **143**, 546–548 (1906)
111. Fatou, P.: Sur les équations fonctionnelles (in French). Bull. Soc. Math. France **47**, 161–271 (1919); **48**, 33–94, 208–314 (1920)
112. Feigenbaum, M.J.: Quantitative universality for a class of nonlinear transformations. J. Stat. Phys. **19**, 25–52 (1978)
113. Felgenhauer, B., Jarvis, A.F.: Mathematics of Sudoku I. Math. Spectrum **39**, 15–22 (2006)
114. Felgenhauer, B., Jarvis, A.F.: There are 6 670 903 752 021 072 936 960 sudoku grids. http://www.shef.ac.uk/~pmlafj/sudoku/
115. Finch, C.E., Jones, L.: A curious connection between Fermat numbers and finite groups. Am. Math. Monthly **109**, 517–524 (2002)
116. Friedlander, J. B., Pomerance, C., Shparlinski, I. E., Period of the power generator and small values of Carmichael's function, Math. Comp. **70**, 1591–1605. Corrigendum ibid **71**(2002), 1803–1806 (2001)
117. Garns, H.: Number Place. Dell Pencil Puzzles and Word Games, No. **16**, 6 (1979)
118. Gauss, C.F.: Disquisitiones arithmeticae. Springer, Berlin (1986)
119. Goldberg, M.: Three infinite families of tetrahedral space-fillers. J. Combin. Theory Ser. A **16**, 348–354 (1974)
120. Golomb, S.W.: Combinatorial proof of Fermat's "little" theorem. Am. Math. Monthly **63**, 718 (1956)
121. Good, I.J.: A reciprocal series of Fibonacci numbers. Fibonacci Quart. **12**, 346 (1974)
122. Gorshkov, A.S.: Method of fast multiplication modulo Fermat primes (in Russian). Dokl. Akad. Nauk SSSR **336**, 175–178. English translation in Soviet Phys. Dokl. **39**, 314–317 (1994)
123. Gorshkov, A. S., On the method of the number-theoretic Mersenne transform (in Russian), Dokl. Akad. Nauk **336**: 33–34. English Transl. Phys. Dokl. **39**, 312–313 (1994)
124. Goto, T., Ohno, Y.: Odd perfect numbers have a prime factor exceeding 10^8. Math. Comp. **77**, 1859–1868 (2008)
125. Granville, A., Tucker, T.J.: It's as easy as abc. Notices Am. Math. Soc. **49**, 1224–1231 (2002)
126. Green, B., Tao, T.: The primes contain arbitrarily long arithmetic progressions. Ann. Math. **167**, 481–547 (2008)
127. Grošek, O., Porubský, Š: Encryption—Algorithms, Methods, Practice (in Slovak). Grada, Prague (1992)
128. Grünbaum, B., Shephard, G.C.: Tilings and Patterns. W. H. Freeman and Company, New York (1987)
129. Gutfreund, H., Little, W.A.: Physicist's proof of Fermat's theorem of primes. Am. J. Phys. **50**, 219–220 (1982)
130. Guy, R.K.: The primes 1093 and 3511. Math. Student **35**, 205–206 (1967)
131. Guy, R.K.: Every number is expressible as the sum of how many polygonal numbers? Am. Math. Monthly **101**, 169–172 (1994)
132. Guy, R.K.: Unsolved Problems in Number Theory, 2nd edn. Springer, Berlin (1994)

133. Hall, M., Jr.: Combinatorial Theory. Blaisdell, Waltham, MA (1967)
134. Halton, J.H.: On Fibonacci residues. Fibonacci Quart. **2**, 217–218 (1964)
135. Hančl, J., Rucki, P.: Increase Your Mathematical Intelligence. PF Ostravské Univerzity, Ostrava (2007)
136. Harary, F.: Graph Theory. Addison-Wesley Publ. Company, London (1972)
137. Hardy, G.H.: A Mathematician's Apology. Cambridge University Press, Cambridge (1940), (1962), (1992)
138. Hardy, G.H., Wright, E.M.: An Introduction to the Theory of Numbers, 4th edn. Clarendon Press, Oxford (1960); other editions: (1938), (1945), (1954), (1979), (2008)
139. Harman, G.: Watt's mean value theorem and Carmichael numbers. Int. J. Number Theory **4**, 241–248 (2008)
140. Hayes, B.: Unwed numbers, the mathematics of Sudoku, a puzzle that boasts, no math. required! Am. Sci. **94**, 12–15 (2006)
141. Hayes, J.H.: On the teeth of wheels. Am. Sci. **88**, 296–300 (2000)
142. Hellegouarch, Y.: Invitation to the Mathematics of Fermat-Wiles. Academic, London (2002)
143. Henrici, P.: Discrete Variable Methods in Ordinary Differential Equations. Wiley, New York (1962)
144. Herzberg, A.L., Murty, M.R.: Sudoku squares and chromatic polynomials. Notices Am. Math. Soc. **54**, 708–717 (2007)
145. Hoggatt, V.E.: Fibonacci and Lucas Numbers. Houghton Mifflin Company, Boston (1969)
146. Holdener, J.: A theorem of Touchard on the form of odd perfect numbers. Am. Math. Monthly **109**, 661–663 (2002)
147. Horský, Z.: Foundation of the Charles Bridge and cosmological symbolism of the Old Town Bridge Tower (in Czech), pp. 197–212. Panorama, Prague, Staletá Praha IX (1979)
148. Horský, Z.: Prague's Horologe (in Czech). Panorama, Prague (1988)
149. Horský, Z., Procházka, E.: Prague horologe (in Czech). Sborník pro dějiny přírodních věd a techniky **9**, 83–146 (1964)
150. Husnine, S.M., Ahmad, U., Somer, L.: On symmetries of power digraphs. Util. Math. **85**, 257–271 (2011)
151. Ireland, K., Rosen, M.: A Classical Introduction to Modern Number Theory, 2nd edn. Springer, New York (1990)
152. Jarden, D.: Existence of an infinitude of composite n for which $2^{n-1} \equiv 1 \pmod{n}$. Riveon Lematematika **4**, 65–67 (1950)
153. Jarden, D.: Recurring Sequences: A Collection of Papers. Riveon Lematematika, Jerusalem (1973)
154. Jeans, J.H.: The converse of Fermat's theorem. Messenger Math. **27**, 174 (1897/1898)
155. Jenkins, P.: Odd perfect numbers have a prime factor exceeding 10^7. Math. Comp. **72**, 1549–1554 (2003)
156. Jensen, K.L.: Om talteoretiske Egenskaber ved de Bernoulliske Tal. Nyt Tidsskr. Mat. B **26**, 73–83 (1915)
157. Jones, J.P.: Diophantine representation of the Fibonacci numbers. Fibonacci Quart. **13**, 84–88 (1975)
158. Joo, I., Phong, B.M.: On super Lehmer pseudoprimes. Studia Sci. Math. Hungar. **25**, 121–124 (1990)
159. Julia, G.: Mémoire sur l'itération des fonctions rationnelles (in French). J. Math. Pures Appl. **4**(7), 47–245 (1918)
160. Kaprekar, D.R.: An interesting property of the number 6174. Scripta Math. **15**, 244–245 (1955)
161. Katrnoška, F.: Latin squares and the genetic code (in Czech). Pokroky Mat. Fyz. Astronom. **52**, 177–187 (2007)
162. Katrnoška, F., Křížek, M.: Genetic code and the theory of monoids or 50 years since the discovery of the structure of DNA (in Czech). Pokroky Mat. Fyz. Astronom. **48**, 207–222 (2003)

163. Katrnoška, F., Křížek, M.: Significant milestones of mathematical genetics (in Czech) Matematika—fyzik—informatika **14**, 1–12 (2004/2005)
164. Katrnoška, F., Křížek, M., Somer, L.: Magic squares and sudoku (in Czech). Pokroky Mat. Fyz. Astronom. **53**, 113–124 (2008)
165. Kiss, E.: Notes on János Bolyai's researches in number theory. Historia Math. **26**, 68–76 (1999)
166. Kiss, P.: A binomial congruence (in Hungarian). Az Egi Ho Si Mihn Tanárképzó Fóiskola füzetei, 457–464 (1978)
167. Kiss, P.: Some results on Lucas pseudoprimes. Ann. Univ. Sci. Budapest. Eötvös, Sect. Math. **28**, 153–159 (1985)
168. Klazar, M.: Primes contain arbitrarily long arithmetical sequences (in Czech). Pokroky Mat. Fyz. Astronom. **49**, 177–188 (2004)
169. Knauer, J., Richstein, J.: The continuing search for Wieferich primes. Math. Comp. **74**, 1559–1593 (2005)
170. Knuth, D.E.: Seminumerical Algorithms. The Art of Computer Programming, 2nd edn., vol. 2. Addison Wesley, Reading (1981)
171. Knuth, D.E.: The Art of Computer Programming. Addison Wesley, Boston (1997)
172. Koblitz, N.: Introduction to Elliptic Curves and Modular Forms. Springer, New York (1984)
173. Koblitz, N.: A Course in Number Theory and Cryptography. Springer, New York (1994)
174. Korselt, A.: Problème chinois. L'Interm. des Math. **6**, 143 (1899)
175. Koshy, T.: Fibonacci and Lucas Numbers with Applications. Wiley, New York (2001); vol. 2 (2019) (2001)
176. Koshy, T.: Elementary Number Theory with Applications. Elsevier, Amsterdam (2002)
177. Kraïtchik, M.: On the factorization of $2^n \pm 1$. Scripta Math. **18**, 39–52 (1952)
178. Kramer-Miller, J.: Structural properties of power digraphs modulo n. In: Proceedings of the 2009 Midstates Conference on Undergraduate Research in Computer Science and Mathematics, pp. 40–49. Oberlin (2009)
179. Křížek, F., Křížek, M.: Astronomical dial of the Prague horologe (in Czech). Rozhledy Mat.-Fyz. **86**, č. 1, 1–6 (2011)
180. Křížek, M.: Ten open problems in number theory (in Czech). Rozhledy Mat.-Fyz. **71**, 4–10 (1994)
181. Křížek, M.: How to divide the space into congruent tetrahedra (in Czech). Rozhledy Mat.-Fyz. **71**, 27–33 (1994)
182. Křížek, M.: On Fermat numbers (in Czech). Pokroky Mat. Fyz. Astronom. **40**, 243–253 (1995)
183. Křížek, M.: Gauss's contribution to Euclidean geometry (in Czech). Rozhledy Mat.-Fyz. **74**, 254–258 (1997)
184. Křížek, M.: The RSA method for encryption by large primes (in Czech). Rozhledy Mat.-Fyz. **75**, 101–107 (1998)
185. Křížek, M.: Does theoretical mathematics have applications in technical practice? (in Czech). Pokroky Mat. Fyz. Astronom. **44**, 14–24 (1999)
186. Křížek, M.: From Fermat numbers to geometry (in Czech). Pokroky Mat. Fyz. Astronom. **46**, 179–191 (2001)
187. Křížek, M.: From Fermat primes to geometry (in Czech). Cahiers du CEFRES **28**, 131–161 (2002)
188. Křížek, M.: On the geometric interpretation of some terms in number theory (in Czech). Pokroky Mat. Fyz. Astronom. **50**, 75–79 (2005)
189. Křížek, M.: From Fermat primes to geometry. In: Proceedings International Conference Presentation of Mathematics '05 (eds. J. Přívratská, J. Příhonská, D. Andrejsová), pp. 45–52. Tech. Univ. Liberec (2006)
190. Křížek, M., Chleboun, J.: A note on factorization of the Fermat numbers and their factors of the form $3h2^n + 1$. Math. Bohem. **119**, 437–445 (1994)
191. Křížek, M., Chleboun, J.: Is any composite Fermat number divisible by the factor $5h2^n + 1$? Tatra Mt. Math. Publ. **11**, 17–21 (1997)

192. Křížek, M., Křížek, P.: The magic pentagonal dodecahedron (in Czech). Rozhledy Mat.-Fyz. **74**, 234–238 (1997)
193. Křížek, M., Křížek, P.: Why has nature invented three stop codons of DNA and only one start codon? J. Theor. Biol. **304**, 183–187 (2012)
194. Křížek, M., Křížek, P., Šolc, J.: Astronomical mistakes accompanying the Prague horologe (in Czech). Čs. Čas. Fyz. **60**, 337–340 (2010)
195. Křížek, M., Liu, L.: Structure of the traditional Chinese calendar (in Czech). Rozhledy Mat.-Fyz. **73**, 270–275 (1996)
196. Křížek, M., Liu, L.: Mathematics in ancient China (in Czech). Pokroky Mat. Fyz. Astronom. **42**, 223–233 (1997)
197. Křížek, M., Liu, L., Šolcová, A.: Fundamental achievements of ancient Chinese mathematicians. Math. Spectrum **38**, 99–107 (2005/2006)
198. Křížek, M., Luca, F., Shparlinski, I., Somer, L.: On the complexity of testing elite primes. J. Integer Seq. **14**, Article 11.1.2, 1–5 (2011)
199. Křížek, M., Luca, F., Somer, L.: 17 Lectures on Fermat Numbers: From Number Theory to Geometry, CMS Books in Mathematics, vol. 9, p. 2011. Springer, New York (2001)
200. Křížek, M., Luca, F., Somer, L.: On the convergence of series of reciprocals of primes related to the Fermat numbers. J. Number Theory **97**, 95–112 (2002)
201. Křížek, M., Luca, F., Somer, L.: Algebraic properties of Fibonacci numbers (in Czech). Pokroky Mat. Fyz. Astronom. **50**, 127–140 (2005)
202. Křížek, M., Luca, F., Somer, L.: From Fermat numbers to geometry. Math. Spectrum **38**, 56–63 (2005/2006)
203. Křížek, M., Luca, F., Somer, L.: Arithmetic properties of Fibonacci numbers. In: Přívratská, J., Příhonská, J., Andres, Z. (eds.) Proceedings of International Conference Presentation of Mathematics '06, pp. 7–18. Tech. Univ. Liberec (2006)
204. Křížek, M., Luca, F., Somer, L.: Desde los números de Fermat hasta la geometría (in Spanish). La Gaceta de la Realm Sociedad Mathematica Espanola **10**(2), 471–483 (2007)
205. Křížek, M., Neittaanmäki, P.: Finite Element Approximation of Variational Problems and Applications, Longman. Harlow; Wiley, New York (1990)
206. Křížek, M., Šolc, J., Šolcová, A.: Is there a crystal lattice possessing five-fold symmetry? Notices Am. Math. Soc. **59**, 22–30 (2012)
207. Křížek, M., Somer, L.: A necessary and sufficient condition for the primality of Fermat numbers. Math. Bohem. **126**, 541–549 (2001)
208. Křížek, M., Somer, L.: Pseudoprimes (in Czech). Pokroky Mat. Fyz. Astronom. **48**, 143–151 (2003)
209. Křížek, M., Somer, L.: 17 necessary and sufficient conditions for the primality of Fermat numbers. Acta Math. Inf. Univ. Ostraviensis **11**, 73–79 (2003)
210. Křížek, M., Somer, L.: Sophie Germain little suns. Math. Slovaca **54**, 433–442 (2004)
211. Křížek, M., Somer, L.: Euclidean primes have the minimum number of primitive roots. JP J. Algebra Number Theory Appl. **12**, 121–127 (2008)
212. Křížek, M., Somer, L.: Architects of symmetry in finite nonabelian groups. Symmetry: Culture Sci. **21**, 333–344 (2010)
213. Křížek, M., Somer, L.: On peculiar Šindel sequences. JP J. Algebra Number Theory Appl. **17**, 129–140 (2010)
214. Křížek, M., Somer, L.: Why quintic polynomial equations are not solvable in radicals. In: Brandts, J. et al. (eds.) Proceedings of Conference Applications of Mathematics, pp. 125–131. Inst. of Math., Acad. Sci. Prague (2015)
215. Křížek, M., Somer, L., Markl, M., Kowalski, O., Pudlák, P., Vrkoč, I.: Abel Prize—The Highest Achievement in Mathematics (in Czech). Academia, Prague (2018)
216. Křížek, M., Somer, L., Šolcová, A.: Magic of numbers. From Large Discoveries to Applications (3rd edition, in Czech). Academia, Prague (2018)
217. Křížek, M., Šolcová, A., Somer, L.: Construction of Šindel sequences. Comment. Math. Univ. Carolin. **48**, 373–388 (2007)

218. Křížek, M., Šolcová, A., Somer, L.: Ten theorems on the astronomical clock of Prague. In: Příhonská, J., Segeth, K., Andrejsová, D. (eds.) Proceedings of International Conference Presentation of Mathematics '07, pp. 53–62. Tech. Univ. Liberec (2007)

219. Kubina, J.M., Wunderlich, M.C.: Extending Waring's conjecture to 471,600,000. Math. Comp. **55**, 815–820 (1990)

220. Kulish, U., Miranker, W.L.: The arithmetic of the digital computer: a new approach. SIAM Rev. **28**, 1–40 (1986)

221. Lagarias, J.C.: The set of primes dividing the Lucas numbers has density 2, 3. Pac. J. Math. **118**, 449–461. Errata ibid. **162**, 393–396 (1985)

222. Lampakis, E.: In Gaussian integers $x^3 + y^3 = z^3$ has only trivial solutions - a new approach. Electron. J. Comb. Number Theory **8**, A32 (2008)

223. Landau, E.: Ueber das Achtdamenproblem und seine Verallgemeinerung. Naturwissenschaftliche Wochenschrift X **I**, 367–371 (1896)

224. Lawther, H.P., Jr.: An application of number theory to the splicing of telephone cables. Am. Math. Monthly **42**, 81–91 (1935)

225. Lehmer, D.H.: An extended theory of Lucas' functions. Ann. Math. **31**, 419–448 (1930)

226. Lehmer, D.H.: On the converse of Fermat's theorem. Am. Math. Monthly **43**, 346–354 (1936)

227. Lenstra, A. K., Lenstra, H. W., Jr., Manasse, M. S., Pollard, J. M., The factorization of the ninth Fermat number, Math. Comp. **61**, 319–349. Addendum ibid. **64**(1995), 1357 (1993)

228. Lenstra, H.W., Jr.: Factoring integers with elliptic curves. Ann. Math. **126**(2), 649–673 (1987)

229. Leopold, J.H.: Almanus Manuscript. Hutchinson and Co., London (1961)

230. Lepka, K.: History of Fermat quotients (in Czech). Dějiny matematiky, sv. 14, Prometheus, Prague (2000)

231. LeVeque, W.J.: Fundamentals of Number Theory. Dover, Mineola, New York (1996)

232. Lidl, R., Neiderreiter, H.: Finite Fields. Addison-Wesley, Reading, MA (1983)

233. Ligh, S., Jones, P.: Generalized Fermat and Mersenne numbers. Fibonacci Quart. **20**, 12–16 (1982)

234. Lind, D.A.: The quadratic field $\mathbb{Q}[\sqrt{5}]$ and a certain diophantine equation. Fibonacci Quart. **6**, 86–93 (1968)

235. Le Lionnais, F.: Les Nombres Remarquables (in French). Hermann, Paris (1983)

236. Livio, M.: The Golden Ratio. Headline Book Publ, London (2002)

237. Ljunggren, W.: On the diophantine equation $x^2 + 4 = Ay^2$, Det. Kgl. Norske Vid.-Selsk. Forh. **24**, 82–84 (1951)

238. London, H., Finkelstein, R.: On Fibonacci Lucas numbers which are perfect powers. Fibonacci Quart. **7**, 476–481, 487. Errata ibid. **8**, 248 (1969)

239. Lovász, L.: Combinatorial Problems and Exercises. North-Holland, Amsterdam (1979)

240. Luca, F.: Products of factorials in binary recurrence sequences. Rocky Mount. J. Math. **29**, 1387–1411 (1999)

241. Luca, F.: Proposed problem H-596. Advanced problem section. Fibonacci Quart. **41**, 187 (2003)

242. Luca, F.: Palindromes in Lucas sequences. Monatsh. Math. **138**, 209–223 (2003)

243. Luca, F.: Números primos y aplicaciones (in Spanish). Soc. Mat. Mexicana, Mexico (2004)

244. Luca, F., Křížek, M.: On the solutions of the congruence $n^2 \equiv 1 \pmod{\phi^2(n)}$. Proc. Am. Math. Soc. **129**, 2191–2196 (2001)

245. Lucas, E.: Question 1180. Nouv. Ann. Math. **14**, 336 (1875)

246. Lucas, E.: Théorèmes d'arithmétique. Atti della Reale Accademia delle Scienze di Torino **13**, 271–284 (1878)

247. Lucheta, C., Miller, E., Reiter, C.: Digraphs from powers modulo p. Fibonacci Quart. **34**, 226–239 (1996)

248. Luo, M.: On triangular Fibonacci numbers. Fibonacci Quart. **27**, 98–108 (1989)

249. Ma, D.G.: An elementary proof of the solution to the Diophantine equation $6y^2 = x(x + 1)(2x + 1)$. Sichuan Daxue Xuebao **4**, 107–116 (1985)

250. van Maanen, J.: Euler and Goldbach on Fermat's numbers: $F_n = 2^{2^n} + 1$ (in Dutch). Euclides (Groningen) **57**, 347–356 (1981/1982)

251. Mahnke, D.: Leibniz auf der Suche nach einer allgemeinen Primzahlgleichung. Bibliotheca Math. **13**, 29–61 (1913)
252. Mahoney, M.S.: The Mathematical Career of Pierre Fermat (1601–1665). Princeton University Press, Princeton, New Jersey (1973), (1994)
253. Mąkowski, A.: Remark on perfect numbers. Elem. Math. **17**, 109 (1962)
254. Mąkowski, A.: On a problem of Rotkiewicz on pseudoprime numbers. Elem. Math. **29**, 13 (1974)
255. Malo, E.: Nombres qui sans être premiers, verifient exceptionnellement une congruence de Fermat. L'Interm. des Math. **10**, 88 (1903)
256. Mandelbrot, B.B.: Les objects fractals: forme, hasard et dimension (in French). Flammarion, Paris (1975)
257. Mann, H.B.: On orthogonal Latin squares. Bull. Am. Math. Soc. **50**, 249–257 (1944)
258. Mann, H.B., Shanks, D.: A necessary and sufficient condition for primality, and its source. J. Combin. Theory Ser. A **13**, 131–134 (1972)
259. Markowsky, G.: Misconceptions about the golden ratio. College Math. J. **23**, 2–19 (1992)
260. Martzloff, J.-C.: The History of Chinese Mathematics. Springer, Berlin (1997)
261. Matiyasevich, Y.: Enumerable sets are diophantine. Soviet Math. Dokl. **11**, 354–358 (1970)
262. Matiyasevich, Y.: My collaboration with Julia Robinson. Math. Intell. **14**(4), 38–45 (1992)
263. Matiyasevich, Y.V., Guy, R.K.: A new formula for π. Am. Math. Monthly **93**, 631–635 (1986)
264. Mauldin, R.D.: A generalization of Fermat's last theorem: the Beal conjecture and prize problem. Notices Am. Math. Soc. **44**, 1436–1437 (1997)
265. Maynard, J.: Small gaps between primes. Ann. Math. **181**, 383–413 (2015)
266. McDaniel, W.: On Fibonacci and Pell numbers of the form kx^2. Fibonacci Quart. **40**, 41–42 (2002)
267. McGuire, G., Tugemann, B., Civario, G.: There is no 16-clue sudoku: solving the sudoku minimum number of clues problem via hitting set enumeration. Exp. Math. **23**, 190–217 (2014)
268. McLaughlin, J.: Small Prime Powers in the Fibonacci Sequence, pp. 1–22 (2002). arXiv:Math/0110150v2
269. Mendel, J.G.: Versuche über Pflanzen-Hybriden (in German, Experiments on plant hybrids). Verh. naturforsch. Verein in Brünn N **4**, 3–47 (1865)
270. Mertens, F.: Ein Beitrag zur analytischen Zahlentheorie. J. Reine Angew. Math. **78**, 46–62 (1874)
271. Meyl, A.-J.-J.: Solution de question 1194. Nouv. Ann. Math. **17**, 464–467 (1878)
272. Mihăilescu, P.: Primary cyclotomic units and the proof of Catalan's conjecture. J. Reine Angew. Math. **572**, 167–195 (2004)
273. Mišoň, K.: Nontraditional application of the dyadic system (in Czech). Pokroky Mat. Fyz. Astronom. **XX**, 332–339 (1975)
274. Mlýnek, J.: Information security (in Czech). Pokroky Mat. Fyz. Astronom. **51**, 89–98 (2006)
275. Mlýnek, J.: Security of Business Data (in Czech). Computer Press, Brno (2007)
276. Mollin, R.A.: Fundamental Number Theory with Applications. CRC Press, New York (1998)
277. Monier, L.: Evaluation and comparison of two efficient probabilistic primality testing algorithms. Theor. Comput. Sci. **12**, 97–108 (1980)
278. Montgomery, P.L.: New solutions of $a^{p-1} \equiv 1 \bmod p^2$. Math. Comp. **61**, 361–363 (1993)
279. Mordell, L.J.: The congruence $((p - 1)/2)! \equiv \pm 1 \pmod p$. Am. Math. Monthly **68**, 145–146 (1961)
280. Muller J.-M. et al.: Handbook of Floating-Point Arithmetic. Birkhäuser (2009)
281. Müller, T.: Searching for large elite primes. Exp. Math. **15**, 183–186 (2006)
282. Müller, T.: On anti-elite prime numbers. J. Integer Seq. **10**, Article 07.9.4 (2007)
283. Müller, T.: A generalization of a theorem by Křížek. Luca and Somer on elite primes. Analysis **28**, 375–382 (2008)
284. Narkiewicz, W.: The Development of Prime Number Theory. From Euclid to Hardy and Littlewood, Springer, Berlin (2000)

285. Nekovář, J.: Modular curves and Fermat's theorem (in Czech). Math. Bohem. **119**, 79–96 (1994)
286. Nemes, I., Pethö, A.: Polynomial values in linear recurrences, II. J. Number Theory **24**, 47–53 (1986)
287. Nicely, T.R.: A new error analysis for Brun's constant. Virginia J. Sci. **52**, 45–55 (2001)
288. Nielsen, P.: Odd perfect numbers have at least nine distinct prime factors. Math. Comp. **76**, 2109–2126 (2007)
289. Nielsen, P.: Odd perfect numbers, Diophantine equations, and upper bounds. Math. Comp. **84**, 2549–2567 (2015)
290. Nirenberg, M.W., Matthaei, J.H.: The dependence of cell-free protein synthesis in *E. coli* upon naturally occurring or synthetic polyribonucleotides. Proc. Natl. Acad. Sci. USA **47**, 1588–1602 (1961)
291. Niven, I., Zuckerman, H.S., Montgomery, H.L.: An Introduction to the Theory of Numbers, 5th edn. Wiley, New York (1991)
292. Ochem, P., Rao, M.: Odd perfect numbers are greater than 10^{1500}. Math. Comp. **81**, 1869–1877 (2012)
293. Ore, O.: Number theory and Its History. Dover, New York (1948)
294. Parker, E.T., Somer, L.: A partial generalization of Mann's theorem concerning orthogonal Latin squares. Canad. Math. Bull. **31**, 409–413 (1988)
295. Peitgen, H.O., Jürgens, H., Saupe, D.: Chaos and Fractals: New Frontiers of Science. Springer, New York (1992)
296. Pépin, P.: (Pépin, J.F.T.), Sur la formule $2^{2^n} + 1$. C. R. Acad. Sci. **85**, 329–331 (1877)
297. Petr, K.: Geometrical proof of Wilson's proposition (in Czech). Čas. Pěst. Mat. Fyz. **34**, 164–166 (1905)
298. Pfender, F., Ziegler, G.M.: Kissing numbers, sphere packing, and some unexpected proofs. Notices Am. Math. Soc. **51**, 873–883 (2004)
299. Phong, B.M.: On super Lucas and super Lehmer pseudoprimes. Studia Sci. Math. Hungar. **23**, 435–442 (1988)
300. Pickover, C.A.: Wonders of Numbers. Oxford University Press, Oxford (2001)
301. Pickover, C.A.: The Zen of Magic Squares, Circles, and Stars. Princeton University Press, Princeton (2002)
302. Pillai, S.S.: On Waring's problem $g(6) = 73$. Proc. Indian Acad. Sci. Sect. A. **12**, 30–40 (1940)
303. Pisano, L.: Fibonacci's Liber abaci A translation into modern English of Leonardo Pisano's Book of calculation, translated by L. E. Sigler, Springer, New York (2002)
304. Polymath, D.H.J.: Variants of the Selberg sieve, and bounded intervals containing many primes. Res. Math. Sci. **1**, Art. 12, 1–83 (2014)
305. Pomerance, C.: On the distribution of pseudoprimes. Math. Comp. **37**, 587–593 (1981)
306. Pomerance, C.: A new lower bound for the pseudoprime counting function. Illinois J. Math. **26**, 4–9 (1982)
307. Pomerance, C.: A tale of two sieves. Notices Am. Math. Soc. **43**, 1473–1485 (1996)
308. Pomerance, C., Selfridge, J.L., Wagstaff, S.S.: The pseudoprimes to $25 \cdot 10^9$. Math. Comp. **35**, 1003–1026 (1980)
309. Porubský, Š: Fermat and theory of numbers or problem of divisors and perfect numbers (in Czech), Matematician Pierre de Fermat (eds. A. Šolcová, M. Křížek, G. Mink). Cahiers du CEFRES **28**, 49–86 (2002)
310. Postnikov, M.M.: Magic Squares. Nauka, Moscow (1964).(in Russian)
311. Poulet, P.: Table des nombres composés vérifiant le théorème de Fermat pour le module 2 jusqu'à 100.000.000, Sphinx **8** (1938), 42–52; Errata in Math. Comp. **25**, 944–945. Math. Comp. **26**(1972), 814 (1971)
312. Pradlová, J., Křížek, M.: Groups around us, I–III (in Czech). Rozhledy Mat.-Fyz. **76**, 209–216, 261–267 (1999); **77**, 5–12 (2000)
313. Proth, F.: Théorèmes sur les nombres premiers. C. R. Acad. Sci. Paris **87**, 926 (1878)

314. Rabin, M.O.: Probabilistic algorithms for testing primality. J. Number. Theory **12**, 128–138 (1980)
315. Rademacher, H., Toeplitz, O.: The Enjoyment of Mathematics. Princeton University Press, New Jersey (1957), (1966) (1970)
316. Rao, T.R.N., Fujiwara, E.: Error-Control Coding for Computer Systems. Prentice-Hall Inc, Englewood Cliffs, New Jersey (1989)
317. Ratliff, L.J.: The dimension of the magic square vector space. Am. Math. Monthly **66**, 793–795 (1959)
318. Reed, I.S., Truong, T.K., Welch, L.R.: The fast decoding of Reed-Solomon codes using Fermat transforms. IEEE Trans. Inform. Theory **24**, 497–499 (1978)
319. Ribenboim, P.: On the square factors of the numbers of Fermat and Ferentinou-Nicolacopoulou. Bull. Greek Math. Soc. **20**, 81–92 (1979)
320. Ribenboim, P.: 13 Lectures on Fermat's Last Theorem. Springer, New York (1979)
321. Ribenboim, P.: The Book of Prime Number Records. Springer, New York (1988), (1989)
322. Ribenboim, P.: The Little Book of Big Primes. Springer, Berlin (1991)
323. Ribenboim, P.: The New Book of Prime Number Records. Springer, New York (1996)
324. Ribenboim, P.: Fermat's Last Theorem for Amateurs. Springer, New York (1999)
325. Ribet, K.A.: From the Taniyama-Shimura conjecture to Fermat's last theorem. Ann. Fac. Sci. Toulouse Math. **11**, 116–139 (1990)
326. te Riele, H.: Four large amicable pairs. Math. Comp. **28**, 309–312 (1974)
327. Riesel, H.: Prime numbers and computer methods for factorization. Birkhäuser, Boston-Basel-Stuttgart (1985), (1994)
328. Ripley, B.D.: Stochastic Simulations. Wiley, New York (1987)
329. Rivest, R.L., Shamir, A., Adleman, L.M.: A method for obtaining digital signatures and public key cryptosystems. Commun. ACM **21**, 120–126 (1978)
330. Robbins, N.: Beginning Number Theory, 2nd edn. Jones and Bartlett Publishers, Sudbury, MA (2006)
331. Robert, F.: Discrete Iterations, vol. 6. Springer Series in Comput. Math. (1986)
332. Rogers, T.D.: The graph of the square mapping on the prime fields. Discrete Math. **148**, 317–324 (1996)
333. Rosen, K.H.: Elementary Number Theory and Its Applications. Addison-Wesley, Reading, MA (2005)
334. Rosenfeld, B.A., Sergeeva, N.D.: Stereographic Projection. Mir Publishers, Moscow (1977)
335. Rotkiewicz, A.: Sur les nombres pseudopremiers de la forme $ax + b$, C. R. Acad. Sci. Paris Sér. I Math. **257**, 2601–2604 (1963)
336. Rotkiewicz, A.: Sur les formules donnant des nombres pseudopremiers. Colloq. Math. **12**, 69–72 (1964)
337. Rotkiewicz, A.: Sur les nombres de Mersenne dépourvus de facteurs carres et sur les nombres naturels n tells que $n^2 \mid 2^n - 2$. Mat. Vesnik **2**(17), 78–80 (1965)
338. Rotkiewicz, A.: On the pseudoprimes of the form $ax + b$. Proc. Cambridge Philos. Soc. **63**, 389–392 (1967)
339. Rotkiewicz, A.: Pseudoprime Numbers and Their Generalizations. Stud. Assoc. Fac. Sci. Univ, Novi Sad (1972)
340. Rump, S.M.: Algorithms for verified inclusions—theory and practice. In: Reliability in Computation (ed. R. E. Moore), pp. 109–126. Academic, New York (1988)
341. Rump, S.M.: Verification methods: Rigorous results using floating-point arithmetic. Acta Numerica **19**, 287–449 (2010)
342. Russell, E., Jarvis, A.F.: Mathematics of Sudoku II. Math. Spectrum **39**, 54–58 (2006)
343. Rybnikov, K.A.: Introduction to Combinatorial Analysis. Izd. Moskovskogo Univ, Moscow (1985).(in Russian)
344. Rychlík, K.: Introduction to Elementary Number Theory (in Czech). JČMF, Prague (1931), (1950)
345. Ryser, H.J.: Combinatorial Mathematics, no. 14, Math. Assoc. of America, New York (1963)
346. Šalát, T.: On perfect numbers (in Slovak). Pokroky Mat. Fyz. Astronom. **IX**, 1–13 (1964)

347. Šalát, T.: Selected Chapters from Elementary Number Theory (in Slovak). SPN, Bratislava (1969)
348. Sandler, K.: Planetary hours of Prague's horologe and angle trisection (in Czech). Pokroky Mat. Fyz. Astronom. **58**, 199–200 (2013)
349. Schönhage, A., Strassen, V.: Fast multiplication of large numbers (in German). Computing **7**, 281–292 (1971)
350. Schroeder, M.R.: Number Theory in Science and Communication. Springer, Berlin (1984), (1986), (2006) (1997)
351. Seibert, J., Trojovský, P.: On factorization of the Fibonacci and Lucas numbers using tridiagonal determinants. Math. Slovaca **62**, 439–450 (2012)
352. Sierpiński, W.: Remarque sur une hypothèse des Chinois concernant les nombres $(2^n - 2)/n$. Colloq. Math. **1**, 9 (1948)
353. Sierpiński, W.: Number Theory (Teoria liczb, in Polish). Warszawa (1950)
354. Sierpiński, W.: 250 Problems in Elementary Number Theory. American Elsevier, New York (1970)
355. Singh, S., Fermat's Last Theorem. Fourth Estate Limited (1998)
356. Sinha, T.N.: Note on perfect numbers. Math. Student **42**, 336 (1974)
357. Skula, L.: Fermat's little theorem (in Czech), Matematician Pierre de Fermat (eds. A. Šolcová, M. Křížek, G. Mink). Cahiers du CEFRES **28**, 163–171 (2002)
358. Sloane, N.J.A.: The on-line encyclopedia of integer sequences, A028355, A028356, see http://www.research.att.com/~njas/sequences/
359. Smyth, C.: The terms in Lucas sequences divisible by their indices. J. Integer. Seq. **13**, Article 10.2.4, 1–18 (2010)
360. Šolcová, A., Křížek, M.: Fermat and Mersenne numbers in Pepin's test. Demonstratio Math. **39**, 737–742 (2006)
361. Šolcová, A., Křížek, M., Mink, G. (eds.): Mathematician Pierre de Fermat (in Czech), Cahiers du CEFRES, No. 28, Prague (2002)
362. Somer, L.: Elementary problems and solutions, Proposed problem B-224. Fibonacci Quart. **9**, 546 (1971)
363. Somer, L.E.: The Fibonacci group and a new proof that $F_p - (5/p) \equiv 0 \pmod{p}$. Fibonacci Quart. **10**, 345–348 (1972)
364. Somer, L.: On Fermat d-pseudoprimes. In: Théorie des nombers (éds. J.-M. De Koninck, C. Levesque (eds.), pp. 841–860. Walter de Gruyter, Berlin, New York (1989)
365. Somer, L.: A note on Parker's paper on mutually orthogonal Latin squares. In: Proceedings of the 24th International Conference on Combinatorics, Graph Theory, and Computing, vol. 98, 188–190. Congr. Numer. (1993)
366. Somer, L.: Divisibility of terms in Lucas sequences by their subscripts, Applications of Fibonacci Numbers, Vol. 5. In: Bergum, G.E. et al. (ed.) Proceedings 5th International Conference on Fibonacci Numbers and Their Applications, pp. 515–525. University of St. Andrews, Scotland (1992); Dordrecht, Kluwer Academic Publishers (1993)
367. Somer, L.: Divisibility of terms in Lucas sequences of the second kind by their subscripts, Applications of Fibonacci numbers. Vol. 6. In: Bergum, G.E. et al. (ed.) Proceedings of the Sixth International Research Conference on Fibonacci Numbers and Their Applications, pp. 473–486. Washington State University, Pullman, WA, USA, 1994, Dordrecht, Kluwer Academic Publishers (1996)
368. Somer, L.: On Lucas d-pseudoprimes, In: Bergum, G.E., Philippou, A.N., Horadam, A.F. (eds.)Applications of Fibonacci numbers, vol. 7, pp. 369–375. Kluwer Academic Publishers, Dordrecht (1998)
369. Somer, L.: Generalization of a theorem of Drobot. Fibonacci Quart. **40**, 435–437 (2002)
370. Somer, L., Křížek, M.: On a connection of number theory with graph theory. Czechoslovak Math. J. **54**, 465–485 (2004)
371. Somer, L., Křížek, M.: Structure of digraphs associated with quadratic congruences with composite moduli. Discrete Math. **306**, 2174–2185 (2006)

372. Somer, L., Křížek, M.: On semiregular digraphs of the congruence $x^k \equiv y \pmod{n}$. Comment. Math. Univ. Carolin. **48**, 41–58 (2007)
373. Somer, L., Křížek, M.: On symmetric digraphs of the congruence $x^k \equiv y \pmod{n}$. Discrete Math. **309**, 1999–2009 (2009)
374. Somer, L., Křížek, M.: The structure of digraphs associated with the congruence $x^k \equiv y \pmod{n}$. Czechoslovak Math. J. **61**, 337–358 (2011)
375. Somer, L., Křížek, M.: Power digraphs modulo n are symmetric of order M if and only if M is square free. Fibonacci Quart. **50**, 196–206 (2012)
376. Somer, L., Křížek, M.: On primes in Lucas sequences. Fibonacci Quart. **53**, 2–23 (2015)
377. Somer, L., Křížek, M.: On moduli for which certain second-order linear recurrences contain a complete system of residues modulo m. Fibonacci Quart. **55**, 209–228 (2017)
378. Somer, L., Křížek, M.: Nondefective integers with respect to certain Lucas sequences of the second kind, Integers 18. Article **A35**, 1–13 (2018)
379. Somer, L., Křížek, M.: Iteration of certain arithmetical functions of particular Lucas sequences. Fibonacci Quart. **58**, 55–69 (2020)
380. Somer, L., Křížek, M.: Second-order linear recurrences having arbitrarily large defect modulo p. Fibonacci Quart. **59**, 108–131 (2021)
381. Sommerville, D.M.Y.: Division of space by congruent triangles and tetrahedra. Proc. R. Soc. Edinburgh **43**, 85–116 (1923)
382. Steuerwald, R.: Über die Kongruenz $2^{n-1} \equiv 1 \bmod n$. S.-B. Math.-Nat. Kl., Bayer. Akad. Wiss., p. 177 (1947)
383. Stewart, C.L.: On the representation of an integer in two different bases. J. Reine Angew. Math. **319**, 63–72 (1980)
384. Stillwell, J.: The story of the 120-cell. Notices Am. Math. Soc. **48**, 17–24 (2001)
385. Strang, G.: Introduction to Linear Algebra, 2nd edn. Wellesley MA, Wellesley-Cambridge (1998)
386. Strassen, V.: Gaussian elimination is not optimal. Numer. Math. **13**, 354–356 (1969)
387. Strauch, O., Porubský, Š: Distribution of Sequences: A Sampler. Peter Lang Publ. Group, Frankfurt a. M. (2005)
388. Suetake, C.: The nonexistence of projective planes of order 12 with a collineation group of order 16. J. Combin. Theory Ser. A **107**, 21–48 (2004)
389. Szalay, L., A discrete iteration in number theory (in Hungarian). In: BDTF Tud. Közl. VIII. Természettudományok 3., pp. 71–91. Szombathely (1992)
390. Szymiczek, K.: Note on Fermat numbers. Elem. Math. **21**, 59 (1966)
391. Szymiczek, K.: On pseudoprimes which are products of distinct primes. Am. Math. Monthly **74**, 35–37 (1967)
392. Táborský, J.: Report on Prague's horologe (in Czech). Josef Teige, Prague (1570), (1901)
393. Tanton, J.: A dozen questions about Fibonacci numbers. Math. Horizons 5–31 (2005)
394. Tate, J.: The higher dimensional cohomology groups of class field theory. Ann. Math. **56**, 294–297 (1952)
395. Tattersall, J.J.: Elementary Number Theory in Nine Chapters. Cambridge University Press, Cambridge (2005)
396. Taylor, R., Wiles, A.: Ring-theoretic properties of certain Hecke algebras. Ann. Math. **141**, 553–572 (1995)
397. Terjanian, G.: Sur l'équation $x^{2p} + y^{2p} = z^{2p}$ (in French), C. R. Acad. Sci. Paris Sér. A-B **285**, A973–A975 (1977)
398. Terjanian, G.: Fermat's last theorem (in Czech). Cahiers du CEFRES **28**, 87–106 (2002)
399. Thompson, T.M.: From Error-Correcting Codes Through Sphere Packing to Simple Groups. Math. Assoc. Amer. (1983)
400. Thompson, A.C.: Odd magic powers. Am. Math. Monthly **101**, 339–342 (1994)
401. Touchard, J.: On prime numbers and perfect numbers. Scripta Math. **19**, 35–39 (1953)
402. Trappe, W., Washington, L.C.: Introduction to Cryptography with Coding Theory. Prentice-Hall, Upper Saddle River, New Jersey (2002)

403. Trenkler, M.: Construction of p-dimensional magic cubes (in Slovak). Obzory Mat. Fyz. Inf. **29**(2), 19–29 (2000)
404. Vajda, S.: Fibonacci and Lucas Numbers, and The Golden Section: Theory and Applications. Wiley, New York (1989)
405. Vorobiev, N.N.: Fibonacci Numbers. Birkhäuser, Basel (2002); Nauka, Moscow (1992)
406. van der Waall, R.W.: Oplossing, Nieuw archief voor wiskunde **XXIV**, 262–263 (1976)
407. Wagstaff, S.S.: Divisors of Mersenne numbers. Math. Comp. **40**, 385–397 (1983)
408. Wagstaff, S.S.: The Joy of Factoring, Student Mathematical Library, vol. 68. AMS, Providence, RI (2013)
409. Waldschmidt, M.: Lecture on the abc conjecture and some of its consequences. In: Cartier, P., Choudary, Waldschmidt, M. (eds.) Mathematics in the 21st Century, pp. 211–230. Springer, Basel (2015)
410. Ward, J.E.: Vector spaces of magic squares. Math. Mag. **53**, 108–111 (1980)
411. Warren, L.R.J., Bray, H.G.: On the square-freeness of Fermat and Mersenne numbers. Pac. J. Math. **22**, 563–564 (1967)
412. Watson, G.N.: The problem of the square pyramid. Messenger Math. **48**, 1–22 (1918/1919)
413. Weil, A.: Number Theory, An Approach Through History. Birkhäuser, Basel (1984)
414. Weintraub, S.H.: Count-wheels. Ars Combinatorica **36**, 241–247 (1993)
415. Weintraub, S.H.: Count-wheels: a mathematical problem arising in Horology. Am. Math. Monthly **102**, 310–316 (1995)
416. Wells, D.: Prime Numbers: The Most Mysterious Figures in Math. Wiley, New Jersey (2005)
417. Wieferich, A.: Beweis des Satzes, dass sich eine jede ganze Zahl als Summe von höchstens neun positiven Kuben darstellen lässt. Math. Ann. **66**, 95–101 (1909)
418. Wiles, A.: Modular elliptic curves and Fermat's last theorem. Ann. Math. **141**, 443–551 (1995)
419. Williams, H.C.: Primality testing on a computer. Ars Combin. **5**, 127–185 (1978)
420. Wilson, B.: Power digraphs modulo n. Fibonacci Quart. **36**, 229–239 (1998)
421. Yan, J.F., Yan, A.K., Yan, B.C.: Prime numbers and the amino acid code: analogy in coding properties. J. Theor. Biol. **151**, 333–341 (1991)
422. Zhang, Y.: Bounded gaps between primes. Ann. Math. **179**, 1121–1178 (2014)
423. Zhu, Z., Cao, L., Liu, X., Zhu, W.: Topological invariance of the Fibonacci sequences of the periodic buds in general Mandelbrot sets. J. Northeast Univ. Nat. Sci. **22**, 497–500 (2001)
424. Znám, Š: Theory of Numbers (in Slovak). ALFA, Bratislava (1987)
425. Zsigmondy, K.: Zur Theorie der Potenzreste. Monatsh. Math. **3**, 265–284 (1892)

References to Webpages

426. http://www.math.cas.cz
427. http://aa.usno.navy.mil/data/docs/easter.php
428. www.en.wikipedia.org/wiki/Primes_in_arithmetic_progressions
429. http://www.claymath.org/millennium
430. http://www.mersenne.org
431. http://primes.utm.edu/mersenne/index.html
432. http://www.isbn.org
433. http://www.issn.org
434. https://en.wikipedia.org/wiki/Fibonacci_prime
435. https://mathworld.wolfram.com/LucasPrime.html
436. http://www.prothsearch.com/fermat.html
437. http://oeis.org

Subject Index

© The Editor(s) (if applicable) and The Author(s), under exclusive license
to Springer Nature Switzerland AG 2021
M. Křížek et al., *From Great Discoveries in Number Theory to Applications*,
https://doi.org/10.1007/978-3-030-83899-7

Author Index